REPRODUCTIVE VERSATILITY IN THE GRASSES

REPRODUCTIVE VERSATILITY IN THE GRASSES

edited by

G. P. Chapman BSc, PhD, FLS

Wye College
University of London

The right of the University of Cambridge to print and sell all manner of books was granted by Henry VIII in 1534. The University has printed and published continuously since 1584.

CAMBRIDGE UNIVERSITY PRESS

Cambridge
New York Port Chester
Melbourne Sydney

CAMBRIDGE UNIVERSITY PRESS
Cambridge, New York, Melbourne, Madrid, Cape Town,
Singapore, São Paulo, Delhi, Tokyo, Mexico City

Cambridge University Press
The Edinburgh Building, Cambridge CB2 8RU, UK

Published in the United States of America by Cambridge University Press, New York

www.cambridge.org
Information on this title: www.cambridge.org/9780521188913

First published 1990
First paperback edition 2011

A catalogue record for this publication is available from the British Library

Library of Congress Cataloguing in Publication data

Reproductive versatility in the grasses/edited by G. P. Chapman.
p. cm.
ISBN 0–521–38060–X
1. Grasses – Reproduction. I. Chapman, G. P. (Geoffrey Peter)
QK495.G74R45 1990
584′.90416–dc20 89–28594 CIP

ISBN 978-0-521-38060-7 Hardback
ISBN 978-0-521-18891-3 Paperback

Additional resources for this publication at www.cambridge.org/9780521188913

All flesh is grass.

Isaiah 40 v. 6

What is the *meaning* of the differences that separate the Gramineae so delicately yet so definitely, from any other order, and that so prevail that a grass remains a grass however freely the type may vary?

Agnes Arber

The Gramineae: a study of cereal, bamboo and grass. Cambridge, 1934

In my opinion, the climax of flowering-plant evolution is represented by the grasses, which, in addition, are the most useful to man of all families.

G. L. Stebbins

Flowering Plants: evolution above the species level. Arnold, 1974

Contents

List of contributors ix

Preface xi

Acknowledgements xiii

1 The grass family, Poaceae 1
L. Watson

2 The spikelet 32
W. D. Clayton

3 Ovule structure and diversity 52
J. Greenham and G. P. Chapman

4 Fertilization and early embryogenesis 76
H. Lloyd Mogensen

5 Apomictic reproduction 100
E. C. Bashaw and Wayne W. Hanna

6 The implications of reproductive versatility for the structure of grass populations 131
A. J. Richards

7 An assessment of grass succession, utilization and development in the arid zone 154
M. D. Kernick

8 *In vitro* technology 182
P. A. Lazzeri, J. Kollmorgen and H. Lörz

9 Reproduction and recognition phenomena in the Poaceae 220
R. Bruce Knox and Mohan B. Singh

10 The widening perspective: reproductive biology of bamboos, some dryland grasses and cereals 240
G. P. Chapman

Appendix: World grass genera 258
L. Watson

Note added in proof on bamboo flowering *in vitro* 266

Author index 267

Organism index 277

Subject index 289

Contributors

Professor E. C. Bashaw
Soil and Crop Sciences
Texas A & M University
College Station
TX 77843, USA

Dr G. P. Chapman
Wye College
University of London
Near Ashford
Kent TN25 5AH, UK

Dr W. D. Clayton
The Herbarium
Royal Botanic Gardens
Kew
Richmond
Surrey TW9 3AB, UK

J. Greenham
Wye College
University of London
Near Ashford
Kent TN25 5AH, UK

Dr W. Hanna
USDA – ARS
Coastal Plain Experimental
Station
Tifton
GA 31793, USA

M. D. Kernick
(Formerly of FAO – Pasture
Improvement Specialist)
Foreland Cottage
St Margaret's Road
St Margaret's Bay
Dover
Kent CT15 6EQ, UK

Professor R. B. Knox
Department of Botany
University of Melbourne
Melbourne
Victoria 3052, Australia

J. Kollmorgen
Victorian Crops Research
Institute
Harshan
Victoria 3400, Australia

P. A. Lazzeri
Institut für Allgemeine
Botanik (AMPII)
Ohnhorststrasse 18
D-2000, Hamburg 52
Germany

Professor H. Lörz
Institut für Allgemeine
Botanik (AMPII)
Ohnhorststrasse 18
D-2000, Hamburg 52
Germany

Professor H. L. Mogensen
Department of Biological Sciences
Northern Arizona University
NAU Box 5640
Flagstaff
AZ 86011, USA

Dr A. J. Richards
Department of Plant Biology
University of Newcastle upon Tyne
Ridley Building
Newcastle upon Tyne NE1 7RU, UK

Dr M. B. Singh
Department of Botany
University of Melbourne
Melbourne
Victoria 3052, Australia

Professor L. Watson
Taxonomy Laboratory
Research School of Biological Sciences
Australian National University
GPO Box 475
Canberra
ACT 2601, Australia

Preface

In 1934, Agnes Arber's *The Gramineae: a study of cereal, bamboo and grass* was published. It is an unusual book, a miscellany of folklore, technical botany and the philosophical reflections of its author. It is not an easy book to read. Detail is relentlessly piled up and the struggling student, one suspects, is more likely to be hindered than helped by the hundreds of line drawings that crowd its pages. None the less, Arber's blend of enthusiasm and scholarship captured something of this extraordinary plant family.

Avdulov's classic paper on grass cytogenetics came late to her notice and is mentioned only in a footnote on page 379. Avdulov's influence was pervasive and led eventually to that landmark in evolutionary biology, G. L. Stebbins' *Variation and evolution in plants*, published in 1950 and in its way an early and exuberant celebration of the Darwinian century. The neo-Darwinian synthesis that Stebbins embodies was a training manual as much for the academic evolutionist as for the plant breeder and must materially have helped prepare the way for the Green Revolution that transformed prospects for agriculture at low latitudes as well as creating surpluses elsewhere.

The years since 1950 have seen the emergence of electron microscopy, molecular biology, tissue culture and a new orthodoxy about how cell organelles might have originated from 'captured' prokaryotes. Even when addressing old questions, biology uses a new language and works with new tools and the presumption is that we therefore have a new understanding, and perhaps we do.

In the preface to her book, Arber remarked that she had written only one of the numberless possible books for which grass might supply material and this latest, even though multi-authored, can contribute to only a limited area, though one of central importance. Part of the secret of the grasses' success must lie in their reproductive versatility, a theme this book seeks to explore in some detail. Increasingly, of course, humankind does not merely study but seeks to direct the evolution of our major cereals, and not surprisingly, therefore, it is around these that much of the newer genetic technology has developed. It is a profound and absorbing question as to whether these cereals have intrinsically more 'responsive' genetic systems than other grasses to traditional plant breeding or the techniques of molecular biology.

Against such a background one obviously needs an awareness of the range of grass types, what appear to be the evolutionary themes, how the reproductive system works and the diversity of which it is capable. From such an understanding the way to new developments becomes more obvious. To these ends, various scientists have contributed to this book, where although cereals are conspicuous, they are not intended to dominate, the aim being to take a family view. Part of this view is in recognizing the increasing importance of perennial grasses in stabilizing the collapsing ecologies of fragile environments and towards which new ideas and techniques must now be directed.

Since Arber wrote, the world has become a more tumultuous place and because of instant visual communication we are almost inured to the sight of dispossessed peoples moving in desolated landscapes. Beside famine relief, there has to develop a strategy for famine avoidance. What science has achieved offers a more optimistic alternative to the gloomy forecast of Thomas Malthus. None the less, we ignore him at our peril and perhaps the shadow of Malthus, quite properly, should be always with us.

It is evident that the long-established interdependence of humankind with grasses is set to continue and that among our biological priorities is the need to understand how grasses can be evolved for our use. What Nature has accomplished we must first comprehend if we are to extend her work in ways that will please rather than dismay us.

G. P. Chapman
Wye College

Acknowledgements

My thanks are due to all the contributors for their ready response to the original invitation and for their willingness to meet various editorial requests. My thanks are due among others to Peggy Pollack, Maxine Rusche, Tong Zhu and Karen Wunderlich.

I should also like to thank the British Council (Australia) and the University of Melbourne for their generosity in arranging for me to visit Australia as the book neared completion. This enabled me to have extended discussions with many agrostologists there including Bruce Knox, Don Marshall, Leslie Watson, Margaret Friedel, David Hayman, Peter Latz, Mike Lazarides, Steve Moreton, Geoff Pickup and Mohan Singh. My hope is that, since two of the book's chapters originate from Australia, these and my own visit will help contribute towards a 'Gondwana' perspective.

Over the years I have appreciated many absorbing discussions with cereal and forage breeders, something of whose approach I have tried to reflect in the penultimate section of the last chapter. Deliberately, however, I have linked this with what I believe must be a shift of emphasis towards perennial grasses in fragile environments.

I have to thank the Wye College Foundation Fund for a contribution towards the cost of publishing the colour illustrations and the College for help with other costs during the preparation of this book. My appreciation is due to Professor Dennis Baker for his encouragement and to Malcolm Kernick for his stimulating awareness of dryland grasses. I also owe particular gratitude to Dr Ted Peat for his friendly but rigorous criticism as this book went through its development.

It is a pleasure to thank, too, the Library staff of Wye College and Mrs Sue Briant, Mrs Margaret Critchley, Mrs Brenda Court and Mrs Terry Dinsdale for their excellent secretarial help and to Jeff Brooks for preparation of diagrams.

The staff of Cambridge University Press have been most helpful throughout.

Finally, I thank my wife, Sheila, for her support and encouragement.

G.P.C.

1

The grass family, Poaceae

L. Watson

Modern taxonomic treatments of the grass family (Poaceae, Gramineae) recognize about 10 000 species and between 650 and 765 genera, figures exceeded among the flowering plants only by those for Compositae, Leguminosae and Orchidaceae. The higher generic estimate reflects an attempt (Watson & Dallwitz, 1988; see Appendix) to account for all those 'genera' accepted by at least some recent authorities, and which seem at least to some extent taxonomically meaningful. It therefore counts many entities which the authors of the lower estimate (Clayton & Renvoize, 1986) refer to the larger genera of which they are obviously satellites, or which might better be recognized at subgeneric rank. A few of the inclusive generic interpretations by Clayton & Renvoize are widely unacceptable, in particular by experimentalists concerned with the Triticeae, and a reasonable compromise would be a figure of around 680 genera. Irrespective of such niceties, the grasses constitute one of the largest families of flowering plants, among which they are unsurpassed in ecological and economic importance. They occur in every habitat available to flowering plants except the sea bed, and dominate the vegetation types which cover about 30% of the earth's land surface. Direct utilization by man includes opportunistic exploitation and consequent modification of natural grasslands, and improvement of these by addition of fertilizers and other management techniques; creation of grasslands, playing fields, lawns, etc., by deforestation; deliberate cultivation, selection for genetic improvement and subsequent distribution and planting of pasture, hay, lawn and cereal crop species; cultivation of assorted species, yielding sugar (*Saccharum*), essential oils and culinary herbs (*Cymbopogon*, *Vetiveria*), raw materials for paper- and rope-making (Esparto grasses – *Lygeum*, *Stipa*, *Ampelodesmos*); exploitation of specialized forms as soil stabilizers and sandbinders (*Ammophila*, × *Ammocalamagrostis*, *Secale*, etc.); and, of course, diverse applications of bamboos as barrier plants, and of their stems, shoots, leaves and fruits for light and heavy constructional work, basketry, simple weaponry, fishing rods, etc., and as food. Besides providing vital food crops for man and livestock, this remarkably adaptable group everywhere contributes significantly to the adventive flora, and many species which seem innocuous may be capable of serving as reservoir hosts for pests and pathogens afflicting crops. Some of the most infamous invasive weeds are also grasses: seven of the 'top ten' according to one estimate (May, 1981), including species of *Cynodon*, *Echinochloa*, *Eleusine*, *Sorghum*, *Panicum* and *Imperata*.

Despite its size, the family Poaceae is by flowering plant standards a very coherent one, whose members exhibit characteristic combinations of unusual morphological and anatomical features. These include basically distichous phyllotaxy, sheathing ligulate leaves with an epidermis containing long-cells, short-cells, silica bodies and stomata with subsidiary cells and peculiar, dumb-bell shaped guard-cells; flowers of reduced and rather uniform construction arranged into 'spikelets'; and caryopses with abundant, starchy endosperm and a laterally placed, very peculiar embryo. Even the most outlandish members (including the overtly moss-like *Micraira*; *Neostapfia*, with non-ligulate leaves lacking most of the usual poaceous anatomical features; *Anomochloa*, with its highly modified spikelets and marantaceous general appearance; the superficially rush-like *Spartochloa*, and its sedge-like counterpart, *Cyperochloa*) are all clearly recognizable as grasses on detailed examination, as indeed are the giant bamboos, which differ significantly only in habit from an assortment of herbaceous relatives. The family is taxonomically isolated among other monocotyledons. There is general agreement, however, that it belongs in the superorder Commeliniflorae (*sensu* Dahlgren, Clifford & Yeo, 1985), and within that alliance there is stronger evidence relating it to the six small southern hemisphere families Anarthriaceae, Centrolepidaceae, Ecdeiocoleaceae, Flagellariaceae, Joinvilleaceae and Restionaceae (cf. Dahlgren *et al.*, 1985: order Poales) than to the Cyperaceae (e.g. Cronquist, 1981). A recent attempt (Campbell & Kellogg, 1987) at reconstructing the phylogeny of the order Poales favoured the monogeneric Joinvilleaceae as the sister group of the grasses, but also emphasized the relative distance of the relationship of the latter to the other families. It is very unlikely, therefore, that there are disguised Poaceae currently resident in any other family.

Some conspicuous features of grass morphology, anatomy, habit and reproductive cycles are readily envisaged as contributing to the relative competitive success of the family, if considered in relation to the evolution of herbivorous mammals and the expansion of human populations. These include intercalary meristems in stems and leaves, and protective leaf sheaths, which together confer unusually high tolerance of grazing, mowing, fire and drought; a basic inflorescence organization lending itself to multitudinous structural variations associated with efficient and opportunistic seed dispersal; and the very widespread herbaceous habit and short life cycles with regular, abundant seed production and a propensity for asexual reproduction. Such obvious attributes are insufficient in themselves, however, to explain the success of the Poaceae relative to other groups enjoying similar advantages. There seems in addition to have been unusual adaptability inherent in earlier grass genotypes, which relatively readily permitted adaptations to wide ranges of environments and facilitated rapid and efficient exploitation of environmental change and new opportunities. The capacity to evolve the C_4 pathway and its different types more than once within the family (cf. Brown, 1977; Hattersley, 1987), permitting very precise climatic adaptation, is a striking example of this; and a still evident high capacity for hybridization and polyploidy, associated with the diversity of reproductive mechanisms discussed in this book, must have constituted an important component of their versatility.

Nobody will need convincing of the economic significance of the Poaceae, which is in itself sufficient to justify massive input of research (including continuing to refine their taxonomy), but it is worth emphasizing that only a tiny proportion of this huge assemblage has been subjected to even the most elementary processes of economic assessment. Thus, fewer than 50 genera are commonly mentioned as sources of cultivated fodder, and fewer than 20 in relation to lawns and playing fields. Only 17 genera are represented in the list of cultivated cereal species, and the potential of several of these (e.g. *Eragrostis tef*) has yet to be fully realized. The total of around 60 genera seen as 'economically important' to the extent of including cultivated representatives takes in some of the largest, encompassing around 4000 species in all; but only about 150 of these seem to be regularly cultivated. On the other hand, most grasses are edible to livestock, many uncultivated forms are seen to be very palatable, and most have much the same nutritional potential in terms of leaf protein amino acid profiles (Yeoh & Watson, 1982). Ethnobotanical references to grain harvesting from wild grasses (e.g. *Loudetia esculentia*, *Hyperthelia edulis*) hint at untapped potential, as does evidence that the grains of many uncultivated species which have not been subjected to selection and genetic improvement (species of *Bromus*, *Stipa* and *Ehrharta*, *Microlaena stipoides*, *Anisopogon avenaceus*, etc.: Yeoh & Watson, 1981) have protein contents, amino acid compositions and chemical scores (i.e. nutritional status) similar to, or better than, many of the cultivated cereals. The assortment of grasses cultivated and subjected to intensive agriculturally orientated research is relatively minute, and often appears to reflect the history of civilization and coincidences of people, times and places rather than scientific sampling; and there is every reason to suppose that the Poaceae afford unlimited opportunities for further agricultural development. From a taxonomist's standpoint, most economically orientated research seems too introverted. In the case of Poaceae, the family *as a whole* should be seen as a magnificent resource, capable not only of supplying many new crops, but of suggesting ideas for improving forms already in cultivation, and of supplying genetic engineers with the material for putting them into effect.

Taxonomy

At their best, taxonomic classifications are highly sophisticated, general-purpose data storage and information retrieval systems. They permit communication, by facilitating generalizations about the facts directly incorporated in them; and more excitingly, they allow predictions about little-known organisms (including those known only as herbarium specimens), with reference to their better-known relatives. Conversely, they should encourage critical evaluation of generalizations based on insufficient evidence: a function of particular importance in relation to experimental sciences, where samples are often unavoidably small. The levels of predictive accuracy available through taxonomy are determined by the reliability with which relationships have been determined, and by the skill with which these are translated into formal systems of names. In opting to work in terms of species and to a lesser extent genera and families, most biologists seem content to allow taxonomists to define their samples and, ultimately, to determine the validity of their scientific generalizations. On the other hand, relatively few take full advantage of taxonomic knowledge to the extent of considering the details of taxonomic hierarchies. In the context of experimental biology, these can play a valuable role in generating testable hypotheses, if given proper consideration in relation to sampling procedures. The intellectual challenge for taxonomists lies in developing hierarchical classifications of demonstrable practical value, and persuading others of the extent of their usefulness (if necessary by practical demonstrations of relevance, involving deliberate approaches by taxonomists to their colleagues in other fields).

There is loud dissent among taxonomists over classificatory methodology, notably in relation to the status of 'phylogenetic' considerations (irrelevant/desirable/necessary/essential), the current expression of this long-continued saga being manifested in arguments about phenetics versus cladistics, and about the relative merits of different methods within these disciplines (especially the latter). For what it is worth, this practitioner subscribes to the view (cf. Sneath, 1988) that valid cladistic inferences from taxonomic data of the traditional kind depend upon prior recognition of phenetic groups. Furthermore, the near universal assumption that existing organisms carry with them information capable of yielding up conclusive phylogenies reflects a faith which may be misplaced. Among some viruses, in fact, it is now suspected that the only source of irrefutable evidence, in the form of nucleic acid sequences, may prove to have been irretrievably scrambled. Hopefully, this would be an extreme case, and similar considerations will not apply very widely: the intention here is not to deny the intellectual excitement of phylogenetic inquiry, its scientific interest or its obvious relevance to taxonomic practice. It should be acknowledged, however, that functional taxonomy does not depend upon the revelation of incontrovertible phylogenies, which may never be generally attainable: even in circumstances where further phylogenetic inquiry seemed futile, few would doubt the necessity for taxonomic classifications or the need to continue refining them.

In any case, here, as in other groups of organisms, neither the phenetic nor the phylogenetic relationships of the hard core of problem taxa will ever be resolved satisfactorily by comparing genotypes indirectly in terms of their morphological, anatomical and physiological, phenotypic manifestations. Nucleic acid and protein structural comparisons should provide more definitive answers for many of them, as well as providing the most reliable evidence about the details of group divergences. As yet, however, comparative molecular studies of angiosperms above generic level are at a very elementary stage. In relation to the main lines of grass classification and evolution, for example, a recent approach aimed at comparing whole genomes (King & Ingrouille, 1987) has yielded bizarre results. These made taxonomic sense within the sample from the subfamily Pooideae, in particular making a clear distinction between the supertribes Poodae and Triticodae and associating *Bromus* with the latter (cf. Macfarlane & Watson, 1982; Fig. 1.1; Appendix); on the other hand, the startling failure of the same procedure to distinguish *Phragmites*, *Spartina*, *Bambusa*, *Oryza*, *Saccharum*, *Sorghum* and *Zea* (representing the subfamilies Arundinoideae, Chloridoideae, Bambusoideae and Panicoideae) from the Pooideae necessitated ignoring them altogther for the purposes of taxonomic analysis and discussion. This tempts the suggestion that grass taxonomy is perhaps being more informative about nucleic

Fig. 1.1. Example of an agglomerative numerical classification of 558 genera of Poaceae, encoded for 253 mainly morphological and anatomical characters (Dissimilarity scale: Gower metric, using Burr's strategy with ISS). (Adapted from Watson, Clifford & Dallwitz, 1985.)

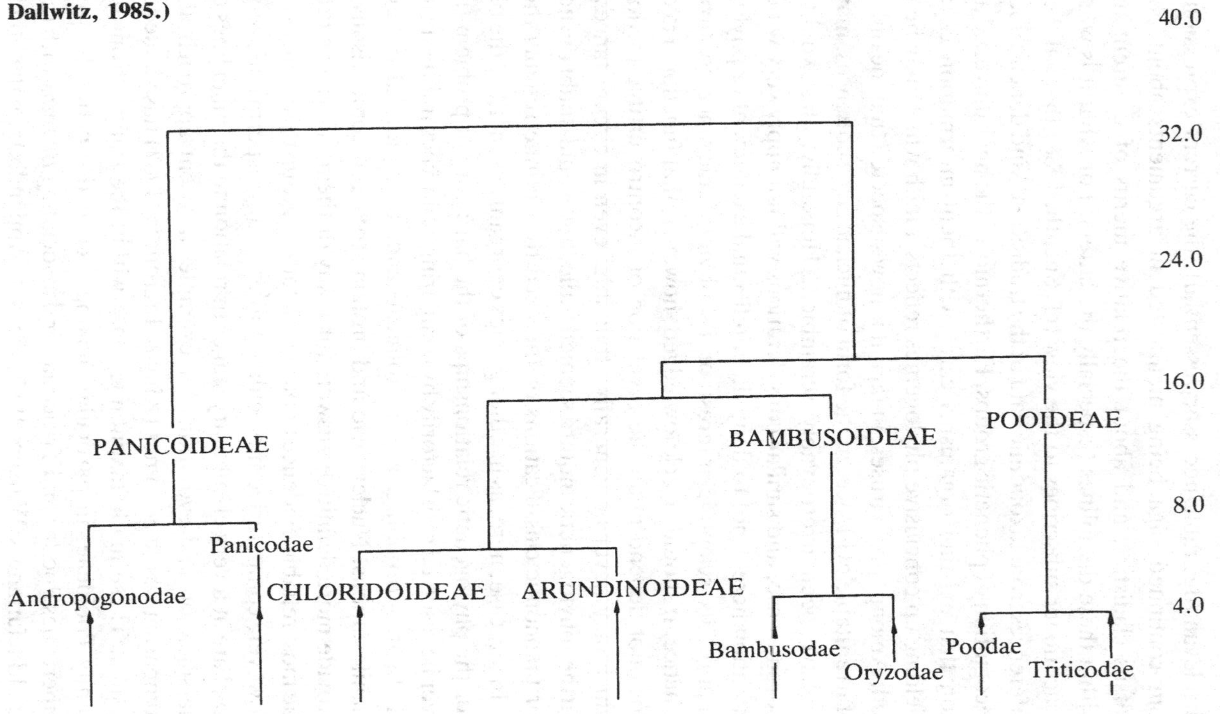

acid organization and the methodology of comparison than *vice versa*. It has been written that 'so-called single-copy DNA can be expected to be composed of innumerable scrambled permutations of repetitive motifs ... which are not separable using standard "single-copy" isolation techniques', and that 'continual fine-scale restructuring of the genome by slippage-like mechanisms and RNA-mediated intermingling of repeats represents a greater source of variation ... than ... point mutations' (Dover, 1987; Joysey, 1988). Interpretive problems associated with repeated DNA could well be significant in the present context, since the Pooideae are characterized as a group among the other grass subfamilies by relatively high per genome DNA values, and among the Pooideae the Triticodae (including *Bromus*) exhibit some of the highest values of all (see Table 1.5). In another recent, near-molecular attempt to address grass taxonomy, Esen & Hilu (1989) have returned (cf. Fairbrothers & Johnson, 1961) to serological comparisons of grass seed proteins, applying the up-to-date techniques of enzyme-linked immunosorbent assays and computer analyses of data to prolamin fractions. They report that the results showed no or few discernible differences between species of *Triticum* or within the tribes Triticeae, Poeae, Chlorideae, Paniceae and Andropogoneae, and they acknowledged distinctions between subfamilies. However, the results are further represented as showing, *inter alia*, lack of support for dividing the Pooideae into the supertribes Triticodae and Poodae and for associating *Bromus* with the latter (by contrast with King & Ingrouille, see above); as indicating close affinity between Chloridoideae and Panicoideae (by contrast, for example, with Fig. 1.1); and as supporting the view that the Oryzeae merit a subfamily separate from the Bambusoideae (though they clustered with the bamboo *Dendrocalamus*, cf. Fig. 1.1). Given the relatively small sample of grasses encompassed (19 species, with only 11 used for antiserum production and in reciprocal assays for taxonomic analyses), however, and taking note of quite serious discrepancies in the alternative hierarchies computed, as well as the alarming failure to persuade *Panicum* to cluster with *Pennisetum* rather than *Tripsacum*, the results in relation to these more subtle questions of grass classification and evolution, though interesting, can scarcely be taken as definitive. Evidence from direct nucleic acid and protein amino-acid sequencing is starting to accumulate (Appels, Scholes & Chapman 1987; Martin & Dowd, 1987; Hamby & Zimmer, 1988), and will presumably be more convincingly interpretable. The phylogeny for Poaceae computed by Hamby & Zimmer from ribosomal RNA sequences, for example, is remarkably sensible considering that it represents only nine genera. In fact, it has proved almost exactly reproducible (Watson, unpublished) via analyses of the same minute sample of genera using the full range of descriptive information in Watson's database, right down to the (unacceptable) clustering of *Hordeum* with *Avena* rather than with *Triticum*. It is clear that acceptable taxonomic and phylogenetic conclusions regarding genuinely contentious matters will require extensive, very laborious and time-consuming sampling across this enormous family, and congruent results from ranges of genes deemed adequate to represent whole genomes.

Fortunately, so far as the consumers of taxonomic output are concerned, the method whereby a classification was produced and the philosophy underlying it will be largely irrelevant, since its worth will depend entirely on its usefulness, and especially on its capacity for generating accurate predictions. There is no way of directly testing the truth of an evolutionary hypothesis, so the same criterion has to suffice even in that context. It is to be hoped that as the data available for classifying increase in quantity and improve in quality, the classificatory implications of phenetic and phylogenetic analyses (including molecular studies) will tend to converge. As it happens, the Poaceae provide a large-scale, comforting pointer to the likelihood of this.

Modern classification of grasses: genera, subfamilies, supertribes and tribes

The majority of the genera recognized in the Appendix are certainly relatively homogeneous, although there remain major problems of generic delimitation, notably in association with *Panicum*, *Ergagrostis*, *Stipa*, and in the Danthonieae, Bambuseae and Triticeae. Description of new species continues at a steady rate, and new genera are proposed every year. Many of the latter suffer the indignity of early reduction to synonymy: the 765 generic names listed in the Appendix are associated with 825 generic synonyms, only a small proportion of which merit recognition even at infra-generic level. Recent discoveries of forms which are undoubtedly sufficiently peculiar to merit new tribes (e.g. *Steyermarkochloeae*: Davidse & Ellis, 1984, 1987) show that the primary taxonomic cataloguing of grasses is far from complete. Nevertheless, the information available for classifying known grasses (or conversely, for which grass classifiers have to account) is better and more extensive than exists for any other large plant group. The difference in this respect from legumes (a family of comparable size), for example, is readily appreciated by comparing two excellent symposium publications (Polhill & Raven, 1981; Söderstrom *et al.*, 1987). In particular, there is an impressive body of comparative anatomical and physiological information on grasses which is unapproached elsewhere.

All classifications based upon comparative morphology are plagued by imponderables over character state definitions and homologies. In the case of Poaceae, this is particularly well illustrated by the controversial and sometimes almost metaphysical questions of spikelet structure (recently reviewed in detail: Clifford, 1987) which have led in the past to wildly divergent taxonomic assignments of the olyroid grasses, and of the Oryzeae. Plant taxonomy everywhere inevitably has a large morphological component; but in the case of Poaceae, the conclusions from comparative morphological studies have been very extensively cross-checked, reassessed and often radically revised with reference to data on relatively inaccessible but less ambiguous features. This process, which commenced about 50 years ago (cf. Prat, 1960), increasingly revealed deficiencies in the standard classification of Hackel (1887), which presented the family under two subfamilies defined mainly on gross morphology of inflorescences and spikelets. It is hard to distinguish many of the grasses now known as the 'chloridoids' from the largely north temperate pooids in terms of spikelet and inflorescence form, and the older scheme presented the former piecemeal among the latter. Variation in karyology was found to correlate with spectacular differences in leaf anatomy, and these associations were subsequently seen to be in accord with differences in embryo structure, photosynthetic pathways and even with patterns of susceptibilities to viruses and fungal pathogens, all fully supporting recognition of the Chloridoideae at subfamily level. The significance of this particular change, in terms of numbers of species and genera involved, is apparent in the Appendix. Similar considerations led to the discovery of other coherent groups, and to proposals for additional subfamiles, and between 1955 and 1962, a plethora of new systems was proposed in which the number of subfamilies ranged from four (Beetle, 1955) through five (Takeoka, 1957), six (Parodi, 1961; Stebbins & Crampton 1961), seven (Prat, 1960), and nine (Pilger, 1954) to twelve (Jacques-Félix, 1962). More recently, Hubbard (1966) distributed the tribes among 19 major groups apparently equivalent to subfamilies, while Tsvelev (1976), by contrast, presented only two subfamilies with informal subordinate groups. During the past 25 years and up to 1985, however, the number of subfamilies recognized tended to stabilize between five (Anon., 1980) and seven (Dahlgren *et al.*, 1985). In the course of all this activity, and in the absence of a comprehensive, up-to-date classification on which to draw, the editors of many floras opted to follow Hitchcock &

Chase (1950) and Bor (1960) in continuing to present the Poaceae under the nineteenth-century scheme of Hackel, or in preserving some of its most misleading aspects, thus denying flora users the benefits of current taxonomic knowledge. The situation has recently been rectified, with publication of a comprehensive classification of the genera into superfamilies and supertribes, with detailed group descriptions and methodological discussions (Watson, Clifford & Dallwitz, 1985); a book with a complete classification to subtribe level, and short descriptions of all the genera (Clayton & Renvoize, 1986); and the aforementioned Symposium publication (Söderstrom *et al.*, 1987), with authoritative discussions of family relationships of the Poaceae, a cladistic 'phylogenetic analysis' of the family, delimitations of subfamilies and generic classifications within them, and taxonomically orientated reviews of a range of other topics, including variation in photosynthetic pathways and in breeding systems. Automated descriptions of all the genera, incorporating comparative data on morphology, anatomy, physiology, cytology and economic aspects, as well as generic synonyms, ecology, geographical distributions, records of rusts and smuts, and classificatory assignments to subfamilies, supertribes and tribes, are now available on floppy disks in the form of an interactive information retrieval and identification system operable from microcomputers (Watson & Dallwitz, 1988). The latter, which is continually being updated and extended, is the direct source of the summarized classification and species counts given in the Appendix and of the data summarized in Tables 1.1 to 1.5 as well as of other figures quoted here, and is furnished with a substantial list of source references. Up-to-date information on grass classification, detailed group decscriptions and extensive data for conducting reassessments and new analyses are now readily available to all who want them. It seems sufficient here, therefore, to present a summarized classification and to discuss its usefulness as a general-purpose guide to sampling, drawing particular attention to salient remaining areas of taxonomic uncertainty.

The Appendix summarizes an extension to tribal level (to subtribal level in the case of Andropogoneae) of the classification detailed by Watson *et al.* (1985), and as given here, both the formal groupings and the sequences in which they appear represent attempts to portray overall similarities and differences between genera in a manner useful for practical purposes. This classification represents extensive and continuing taxonomic analyses of a large set of descriptive data, employing a variety of mainly phenetic numerical methods. Fig. 1.1 exemplifies the clusters and the pattern of relationships among high-level groups, which such analyses tend to depict, the relative positions of the subordinate groupings in adjoining higher-level groups being ordered horizontally as suggested by group ordinations (e.g. with Chloridoideae adjacent to Panicoideae, Arundinoideae adjacent to Bambusoideae). The strategy of this particular analysis enhances the intensity of clustering, and oversimplifies the situation by exaggerating the discreteness of groups. The result is not claimed to be a phylogenetic tree, and while the named groupings are for the most part presented with some confidence, neither the vertical scale nor details of the pattern of fusions in any one analysis of this kind should be regarded as definitive: practical taxonomic implementation of such work involves conducting a range of analyses, applying a variety of classificatory strategies to alternative interpretations of the descriptive data, and attempting to steer a sensible course between conflicting results. By contrast with this kind of product, the classification of Clayton & Renvoize (1986) is avowedly 'phylogenetic'. The difference in approach seems rather fundamental, but allowing for the possibilities which exist for translating similar views on relationships into superficially different formal schemes, the resulting classifications are remarkably similar. Thus, the five subfamilies Pooideae, Bambusoideae, Arundinoideae, Chloridoideae and Panicoideae are the same in name and, for the most part, in content. There is close correspondence regarding tribal delimitations; and the supertribes of Watson *et al.* (see Appendix) are detectable in sequences of genera and tribes as presented by Clayton & Renvoize (Bromeae adjoining and *Brachypodium* within the Triticeae; herbaceous bambusoids together; the eupanicoids and andropogonoids constituting well-separated groups within the Panicoideae, with the Arundinelleae between them). The methodological exposition of Clayton & Renvoize is likely to persuade cladists that their classification has rather dubious phylogenetic credentials. However, the overtly cladistic analyses of Kellogg & Campbell (1987) have produced an extensively compatible phylogenetic interpretation; and further confirmation of overall agreement among grass taxonomists will be seen in comparisons between Fig. 1.1 and recent treatments of the subfamilies by specialists (Conert, 1987; Jacobs, 1987; Macfarlane, 1987). In short, there is at present something approaching consensus regarding the main outlines of grass classification, of which the Appendix is reasonably representative for practical purposes.

The classification summarized there assigns all the genera to a subfamily, a tribe and (where appropriate) to a supertribe, and indicates the large genera. The supertribes of Pooideae probably represent the most useful subdivision in that subfamily where, with the exceptions of Triticeae, Bromeae and Brachypodieae, satisfactory tribes have proved hard to define. Those in the Panicoideae acknowledge a rather clear distinction between the eu-panicoid and andropogonoid tribes, which is to some extent blurred by the Arundinelleae (a mostly homogeneous tribe showing an interesting mixture of eu-panicoid, andropogonoid and danthonioid features). Presentation of two subtribes within the Andropogoneae acknowledges the existence of a rather strong distinction among the members of that large and important tribe.

Naturally, there are some genera whose position is uncertain, through lack of information (e.g. *Euthryptochloa*, *Chandrasekharania*, *Danthoniastrum*) or because they exhibit genuinely intermediate or highly peculiar features (e.g. *Steyermarkochloa*, *Sartidia*, *Brachyelytrum*, *Ehrharta*, *Lygeum*, *Nardus*). A greater concern is that at subfamily level the Arundinoideae constitute an unsatisfactory assemblage of convenience, which is not amenable to anything approaching a diagnostic description and is probably polyphyletic (see Watson *et al.*, 1985, and Kellogg & Campbell, 1987, for detailed analyses counched in phenetic and phylogenetic terms, respectively). The individual arundinoid tribes are coherent and useful, but their relationships to one another and to the other subfamilies remain controversial. Kellogg & Campbell interpret their results as showing that 'the Pooideae, Bambusoideae, Panicoideae and perhaps the Chloridoideae are monophyletic and are derived from a highly polyphyletic Arundinoideae'. Elsewhere in the system, Söderstrom & Ellis (1987) have excluded from the Bambusoideae a number of isolated genera and small groups presented here among the herbaceous bambusoids (supertribe Oryzodae) without, however, suggesting alternative locations: i.e. *Brachyelytrum*, *Diarrhena*, *Phaenosperma*, the Ehrharteae and the Centotheceae. The latter are the least convincingly bambusoid of all these: they are sometimes presented as a separate subfamily (Söderstrom, 1981; Clayton & Renvoize, 1986), but they might be better transferred to the rag-bag of the Arundinoideae. In any event, only about 12% of grass species fall into these areas of uncertainty at the highest hierarchical level, and most of them are contributed by the constellations around *Aristida*, *Stipa* and *Danthonia*. Regarding the tribes as portrayed in the Appendix, the most unsatisfactory aspect is the *sensu lato* version of the Chlorideae. It is evidently desirable that this unwieldy assemblage be effectively broken down, but the time-honoured division into two weakly distinguished large tribes (Cynodonteae or Chlorideae *sensu stricto* and Eragrosteae) and a number of small tribes of dubious worth seems hardly to be recommended.

Persons wanting a more detailed breakdown of this group are referred to Clayton & Renvoize (1986), who present many sequences of convincingly related genera as subtribes.

Detailed generic and higher group descriptions are available from the automated database mentioned earlier, and on associated microfiches (see also Watson *et al.*, 1985; Clayton & Renvoize, 1986; Watson, Dallwitz & Johnson, 1986). Taxonomic bias at subtribal, tribal, supertribal or superfamily level is shown by most of the 350 morphological and anatomical characters represented there, and is also evident in the compiled data on physiology, cytology, ecology, geography, and for host ranges of rusts, smuts and viruses. It is apparent even in terms of quantitative DNA estimations, exemplified by the diploid 2*c* values in Table 1.5 (the relatively scanty data for tetraploids, hexaploids and octoploids show the same trend, with values for a few *polyploid* chloridoids and panicoids lower than those of many *diploid* Pooideae; cf. Appels *et al.*, 1987). The difference between Pooideae and the other subfamilies in this respect is evidently related to possession of large versus small chromosomes, an aspect of variation first noted in an agrostological context by Avdulov in 1931. Information on the more esoteric and inaccessible features naturally represents extrapolation from small samples, the extent of which has been discussed elsewhere (Watson, 1987) and is also ascertainable from the automated database. The brief descriptions of the larger groups given in the Appendix are intended not only to indicate how they are distinguished, but also to illustrate the kinds of features upon which the taxonomic arrangement imposes patterns, some of which are directly related to the subject matter of the present publication. These include many details of spikelet and infloresence organization, where grass evolution has developed and elaborated on the multitude of themes exploited by classical taxonomists, while conserving a relatively uniform flower.

A notable recurrent tendency is towards increasing the preponderance of male structures and pollen over female structures and ovules, achieved by restricting proximal florets and/or complete spikelets to the production of stamens. It is tempting to envisage this as bestowing advantages in relation to anemophily, but by contrast with the Panicoideae, the equally successful Pooideae have largely ignored it (nothwithstanding that proximal male-only florets fall within the capabilities of the pooid genotype: witness *Hierochloë*). Three of the features included in Tables 1.1 and 1.2 concern the related tendency for grasses to separate their male and

Table 1.1. *Grass genera known to exhibit inbreeding (autogamy), cleistogamy, apomixis, hidden cleistogenes, subterranean cleistogenes (rhizanthogenes), fertile florets all unisexual, monoecism (fertile spikelets unisexual, males and females on the same plant), and dioecism (separate male and female-fertile plants)*

Note that the presence of the first five of these features is often variable in the genera listed as exhibiting them.

Inbreeding:
Aira Australopyrum Avena Bouteloua Brachypodium Briza Bromus Catapodium Chionochloa Chloris Cockaynea Cortaderia Danthonia Deschampsia Deyeuxia Dichelachne Digitaria Echinochloa Eleusine Elymus Eragrostis Festuca Glyceria Hordeum Hubbardochloa Hyparrhenia Koeleria Lagurus Lolium Microlaena Oryza Panicum Paspalum Phalaris Phleum Poa Polypogon Puccinellia Rottboellia Rytidosperma Secale Setaria Sieglingia Sorghum Sporobolus Stipa Thinopyrum Trisetum Triticum Vulpia Zizania

Cleistogamy:
Aciachne Acrachne Agrostis Amphibromus Amphicarpum Aniselytron Arctagrostis Aristida Astrebla Avena Bothriochloa Bouteloua Brachyachne Briza Bromus Calamagrostis Calosteca Calyptochloa Catapodium Chasmanthium Chikusichloa Chloris Cleistochloa Coelorhachis Cottea Dactyloctenium Danthonia Deschampsia Diandrochloa Dichanthelium Dichanthium Dichelachne Digitaria Dimorphochloa Diplachne Echinochloa Ectrosia Eleusine Enneapogon Eragrostis Eriachne Erianthecium Erianthus Festuca Garnotia Gymnachne Gymnopogon Habrochloa Helictotrichon Hemarthria Heterachne Hordeum Hubbardochloa Hypseochloa Kengia Leersia Leptochloa Lombardochloa Melica Microbriza Microchloa Microlaena Microstegium Muhlenbergia Nassella Oryza Oryzopsis Panicum Pappophorum Paspalum Pheidochloa Phippsia Piptochaetium Poa Poidium Puccinellia Relchela Rhomboelytrum Rottboellia Rytidosperma Schizachyrium Secale Setaria Sieglingia Sorghum Spartina Spathia Sporobolus Stipa Tetrapogon Thellungia Thyridolepis Tridens Triplasis Trisetum Uniola Vulpia Willkommia

Apomixis:
Anthephora Apluda Bothriochloa Bouteloua Brachiaria Calamagrostis Capillipedium Cenchrus Chloris Coix Cortaderia Dichanthelium Elymus Eragrostis Eriochloa Fingerhuthia Heteropogon Hierochloë Hilaria Hyparrhenia Lamprothyrsus Nardus Panicum Paspalum Pennisetum Poa Saccharum Setaria Sorghum Themeda Tricholaena Tripsacum Urochloa

Hidden cleistogenes:
Amphibromus Amphicarpum Andropogon Aristida Calyptochloa Chloris Cleistochloa Cottea Danthonia Dichanthelium Digitaria Diplachne Enneapogon Garnotia Humbertochloa Kengia Leersia Libyella Microlaena Muhlenbergia Pappophorum Paratheria Pariana Paspalum Pennisetum Phippsia Rytidosperma Sieglingia Stipa Triplasis

Subterranean cleistogenes:
Amphicarpum Chloris Libyella Pariana Paspalum

Table 1.1. (*cont.*)

Only unisexual florets:
Allolepis Apocopis Arberella Arrhenatherum Arundoclaytonia Buchloë Buchlomimus Buergersiochloa Calderonella Cathestechum Chamaeraphis Chionachne Chumsriella Coix Cortaderia Cryptochloa Cyclostachya Diandrolyra Distichlis Ekmanochloa Euchlaena Germainia Griffithsochloa Gynerium Heteropogon Humbertochloa Hydrochloa Hygrochloa Hypogynium Isachne Iseilema Ixophorus Jouvea Lamprothyrsus Leptaspis Lepturidium Limnopoa Lithachne Luziola Maclurolyra Mniochloa Monanthochloë Neeragrostis Olyra Opizia Oryzidium Otachyrium Pariana Parodiolyra Pharus Phyllorhachis Piresia Poa Pohlidium Polytoca Pringleochloa Pseudochaetochloa Pseudoraphis Puelia Raddia Raddiella Reederochloa Rehia Reitzia Sclerachne Scleropogon Scrotochloa Soderstromia Sohnsia Spinifex Sucrea Thuarea Trilobachne Triplopogon Tripsacum Xerochloa Zea Zeugites Zizania Zizaniopsis Zygochloa

Monoecious:
Arberella Arundoclaytonia Buchloë Buergersiochloa Chionachne Chumsriella Coix Cryptochloa Diandrolyra Distichlis Ekmanochloa Euchlaena Germainia Humbertochloa Hydrochloa Hygrochloa Hypogynium Jouvea Leptaspis Lithachne Luziola Maclurolyra Mniochloa Olyra Pariana Parodiolyra Pharus Phyllorhachis Piresia Pohlidium Polytoca Pringleochloa Raddia Raddiella Rehia Reitzia Sclerachne Scleropogon Scrotochloa Soderstromia Sucrea Trilobachne Tripsacum Zea Zizania Zizaniopsis

Dioecious:
Allolepis Buchloë Buchlomimus Cortaderia Cyclostachya Distichlis Gynerium Jouvea Lamprothyrsus Lepturidium Monanthochloë Neeragrostis Opizia Poa Pseudochaetochloa Reederochloa Scleropogon Sohnsia Spinifex Zygochloa

Extracted from Watson's automated generic descriptions (1989: cf. Watson & Dallwitz, 1988); main primary sources for features 1–5 Connor (1979, 1981, 1987) and Campbell *et al.* (1983).

female components, which is represented there by comprehensive figures for genera including species where:

1 the fertile *florets* are exclusively unisexual;
2 the plants are monoecious, with unisexual *spikelets* on the same plant; and
3 the plants are dioecious, with separate male and female-fertile individuals.

Category 1 is not exclusive of the other two, but includes in addition forms where the individual spikelets bear both male and female florets, but no hermaphrodites. Table 1.2 illustrates how this kind of reproductive specialization, though occurring in relatively few genera, is very widespread in the family. It also emphasizes, however, that it is largely a non-pooid tendency, the only exceptions being a few species of *Poa* from extreme environments, and *Arrhenatherum* where the upper floret is

Table 1.2. *Taxonomic distribution, by subfamilies and supertribes, of the grass breeding system features recorded generically in Table 1.1. Classification as in the Appendix*

	Pooideae		Bambusoideae				Panicoideae	
Feature	Triticodae	Poodae	Oryzodae	Bambusodae	Arundinoideae	Chloridoideae	Panicodae	Andropogonodae
Inbreeding	9	19	3	0	6	6	5	3
Cleistogamy	3	27	5	0	12	29	12	10
Apomixis	1	3	0	0	3	5	11	10
Hidden cleistogenes	0	3	4	0	5	8	9	1
Subterranean cleistogenes	0	1	1	0	0	1	2	0
Unisexual florets	0	2	29	1	4	16	13	15
Monoecism	0	0	28	0	1	6	1	11
Dioecism	0	1	0	0	3	13	3	0

Table 1.3. *World geographical distributions of grass genera which include species having only unisexual florets, and those which include monoecious and/or dioecious species (cf. Table 1.1)*

'Eastern Asia' = Japan, China to India; 'Africa' includes Saudi Arabia and Madagascar; 'Pacific' = Malaysia, Indonesia, Australasia and Pacific Islands; 'North America' = Canada, Alaska, USA and Mexico.

	W. Eurasia, USSR	Mediterranean	Eastern Asia	Africa	Pacific	North America	S. and C. America, West Indies
Total genera	191	165	299	369	284	220	264
% unisexual florets	2.6	2.4	5.7	3.5	8.8	13.6	16.0
% monoecious and/or dioecious	1.6	1.2	3.3	1.9	5.6	10.9	14.0

Data from Watson's automated generic descriptions (1989: cf. Watson & Dallwitz, 1988).

occasionally female-only (*Hierochloë* has hermaphrodite florets as well as proximal males, and is exluded here). The woody bamboos (Bambusoideae–Bambusodae) are relatively unenterprising in this connection, furnishing no examples of monoecism or dioecism, and only *Puelia* (a monogeneric tribe from Africa) with exclusively unisexual florets.

Table 1.3 draws attention to a curious fact of geographical distribution associated with these features of reproductive organization, for which there is no obvious adaptive, phylogenetic or taxonomic explanation; namely, the conspicuous concentration of unisexual florets, monoecism and dioecism in the Americas, by contrast with their scarcity in Eurasia and Africa. The trend is expressed regardless of subfamily: for example, there are almost twice as many genera of Panicoideae in Africa (144) as in the Americas (90), but only 5 of them (as opposed to 9 from the Americas) lack hermaphrodite florets; more impressively, there are more Chloridoideae in Africa (86) than in the Americas (72), yet 16 of the unisexual-floreted, monoecious and dioecious genera come from the Americas (14 being represented in North America and Mexico, 8 in South America), as opposed to none from Africa. Naturally, the figures reflect the distributions of some tribes and lower-level taxonomic assemblages in which the tendency to separate the sexes is seen most strongly, in particular the exclusively monoecious Olyreae and Maydeae and the commonly dioecious relatives of *Bouteloua*; but given the capacity of all the subfamily genotypes to follow this particular trend, it is not at all obvious why they should have done so on a geographical basis, rather than a climatic or ecological one. Nor is it easy to envisage appropriate climatic or ecological factors which might be operative, in connection with Poaceae, on a continental scale. Significantly, the trend is somewhat obscured in figures extracted from the database in terms of Takhtajan's *floristic Kingdoms* (Takhtajan, 1969), though they re-emerge in comparisons conducted at lower hierarchical levels in his floristic-regional classification. The most commonly invoked source of geographically orientated selection pressures in connection with floral and inflorescence organization, in the form of animal pollinators, is of course inapplicable to this almost exclusively wind-pollinated family. For further spectacular examples of angiosperm structural features posing similar problems, and an imaginative discussion, the reader is referred to a paper by Went (1971).

Information on inbreeding, cleistogamy and apomixis, where the data are inevitably very incomplete, is also summarized at generic level in Table 1.1 and in terms of major taxonomic groups in Table 1.2. Absence of records for cleistogamy (the most widely documented of these) from the tribes Bambuseae, Arundineae, Orcuttieae (by contrast with Pappophoreae) and Isachneae may reflect genuine taxonomic bias in the propensity to produce them; but it is clear from the figures that all three phenomena are very widespread accomplishments of the grasses as a family. 'Hidden cleistogenes' involve production, in addition to the 'normal' inflorescences and spikelets, of more or less modified versions which are usually to be found in the leaf axils and which probably have survival value in relation to fire and drought (Campbell *et al.*, 1983). The low figures of Pooideae here are probably a genuine indicator of relative scarcity in that comparatively well-studied group; but Table 1.2 shows that hidden cleistogenes – and even rhizanthogenes, their rare and highly specialized subterranean manifestation – are also represented across the system. Clearly, none of these features of grass breeding systems contributes much towards helping taxonomists recognize useful groups, hence their omission from the Appendix. Rather, with the apparent exception of the woody bamboos, they are significant components of the evolutionary and ecological versatility of the Poaceae as a whole. They are also of singular interest to plant breeders, as is the near universal capacity of grasses to generate viable polyploids (a feature which is noticeably unusual, however, in the Bambusoideae–Oryzodae).

Hybridization is another topic of particular interest to plant breeders, and contributes important information to taxonomy. It is a pity that no agricultural or agronomic institution has yet recognized the need to organize and maintain a comprehensive database of authenticated grass hybrids, summarizing the evidence available and carrying references. The compilation by Knobloch (1968) is a valuable source of references up to 1967, but it needs updating and in any case does not provide information for critical assessment of records. Table 1.4 lists the brief information on *intergeneric* hybrids currently summarized in Watson's automated generic descriptions, which are accompanied by a list of source references. Here, an attempt has been made to include all convincing cases of hybridization, as well as all formally described hybrid genera. The data are no doubt incomplete. So far as they go, however, they are nearly all compatible with taxonomy (cf. Appendix), being almost entirely concerned with genera from the same tribe, and often with those which are sometimes and with some justification reduced to synonymy (e.g. around *Saccharum*; but see *Oryza*). In contemplating the numerous hybrids involving genera of the Triticeae (e.g. *Elytrigia*, *Triticum*), it should be borne in mind that several of the genera themselves have well-documented hybrid origins. As a result of this demonstrated (presumably relatively recent) reticulate evolution, satisfactory generic definitions have proved very elusive indeed, and the Triticeae are recognized as posing very serious difficulties for phylogenetic reconstruction and taxonomic interpretation (see Löve, 1984; Baum, Estes & Gupta, 1987; Jauhar & Crane, 1989; Kellogg, 1989). They have been studied far more intensively in cytogenetic terms than any other sizeable group of grasses (or any other group of plants, for that matter); and it would be unwarranted to assume that such complications do not apply in other grass groups which have had less attention lavished upon them at all levels, and have as yet been subjected to no 'genomic analysis' at all.

Of the 765 genera listed in the Appendix, 390 are monotypic or ditypic; 23 have about 100 or more species, and contain almost half of all grass species; and 31% of the species are concentrated in the 10 largest genera (*Agrostis*, *Aristida*, *Calamagrostis*, *Digitaria*, *Eragrostis*, *Festuca*, *Panicum*, *Paspalum*, *Poa* and *Stipa*), each with 200 or more species. Inspection of the classification will show that species numbers are fairly evenly distributed across the subfamilies and large tribes, which have distributions that are broadly predictable in climatic and latitudinal terms.

Table 1.4. *Summarized data on intergeneric hybrids*

Genus	Genera with which hybrids formed	Notes/references
Aegilops L.	*Triticum* (× *Aegilotriticum* Wagner ex Tschermak) *Secale* (× *Aegilosecale* Ciferri & Giacom.) *Dasypyrum* *Elytrigia*	
Agropyron Gaertn.	*Hordeum* (× *Agrohordeum* A. Camus) *Leymus* (× *Leymopyron* Tsvelev) *Elytrigia* (× *Agrotrigia* Tsvelev) *Secale* *Triticum* (× *Agrotrisecale* Ciferri & Giacom., × *Agrotriticum* Ciferri & Giacom.) *Sitanion* (× *Agrositanion* Bowden)	See also × *Agroelymus* A. Camus
Agrostis L.	*Polypogon* (× *Agropogon* P. Fourn.) *Calamagrostis*	
Ammophila Host	*A. arenaria* with *Calamagrostis epigejos* (× *Ammocalamagrostis* P. Fourn.)	
Arctophila Rupr. ex Andersson	*Dupontia* (× *Arctodupontia* Tsvelev)	
Arrhenatherum P. Beauv.	*Avena*	
Avena L.	*Arrhenatherum*	
Bambusa Schreber		Claimed hybrids with *Saccharum* represent apomixis
Bothriochloa Kuntze	*Capillipedium* *Dichanthium*	
Bromus L.	? *Festuca* (× *Bromofestuca* Prodan)	*Bull. Grad. Bot. Univ. Cluj.* **16**, 93 (1936)

Calamagrostis Adans	*Agrostis* *C. epigejos* hybridizes with *Ammophila arenaria* (× *Ammocalamagrostis* P. Fourn.)	See also × *Calamophila* O. Schwartz
Capillipedium Stapf	*Bothriochloa*	
Chloris O. Schwartz	*Cynodon* (× *Cynochloris* Clifford & Everist)	Several species involved
Cynodon Rich.	*Chloris* (× *Cynochloris* Clifford & Everist)	Several species involved
Danthonia DC.	*Sieglingia* (× *Danthosieglingia* Domin.)	
Dasypyrum (Cosson & Durieu) Durand	*Aegilops*	
Dichanthium Willem.	*Bothriochloa*	
Dupontia R. Br.	*Arctophila* (× *Arctodupontia* Tsvelev)	
Elymus L.	*Sitanion* *Triticum*	
Elytrigia Desv.	*Agropyron* (× *Agrotrigia* Tsvelev) *Hordeum* (× *Elytrohordeum* Hylander) *Aegilops* *Leymus* (× *Leymotrigia* Tsvelev) *Lophopyrum* *Secale* *Triticum* (× *Trititrigia* Tsvelev) *Thinopyrum*	
Erianthus Michx.	*Saccharum*	
Euchlaena Schrad.	*Zea* (× *Euchlaezea* Janaki ex Bor)	
Festuca L.	*Vulpia* (× *Festulpia* Melderis ex Stace & Cotton) *Lolium* (× *Festulolium* Aschers. & Graebn.) ? *Bromus* (× *Bromofestuca* Prodan)	

Table 1.4 (*cont.*)

Genus	Genera with which hybrids formed	Notes/references
Hordeum L.	*Elytrigia* (× *Elytrohordeum* Hylander)	See also × *Elyhordeum* Zizan & Petrowa
	Agropyron (× *Agrohordeum* A. Camus)	
	Secale (× *Hordale* Ciferri & Giacom.)	
	Sitanion (× *Sitordeum* Bowden)	
	Triticum (× *Tritordeum* Aschers. & Graebn.)	
Imperata Cirillo	*Saccharum*	
Koeleria Pers.	*Trisetum* (× *Trisetokoeleria* Tsvelev)	
Lepturopetium Morat		All possible hybrids involving *Lepturus* as one parent
Leymus Hochst	*Agropyron* (× *Leymopyron* Tsvelev)	See also × *Elyelymus* Baum.
	Elytrigia (× *Leymotrigia* Tsvelev)	
	Psathyrostachys (× *Leymostachys* Tsvelev)	
	Thinopyrum (several species involved)	
Lolium L.	*Festuca* (× *Festulolium* Anschers. & Graebn.)	Several species of each genus involved
Lophopyrum A. Löve	*Elytrigia*	
	Leymus	
	Sitanion	
	Triticum	
Miscanthidium Stapf	*Saccharum*	
Miscanthus Anderss.	*Saccharum*	
Narenga Bor	*Saccharum*	
	Sclerostachya	
Oryza L.	? *Triticum* (× *Oryticum* Wang & Tang)	*Acta Phytotax. Sin.* **20**, 179 (1982)
Oryzopsis Michx.	*Stipa* (× *Stiporyzopsis* B. L. Johnson & Rogler)	

Phippsia R. Br.	*Puccinellia* (× *Pucciphippsia* Tsvelev)	
Polypogon Desf.	*Agrostis* (× *Agropogon* P. Fourn.)	
Psathyrostachys Nevski	*Leymus* (× *Leymostachys* Tsvelev)	See also × *Elymostachys* Tsvelev
Puccinellia Parl.	*Phippsia* (× *Pucciphippsia* Tsvelev)	
Saccharum L.	*Erianthus* *Imperata* *Miscanthidium* *Miscanthus* *Narenga* *Sclerostachya* *Sorghum*	A claim involving *Zea* is probably erroneous, and one involving *Bambusa* has been proven so
Sasa Makino & Shibata	? *Semiarundinaria* (× *Hibanobambusa* Maruyama & Okamura)	
Sclerostachya A. Camus	*Narenga* *Saccharum*	
Secale L.	*Triticum* (× *Triticosecale* Wittmack) *Agropyron* *Aegilops* (× *Aegilosecale* Ciferri & Giacom.) *Elytrigia*	× *Agrotrisecale* Ciferri & Giacom. = *Agropyron* × *Secale* × *Triticum*
Semiarundinaria Makino	? *Sasa* (× *Hibanobambusa* Maruyama & Okamura)	
Sieglingia Bernh.	*Danthonia* (× *Danthosieglingia* Domin)	
Sitanion Raf.	*Agropyron* (× *Agrositanion* Bowden) *Elymus* *Hordeum* (× *Sitordeum* Bowden) *Lophopyrum*	See also × *Elysitanion* Bowden

Table 1.4 (*cont.*)

Genus	Genera with which hybrids formed	Notes/references
Sorghum Moench	*Saccharum*	
Sphenopholis Scribner	*Trisetum*	
Stipa L.	*Oryzopsis* (× *Stiporyzopsis* B. L. Johnson & Rogler)	
Tarigidia Stent		Sharing features of *Digitaria* and *Anthephora*, originally suggested as an intergeneric hybrid – which accords with absence of fruit in material seen, and the scattered distribution
Thinopyrum A. Löve	*Leymus* *Elytrigia*	
Tripsacum L.	*Zea*	
Trisetum Pers.	*Koeleria* (× *Trisetokoeleria* Tsvelev) *Sphenopholis*	
Triticum L.	*Aegilops* (× *Aegilotriticum* Wagner ex Tschermak) *Agropyron* (× *Agrotriticum* Ciferri & Giacom.) *Elymus* *Elytrigia* (× *Trititrigia* Tsvelev)	See also × *Elymotriticum* P. Fourn., × *Agrotrisecale* Ciferri & Giacom. (*Agropyron* × *Secale* × *Triticum*), × *Oryticum* Wang & Tang (supposedly *Oryza* × *Triticum*)

	Hordeum (× *Tritordeum* Aschers. & Graebn.)	
	Lophopyrum	
	Secale (× *Triticosecale* Wittmack)	
Vulpia C. Gmelin	*Festuca* (× *Festulpia* Melderis ex Stace & R. Cotton)	Several species involved
Zea L.	*Euchlaena* (× *Euchlaezea* Janaki ex Bor)	
	Saccharum	
	Tripsacum	

From Watson's automated generic descriptions (1989: cf. Watson & Dallwitz, 1988).

Table 1.5. *Generic means and ranges of diploid 2c DNA values, with genera arranged by subfamilies and supertribes (cf. Appendix)*

Genus (arranged by subfamily and supertribe)	No. of species investigated	Range of diploid 2*c* DNA values (pg)	Generic mean 2*c* DNA value (pg)
Pooideae			
Triticodae			
Aegilops	11	7.2–14.3	11.2
Boissiera	1	—	3.7
Bromus	11	6.5–12.5	9.8
Eremopyrum	1	—	11.0
Hordeum	11	10.8–11.1	11.0
Lophopyrum	1	—	11.2
Secale	6	14.8–19.0	16.8
Taeniatherum	1	—	8.8
Triticum	3	9.8–13.8	12.2
Poodae			
Agrostis	1	—	6.9
Aira	2	—	6.0
Anthoxanthum	1	—	11.8
Avena	13	8.8–11.0	9.9
Briza	3	14.6–21.6	17.7
Catapodium	1	—	9.6
Corynephorus	1	—	2.3
Dactylis	1	—	9.8
Deschampsia	1	—	10.0
Festuca	9	3.4–9.5	6.5
Helictotrichon	1	—	10.1
Lolium	5	(4.3–)6.4–13.6	9.9
Mibora	1	—	5.5
Periballia	1	—	6.6
Phalaris	9	(2.8–)3.5–9.0	5.8
Phleum	1	—	3.4
Poa	3	2.4–5.6	3.6
Bambusoideae			
Oryzodae			
Oryza	10	1.1–2.5	1.7
Zizania	1	—	4.4
Chloridoideae			
Chloris	1	—	0.7
Eleusine	1	—	1.4
Panicoideae			
Cenchrus	1	Ploidy unknown	?2.6
Echinochloa	1	Ploidy unknown	?2.7
Paspalum	2	Ploidy unknown	1.6
Zea		4.4–11.0	5.2

As summarized in Watson's automated generic descriptions (1989: cf. Watson & Dallwitz, 1988); primary data from Bennett & Smith (1976); Bennett, Smith & Heslop-Harrison (1982); Grime, Shacklock & Band (1985).

Genera and species, on the other hand, tend to be confined to single land masses, though there is a significant proportion of disjunct genera, and the picture is everywhere confused by introductions reflecting human activities. These facts combine with fossil evidence to suggest that the family was in existence by the Late Cretaceous, and that the major tribes were already established before the continents became widely separated in the late Tertiary period (Clayton, 1981; Stebbins, 1987; Thomasson, 1987).

The relatively high concentration of experimental biologists in the North Temperate zone, combined with the facts of grass group distribution, has had the unfortunate effect of biasing their activities in favour of the Pooideae. Table 1.5 is a fairly typical example of such bias; it will be observed that the data for non-pooids are both fewer and poorer in quality (e.g. in the Paniceae, where low values could represent polyploids, and therefore be misleadingly high). Errors consequent on taxonomically unbalanced sampling have ranged from the relatively harmless textbook myth that Poaceae (as opposed to the subfamily Pooideae) are characterized by hollow culms, to more far-reaching misconceptions about selective herbicides and pollen allergens. Clearly, people trying to discover variation in grasses, or wanting to generalize about the family from minimal observations, should deliberately seek to maximize the diversity of their samples by arranging to include (say) a bamboo (or rice), wheat and maize. Likewise, those wishing to generalize about Pooideae, Bambusoideae or Panicoideae, respectively, should arrange sampling with reference to the supertribes and tribes. The best strategy when hoping to uncover taxonomic predictability is to start by bracketing known diversity as represented by the subfamilies, avoiding the Stipeae, Aristideae and other contentious taxa. Those desirous of contributing taxonomically useful feedback, on the other hand, might deliberately choose to include some of the latter, while covering enough of the rest to lend perspective. With luck, judicious initial sampling will help concentrate further work on those parts of the system most likely to repay the effort.

References

Appels, R., Scholes, G. & Chapman, G. D. (1987). The nature of change in nuclear DNA in the evolution of grasses. In *Grass Systematics and Evolution*, ed. T. R. Söderstrom, K. W. Hilu, C. S. Campbell & M. E. Barkworth. Smithsonian Institution Press: Washington DC.

Avdulov, N. P. (1931). Karyo-systematische Untersuchungen der Familie Gramineen. *Bull. Appl. Bot. Suppl.* **44**, 428 pp.

Baum, B. R., Estes, J. R. & Gupta, P. K. (1987). Assessment of the genomic system of classification in the Triticeae. *Amer. J. Bot.* **74**, 1388–95.

Beetle, A. A. (1955). The four subfamilies of the Gramineae. *Bull. Torrey Bot. Club*, **82**, 196–7.

Bennett, M. D. & Smith, J. B. (1976). Nuclear DNA amounts in angiosperms. *Phil. Trans. R. Soc. Lond. B*, **274**, 227–74.

Bennett, M. D., Smith, J. B. & Heslop-Harrison, J. S. (1982). Nuclear DNA amounts in angiosperms. *Phil. Trans. R. Soc. Lond. B*, **216**, 179–99.

Bor, N. L. (1960). *Grasses of Burma, Ceylon, India and Pakistan* (excluding Bambuseae). Pergamon Press: Oxford.

Brown, W. V. (1977). The Kranz syndrome and its subtypes in grass systematics. *Mem. Torrey Bot. Club*, **23**, 1–97, 126–30.

Campbell, C. S. & Kellogg, E. A. (1987). Sister group relationships of the Poaceae. In *Grass Systematics and Evolution*, ed. T. R. Söderstrom, K. W. Hilu, C. S. Campbell & M. E. Barkworth. Smithsonian Institution Press: Washington DC.

Campbell, S. C., Quinn, J. A., Cheplick, G. P. & Bell, T. J. (1983). Cleistogamy in grasses. *Ann. Rev. Ecol. Syst.* **14**, 411–41.

Clayton, W. D. (1981). Evolution and distribution of grasses. *Ann. Missouri Bot. Gard.* **68**, 5–14.

Clayton, W. D. & Renvoize, S. A. (1986). *Genera Graminum: grasses of the world. Kew Bull. Addit. Ser.* 13. Royal Botanic Gardens, Kew: London.

Clifford, H. T. (1987). Spikelet and floral morphology. In *Grass Systematics and Evolution*, ed. T. R. Söderstrom, K. W. Hilu, C. S. Campbell & M. E. Barkworth. Smithsonian Institution Press: Washington DC.

Conert, H. J. (1987). Current concepts in the systematics of the Arundinoideae. In *Grass Systematics and Evolution*, ed. T. R. Söderstrom, K. W. Hilu, C. S. Campbell & M. E. Barkworth. Smithsonian Institution Press: Washington DC.

Connor, H. E. (1979). Breeding systems in grasses: a survey. *N.Z. J. Bot.* **17**, 547–74.

(1981). Evolution of reproductive systems in the Gramineae. *Ann. Missouri Bot. Gard.* **68**, 48–74.

(1987). Reproductive biology in the grasses. In *Grass Systematics and Evolution*, ed. T. R. Söderstrom, K. W. Hilu, C. S. Campbell & M. E. Barkworth. Smithsonian Institution Press: Washington DC.

Cronquist, A. (1981). *An Integrated System of Classification of Flowering Plants*. 1262 pp. Columbia University Press: New York.

Dahlgren, R. M. T., Clifford, H. T. & Yeo, P. F. (1985). *The Families of the Monocotyledons*. 520 pp. Springer Verlag: Heidelberg.

Davidse, G. & Ellis, R. P. (1984). *Steyermarkochloa unifolia*, a new genus from Venezuela and Colombia (Poaceae: Arundinoideae: Steyermarkochloeae). *Ann. Missouri Bot. Gard.* **71**, 994–1012.

(1987). *Arundoclaytonia*, a new genus of the Steyermarkochloeae (Poaceae: Arundinoideae) from Brazil. *Ann. Missouri Bot. Gard.* **74**, 479–90.

Dover, G. A. (1987). DNA turnover and the molecular clock. *J. Mol. Evol.* **26**, 47–58.

Esen, A. & Hilu, K. W. (1989). Immunological affinities among subfamilies of the Poaceae. *Amer. J. Bot.* **76**, 196–203.

Fairbrothers, D. E. & Johnson, M. A. (1961). The precipitation reaction as indication of relationships in some grasses. *Rec. Adv. Bot.* **1**, 116–20.

Grime, J. P., Shacklock, J. M. L. & Band, S. R. (1985). Nuclear DNA contents, shoot phenology and species co-existence in a limestone grassland community. *New Phytol.* **100**, 435–45.

Hackel, E. (1887). Gramineae. In *Die Naturlichen Pflanzenfamilien*, ed. A. Engler & K. Prantl. Teil II, Abteilung 2. Engelmann: Leipzig.

Hamby, K. R. & Zimmer, E. A. (1988). Ribosomal RNA sequences for inferring phylogeny within the grass family (Poaceae). *Plant System. Evol.* **160**, 29–37.

Hattersley, P. W. (1987). Variations in photosynthetic pathway. In *Grass Systematics and Evolution*, ed. T. R. Söderstrom, K. W. Hilu, C. S. Campbell & M. E. Barkworth. Smithsonian Institution Press: Washington DC.

Hitchcock, A. S. & Chase, A. (1950). *Manual of the Grasses of the United States. U.S. Dept. of Agric., Misc. Publ.* No. 200. Govt Printing Office: Washington DC.

Hubbard, C. E. (1966). Gramineae. In *A Dictionary of Flowering Plants and Ferns*, ed. J. C. Willis. 8th edn, revised by H. K. Airy Shaw. Cambridge University Press.

Jacobs, S. W. L. (1987). Systematics of the chloridoid grasses. In *Grass Systematics and Evolution*, ed. T. R. Söderstrom, K. W. Hilu, C. S. Campbell & M. E. Barkworth. Smithsonian Institution Press: Washington DC.

Jacques-Félix, H. (1962). *Les Graminées (Poaceae) d'Afrique tropicale. I. Généralités, classification, description des genres. Bull. Sci. Inst. des Recherches Agronomiques Tropicales*, **8**. I.R.A.T.: Paris.

Jauhar, P. P. & Crane, C. F. (1989). An evaluation of Baum *et al.*'s assessment of the genomic system of classification in the Triticeae. *Amer. J. Bot.* **76**, 571–6.

Joysey, K. A. (1988). Some implications of the revolution in molecular biology. In *Prospects in Systematics*, ed. D. L. Hawksworth. Systematics Association Special Volume No. 36. Clarendon Press: Oxford.

Kellogg, E. A. (1989). Comments on genomic genera in the Triticeae (Poaceae). *Amer. J. Bot.* **76**, 796–805.

Kellogg, E. A. & Campbell, C. S. (1987). Phylogenetic analyses of the Gramineae. In *Grass Systematics and Evolution*, ed. T. R. Söderstrom, K. W. Hilu, C. S. Campbell & M. E. Barkworth. Smithsonian Institution Press: Washington DC.

King, G. J. & Ingrouille, M. J. (1987). Genome heterogeneity and classification of the Poaceae. *New Phytol.* **107**, 633–44.

Knobloch, I. W. (1968). *A Check List of Crosses in the Gramineae.* Dept of Botany and Plant Pathology, Michigan State University: E. Lansing.

Löve, A. (1984). Conspectus of the Triticeae. *Feddes Repert.* **95**, 425–521.

Macfarlane, T. D. (1987). Poaceae subfamily Pooideae. In *Grass Systematics and Evolution*, ed. T. R. Söderstrom, K. W. Hilu, C. S. Campbell & M. E. Barkworth. Smithsonian Institution Press: Washington DC.

Macfarlane, T. D. & Watson, L. (1982). The classification of Poaceae subfamily Pooideae. *Taxon*, **31**(2), 178–203.

May, R. M. (1981). The world's worst weeds. *New Scientist*, **290**, 85–6.

Martin, P. G. & Dowd, J. M. (1987). A phylogenetic sequence for some monocotyledons and gymnosperms derived from protein sequences. *Taxon*, **35**, 469–75.

Parodi, L. R. (1961). La taxonomia de las Gramineas Argentinas a la luz de las investigaciones mas recentes. *Rec. Adv. Bot.* **1**, 125–9.

Pilger, R. (1954). Das System der Gramineae unter Ausschlus der Bambusoideae. *Bot. Jahrb. Syst., Pflanzengeschichte und Pflanzengeographie*, **76**(3), 281–4.

Polhill, R. M. & Raven, P. H. (1981). *Advances in Legume Systematics*. Royal Botanic Gardens, Kew. Ministry of Agriculture, Fisheries and Food: London.

Prat, H. (1960). Revue d'Agrostologie; vers une classification naturelle des Graminées. *Bull. Soc. Bot. Fr.* **107**, 32–79.

Sneath, P. H. (1988). The phenetic and cladistic approaches. In *Prospects in Systematics*, ed. D. L. Hawksworth. Clarendon Press: Oxford.

Söderstrom, T. R. (1981). The grass subfamily Centostecoideae. *Taxon*, **30**, 614–16.

Söderstrom, T. R. & Ellis, R. P. (1987). The position of bamboo genera and allies in a system of grass classification. In *Grass Systematics and Evolution*, ed. T. R. Söderstrom, K. W. Hilu, C. S. Campbell & M. E. Barkworth. Smithsonian Institution Press: Washington DC.

Söderstrom, T. R., Hilu, K. W., Campbell, C. S. & Barkworth, M. E. (ed.) (1987). *Grass Systematics and Evolution*. Smithsonian Institution Press: Washington DC.

Stebbins, G. L. (1987). Grass systematics and evolution: past, present and future. In *Grass Systematics and Evolution*, ed. T. R. Söderstrom, K. W. Hilu, C. S. Campbell & M. E. Barkworth. Smithsonian Institution Press: Washington DC.

Stebbins, G. L. & Crampton, B. (1961). A suggested revision of the grass genera of temperate North America. *Adv. in Bot. (Lectures and Symposia, IX Int. Bot. Congr.)* **1**, 133–45.

Takeoka, T. (1957). Proposition of a new phylogenetic system of Poaceae. *J. Jap. Bot.* **32**, 275–87.

Takhtajan, A. (1969). *Flowering Plants: Origin and Dispersal*. Transl. C. A. Jeffrey. Oliver & Boyd: Edinburgh.

Thomasson, J. R. (1987). Fossil grasses: 1820–1986 and beyond. In *Grass Systematics and Evolution*, ed. T. R. Söderstrom, K. W. Hilu, C. S. Campbell & M. E. Barkworth. Smithsonian Institution Press: Washington DC.

Tsvelev, N. N. (1976). *Grasses of the Soviet Union*, Vols I and II. Nanka: Leningrad. (English translation 1983, Oxonion Press: New Delhi.)

Watson, L. (1987). Automated taxonomic descriptions. In *Grass Systematics and Evolution*, ed. T. R. Söderstrom, K. W. Hilu, C. S. Campbell & M. E. Barkworth. Smithsonian Institution Press: Washington DC.

Watson, L. & Dallwitz, M. J. (1988). *Grass Genera of the World: illustrations of characters, descriptions, classification, interactive identification, information retrieval.* Research School of Biological Sciences, Australian National University, Canberra.

Watson, L., Clifford, H. T. & Dallwitz, M. J. (1985). The classification of Poaceae: subfamilies and supertribes. *Aust. J. Bot.* **33**, 433–84.

Watson, L., Dallwitz, M. J. & Johnson, C. R. (1986). Grass genera of the world: 728 detailed descriptions from an automated database. *Aust. J. Bot.* **34**, 223–30 (with 3 microfiches).
Went, F. W. (1971). Parallel evolution. *Taxon*, **20**, 197–226.
Yeoh, Hock-Hin & Watson, L. (1981). Systematic variation in amino acid compositions of grass caryopses. *Phytochem.* **20**, 1041–51.
(1982). Taxonomic variation in total leaf protein amino acid compositions of grasses. *Phytochem.* **21**, 615–26.

Addendum

A recent version of Tsvelev's classification (Tsvelev 1989: with no references beyond 1984) continues to recognize only two subfamilies, Bambusoideae and Pooideae ('true grasses'), with the latter including some tribes more usually regarded as bambusoid (notably Oryzeae and Phyllorachidae). Even here, however, the tribes and the relationships depicted among them are extensively conventional.

Tsvelev, N. N. (1989). The system of Grasses (Poaceae) and their evolution. *Bot. Rev.* **55**, 141–204.

2

The spikelet

W. D. Clayton

Introduction

The grass flower is reduced to little more than the reproductive organs, typically comprising two tiny lodicules, three stamens and a unilocular ovary bearing two stigmas. It is widely believed that the lodicules are homologous with perianth members, a contention which is open to argument (Clifford, 1987), though most of the conflicting evidence comes from special cases of ontogeny or teratology whose validity as a basis for generalization is questionable.

The flower is enclosed by two bracts, a lemma on the outer side and a palea on the inner. By functional analogy with a perianth, the whole organ is called a floret, although it is now considered that the lemma is a modified leaf subtending the floral branchlet (Tran, 1973), and that the palea is a prophyll. The latter is the first leaf of an axillary branch which, in monocotyledons, is commonly squeezed between branch and stem, taking the form of a little two-keeled bract (Blaser, 1944). Typically the florets are borne distichously along a rhachilla, at the base of which are two empty scales called glumes. Glumes, rhachilla and florets together comprise a spikelet.

The spikelets are borne in an inflorescence whose branches are devoid of subtending leaves or bracts. There are two basic inflorescence types, a many-branched panicle and a single-axis raceme (sometimes called a spike when the spikelets are sessile, but the difference is often so ambiguous that the distinction is unhelpful). The basic types are but the two ends of a continuum spanning a wide range of transitional forms. These include a spiciform panicle whose branches are more or less accrescent into a single axis, multiple racemes disposed digitately or along a central axis, and composite racemes when secondary clusters replace single spikelets. The terminology rests upon another analogy, this time between spikelet and petaloid flower. Attempts have been made to correct some of its inconsistencies, such as Allred's (1982) critical review and Stapf's (1904 and subsequently) use of anthoecium instead of floret, but they have not had much impact on the well-established, though inexact, convention of floret, raceme and panicle.

The linchpin of the reproductive structures described above is the spikelet. It is a fascinating organ whose groundplan is remarkably consistent throughout the family; in only two genera (*Anomochloa* and *Streptochaeta*) is there any real difficulty in relating spikelet structure to the standard form. Yet it contrives an extraordinary wealth of variation, with bewildering changes in evolutionary direction which make it difficult to present in any rational sequence. To some extent this reflects inadequate knowledge of the selection pressures that relate form to function, but three other factors make the biology of the grass spikelet particularly difficult to unravel.

In the first place, spikelet structure is only loosely correlated with the primary phylogenetic divisions based on photosynthetic metabolism, so there is much parallel evolution. Secondly, grasses are opportunists of fluctuating or disturbed ecosystems, conditioned by climatic stress, fire, grazing and farming. Much of their success in these unstable environments is due to a versatile reproductive system, readily responsive to change and driving complementary adaptations in spikelet morphology. Finally, spikelets serve as both flower and fruit, so that pressures on the various functions are likely to be different and sometimes in conflict.

The upshot is that spikelet evolution proceeds by disjointed parallel sequences rather than a single coherent sweep. Nevertheless, if weakly expressed or capriciously isolated occurrences of a character are disregarded for a first approximation, some useful generalizations can be made. The process of winnowing out significant data inevitably entails some arbitrary judgements, therefore the examples and counts given indicate the relative frequency and principal sites of variation, but make no pretence of exact demarcation. Further details of the taxa cited, and the classification adopted here, may be found in Clayton & Renvoize (1986).

The spikelet as pollinator

The standard floral formula – P2A3$\underline{G}$(2) – is by far the commonest, and is surprisingly stable considering the variation shown by spikelets. However, the number of parts can vary and Clifford (1961) has discussed the combinations that occur. The trimerous formula – P3A6$\underline{G}$(3) – recalls the basic monocotyledonous floral symmetry and presumably represents a legacy from some ancestral form. This pattern and its immediate derivatives are almost confined to Bambusoideae (*Sasa* is a good example) which are thereby deemed to have primitive flowers although, paradoxically, their vegetative structure is highly advanced. Indeed, the few non-bambusoid examples (Stipeae, *Ampelodesmos*, *Metcalfia* and *Anisopogon* with three lodicules: *Megalachne*, *Pseudodanthonia* and *Neurachne* with three stigmas) have provoked much debate concerning their taxonomic affinity. In about 10 genera, all bambusoid, the filaments of the stamens are united. Reduction to one or two stamens occurs sporadically and unpredictably across the family. Reduction to one stigma is rare (*Oreobambos*, *Gigantochloa*, *Dendrocalamus*, *Nardus*, *Lygeum*, *Odontelytrum* and *Zea*); the two stigmas are connate in some *Pennisetum*.

Grasses are wind-pollinated and their flowers devoid of insect lures. The display of purple pigments in the inflorescence and the occasional presence of aromatic oils in the leaves seem unrelated to pollination. Bees have been observed visiting grasses to collect pollen (Bogdan, 1962), but only a few specialized grasses of deep forest shade, such as *Pariana* and Olyreae, have acquired any significant reliance upon entomophily (Söderstrom & Calderon, 1971).

Pollen release has engendered little structural diversity, as witnessed by the unspecialized male spikelets of monoecious and dioecious genera. There is, however, a widespread tendency to suppress sexuality in some florets or spikelets. They are commonly referred to as 'sterile', meaning no ovary, but in fact they often vary from male to barren even within the same species; when male, the effect is to increase the stamen–ovary ratio. It may be naïve to explain this solely as a response to the uncertainties of anemophily (see below under andromonoecism), because retaining the full number of six stamens per floret would seem to be a simpler strategy. In fact, this option is exercised only by bambusoid genera, three of which actually increase the stamen number. *Luziola* has up to 16 stamens per floret, *Pariana* up to 30 and *Ochlandra* verges on the ridiculous with up to 120 stamens accompanied by 3–15 lodicules and 4–6 stigmas. The implication is that the 'primitive' bamboo flower has been preserved, not as an anachronism, but as well adapted to the low wind speeds of its forest environment, and that stamen reduction became advantageous when grasses moved on to the windy plains.

Pollen entrapment is governed by aerodynamic considerations which undoubtedly influence the shape of inflorescence and spikelet, though studies on this aspect are in their infancy (Niklas, 1985). Exposure of the flower during anthesis is very brief, 10 minutes to an hour or so, possibly to reduce the risk of catching fungal spores. To achieve this grasses have evolved a very precise timing system in which the lodicules play a part. They swell to force the lemma and palea apart for the anthers to emerge, then subside so that the scales can spring back until the stigmas are ready for exsertion (Pissarek, 1971). Lodicules are absent from about 40 genera, a condition strongly, but not invariably, associated with protogyny, apical extrusion of the stigmas and cleistogamy.

Timing assumes a more extravagant dimension in many bamboos, whose flowering occurs in cycles of 20–120 years. The mechanism is unknown, but the phenomenon is thought to be associated with ensuring regeneration by temporary satiation of grain predators (Janzen, 1976).

Outbreeding is determined by a complex incompatibility system at the pollen–stigma interface, but some grasses have reinforced it by dicliny, a topic extensively reviewed by Connor (1980, 1981 and 1987). Andromonoecism (male and bisexual flowers) is quite common in the family, becoming usual in Panicoideae. There is much variety in the arrangement of the sexes and andromonoecism often seems to be not so much a response to the demands of pollination, as consequential to modification in lemma function after fertilization. However, it offers a route to full monoecism with all the spikelets unisexual. Here again, the disposition of the sexes within the inflorescence, or more rarely in different inflorescences, displays considerable variability and the genera are distributed across a wide range of tribes: Olyreae (all 16 genera), Parianeae (both genera), Phareae (both genera), Oryzeae (*Zizania*, *Zizaniopsis*, *Luziola*), Phyllorachideae (both genera), Centotheceae (*Pohlidium*), Paniceae (*Hygrochloa*) and Andropogoneae (*Tripsacum*, *Zea*, *Chionachne* and three allies, *Coix*). The final step is to dioecy, which is also erratically scattered across the tribes: Poeae (some *Poa* and *Festuca* or the latter dichogamous, meaning the sexes borne on the same plant at different times), Arundineae (*Cortaderia*, *Lamprothyrsus*, *Gynerium*), Eragrostideae (*Jouvea*, *Distichlis* and three allies, *Eragrostis reptans*, *Sohnsia*, *Scleropogon*), Cynodonteae (*Buchloe* and five allies) and Paniceae (*Zygochloa*, *Spinifex*). The spikelets of dioecious plants are often quite similar, but the two sexes in *Jouvea*, the *Buchloe* group, *Scleropogon* and *Spinifex* are strongly heteromorphic.

Self-compatibility is an alternative strategy pursued by many grasses, trading a reduction in genetic flux against greater certainty of fertilization. It is taken a step further in cleistogamy – self-pollination within the closed spikelet – which Campbell *et al.* (1983) have divided into four levels of spikelet differentation:

1 Aerial fertilization (6 genera). Cleistogamous and chasmogamous inflorescences virtually identical. Probably more frequent but unrecognized.
2 Sheath fertilization (41 genera). Spikelets remain within the

upper leaf-sheaths; often smaller than chasmogamous and may be mistaken for immature.

3 Cleistogenes (*Stipa*, *Danthonia*, *Cottea*, *Enneapogan*, *Pappophorum*, *Muhlenbergia*, *Triplasis*, *Calyptochloa* and *Cleistochloa*). Modified spikelets within basal leaf-sheaths; usually the spikelet scales reduced and grain enlarged; often dispersed with the sheath when culms break up at the end of the season.

4 Rhizanthogenes (*Eremitis*, *Enteropogon chlorideus*, *Amphicarpum* and *Paspalum amphicarpon*). Spikelets highly modified and borne on specialized underground rhizomes.

Cleistogamy seems never to be obligate but is mediated by the environment, taking precedence over chasmogamy in times of stress. It is odd that *Pennisetum clandestinum* has evolved a form of chasmogamous sheath fertilization using very long filaments and stigmas; it reduces damage by grazing and, speculatively, could have arisen via cleistogamy. The ultimate expression of selfing – apomixis – is not reflected in spikelet morphology apart from some reduction in lodicules and stamens.

The final denial of sexuality is shown by those species which regularly propagate by means of plantlets formed from vegetative proliferation of the spikelets (Langer & Ryle, 1958; Beetle, 1980). They are mainly Arctic–alpine Pooideae growing at the limits of tolerance, but it can occur as a teratological condition elsewhere.

The spikelet as inflorescence

The spikelet is the basic unit of the inflorescence, but the homologies of its parts show that it is itself derived from a condensed inflorescence. The distinction between spikelet and inflorescence is thus one of degree, presenting no conceptual obstacle to further condensations which may sometimes transgress the spikelet–inflorescence boundary.

The commonest condensations are the reduction of the inflorescence from paniculate to racemose, and of the spikelet from many fertile florets to only one. These set the stage for a repositioning of the abscission level whose effect is to shift the floral analogue along the transformation series: floret–spikelet–raceme segment–spikelet cluster. The possibilities are numerous, but the frequency of the principal combinations and their overall trend can be judged from Table 2.1.

At its simplest, shifting the abscission level into the inflorescence merely enlarges the diaspore to a group of florets, but it opens the way for the incorporation of branch elements into the spikelet structure. These

Table 2.1. *Numbers of genera according to their floret number, inflorescence type and abscission mode*

	Abscission	
	above glumes	below glumes
Two or more fertile florets		
Paniculate	134	8
Racemose	71	24
One fertile floret		
Paniculate	92	76
Racemose	47	184

amalgamations of spikelet and inflorescence can be rather complex, but they fall into six main types listed below with examples:

1 Fragile raceme with a spikelet (occasionally more) at each node. Common in Triticeae, Hainardieae and Parianeae; also *Gaudinia*, *Oropetium*, *Lepturus* and *Uranthoecium*.
2 Fragile raceme with a heteromorphic pair of spikelets at each node. A special case of the above, characteristic of Andropogoneae. Tough racemes with homomorphic pairs are common in the neighbouring tribe, Paniceae, and could be regarded as a transitional condition, though the pressures selecting the paired state are unknown.
3 Deciduous racemelets of one to five spikelets. Common in Cynodonteae (Boutelouinae, Zoysiinae) and Phyllorachideae; also *Lycurus* and a few Eragrostideae (*Dinebra* and allies).
4 Whole raceme. A variety of structures in which the spikelet becomes integral with a raceme: *Aegilops*, *Stenotaphrum*, *Thuarea*, *Iseilema* amd *Thaumastochloa*.
5 Combined wtih sterile panicle branches. Paniceae (*Setaria* and allies, Cenchrinae). Precursory forms with branch tips or lowest whorl of the inflorescence sterile occur in Paniceae and elsewhere.
6 Triads. Panicle branches tipped by three spikelets with connate pedicels, thus imitating a single spikelet: Arundinelleae (*Loudetiopsis*, *Tristachya* and *Zonotriche*). Clusters of three also occur as special cases of type 1 (*Hordeum*), type 2 (*Chrysopogon*, *Lasiurus*) and type 3 (*Aegopogon* and allies).

The reduction of the spikelet to one floret followed by reinstatement of a many-flowered unit by aggregating spikelets seems paradoxical, but its effect is to bring novel structures into play. Sometimes these augment existing trends, but sometimes they seem to be restoring functions discarded by earlier adaptations.

The further development of the inflorescence is equally paradoxical. Certain genera of Bambuseae (such as *Phyllostachys*) and Andropogoneae (such as *Hyparrhenia*), having reduced the inflorescence to a short raceme or raceme-pair, multiply these by axillary branching and place the whole system at the top of the culm. This aggregate of racemose inflorescences is called a compound inflorescence. It imitates the simple panicle except that each of its elements (the original racemes) has gained an enveloping 'spatheole' (inflated sheath of the subtending leaf) in the process. *Bambusa* and 10 other bamboo genera take yet another step, contracting the compound inflorescence on to the nodes to form a globular 'iterauctant' inflorescence. The process culminates in the loss of all the leaves, and the conversion of the whole culm into an inflorescence (*Dinochloa*). These extraordinary oscillations between reduction and proliferation testify to the family's adaptability, and would seem to imply periodic reversals in evolutionary direction though the causal factors are largely unknown.

The spikelet as seed-bearer

Grasses place a high premium on enclosure of the flower. This is evident from the beginning, when the inflorescence remains within the uppermost leaf-sheath until the spikelets are close to anthesis. In fact, those few genera (*Phragmites*, *Thysanolaena*, *Pogonarthria* and Arundinelleae) which emerge with juvenile spikelets regularly confuse the novice. In diclinous genera enclosure is better developed around the female flower, suggesting that it is primarily concerned with the ovary. The function of enclosing organs is largely speculative. They are obviously protective, though whether against infection, predation or desiccation is far from clear, but that is no more than a partial explanation because spikelet morphology plays a significant role in reproductive biology from pollination to germination. Indeed, Redman & Reekie (1982) report that most of the assimilate transported to the seed is produced in the inflorescence or flag leaf, so that spikelet parts – such as the awns of *Triticum* (Grandbacher, 1963) – include photosynthesis among their many functions.

Variations on the enclosure theme are multifarious, with functions permuting freely among the different parts, but they are of two main kinds: induration of scales, and provision of additional envelopes. Rarely, redundant parts are suppressed.

Glumes

Glumes are normally small and apparently function as bud scales, but in a number of tribes (Olyreae, Stipeae, Hainardieae, Aveneae, Arundineae, Aristideae, Pappophoreae, Cynodonteae and Arundinelleae) they are long enough to enclose the mature florets, becoming fused at the base into a cup in *Alopecurus*. In Cynodonteae (Zoysiinae) and Andropogoneae they are also indurated.

The lower glume is often suppressed when its function is usurped by some other organ. In Hainardieae and Triticeae, where the spikelet is appressed to a rhachis internode, the lower glume is not suppressed but swivels to an abaxial position beside the upper glume. Both glumes are suppressed or rudimentary in Oryzeae and a scattering of other genera (*Ehrharta*, *Brachyelytrum*, *Lygeum*, *Nardus*, *Psilurus*, *Aphanelytrum*, *Coleanthus*, *Phippsia*, *Libyella*, *Pohlidium*, *Monanthochloe* and *Jouvea*).

Lemma

The glumes are supplemented by one or two sterile or male lemmas at the base of the spikelet in a wide range of genera including: *Chusquea*, *Nastus*, *Ehrharta*, *Brylkinia*, *Anthoxanthum*, *Hierochloe*, *Bromuniola*, *Chasmanthium*, *Alloeochaete*, *Phragmites*, *Uniola*, *Tetrachne*, *Fingerhuthia*, *Blepharidachne*, some *Eragrostis* and *Ctenium*. There are also two vestigial sterile lemmas in Oryzeae and *Phalaris*, perhaps functioning as elaiosomes in the latter.

Panicoideae are a special case whose spikelets comprise a lower male or sterile floret and an upper fertile floret. It is not clear why this arrangement, occurring in 202 genera, should be so successful. Spikelets composed of two fertile florets are not uncommon (occurring in about 30 genera) but shed no light on the problem. In fact, of these, Isachneae, Eriachneae and *Dissochondrus* seem to be derived from Paniceae by restoration of fertility in the lower floret, affording further evidence of evolutionary reversal.

Although not strictly additional envelopes, convolute lemmas, completely enfolding the flower and palea, should be mentioned (Stipeae, Aristideae, the *Amphipogon* group and a few others). More extreme are the tubular lemmas, open only at an apical pore, of *Leptaspis* and *Jouvea*. *Lygeum* achieves a similar effect by fusing two opposite lemmas edge to edge with the accrescent paleas as a septum between them.

Induration is particularly well developed in Olyreae, Stipeae and the upper floret of Paniceae; but when the upper floret is embedded in the rhachis, the lower lemma may be indurated instead (*Stenotaphrum*, *Trachys* and Phyllorachideae).

Palea

In Olyreae and Paniceae the indurated lemma is accompanied by an equally tough palea to form a cowrie-like box around the flower. This may be further reinforced by induration of the flanks and keels of the palea of the lower floret: *Steinchisma*, *Plagiantha*, *Otachyrium*, *Thrasya*, *Thrasyopsis*, *Holcolemma* and *Ixophorus*. The lower floret of *Gilgiochloa* has a curious palea like a little chip of wood. In the glumeless single-flowered spikelets of Oryzeae the palea is fully exposed, resembling and functioning like a lemma.

Reduction of the palea in Stipeae and Aristideae is associated with a convolute lemma, and in *Alopecurus* with basal fusion of the lemma margins. In Andropogoneae the function of the palea of the upper floret is often usurped by the lemma of the sterile floret (or, paradoxically, by the palea of the sterile floret in *Microstegium vagans*), leaving each floret with only one scale apiece. Loss of the palea in genera such as *Agrostis* and *Hubbardia* is less easy to understand.

Rhachilla

An enlarged spongy rhachilla, fused to the indurated lemmas, is a prominent component of the peculiar thorn-like female spikelets of *Jouvea*. The S-shaped rhachilla of *Eccoptocarpha* straightens at maturity to extrude the fertile floret.

Pedicel

Expanded pedicels bearing male or sterile spikelets embrace the fertile spikelet in *Pariana* and *Apluda*. In *Eremitis* they form a long tube completely enclosing the fertile spikelet.

In a few genera abscission occurs at the base of the pedicel, which then imitates a spikelet callus (*Brylkinia*, *Polypogon*, *Chaetopogon*, *Thysanolaena*, *Harpachne*, *Viguierella* and *Myriostachya*).

Rhachis

The raceme rhachis is one of the most important accessory spikelet structures, with great capacity for evolutionary transformation. The first step is to cover the adaxial side of the spikelet, as in *Lolium* where it supplants the lower glume. The second step is either to become thickened so that spikelets sink bodily into it (*Oropetium*,

Stenotaphrum), or to become foliaceous and wrapped around the spikelets (Phyllorachideae, *Thuarea*). Finally, abscission shifts to the rhachis, and its internode becomes an integral part of the spikelet, as in *Lepturus*. The spikelet is normally attached to the internode alongside it, but in *Gaudinia* and Triticeae it is attached to the internode below, which functions as a callus.

The most advanced forms occur in Andropogoneae where fragile racemes bear spikelets in pairs, the one sessile and fertile, the other pedicelled and male or sterile (rarely the sexes are reversed as in *Trachypogon*). The fertile spikelet is therefore amalgamated with an internode, pedicel and sterile spikelet, affording great scope for elaborate modification. *Rottboellia* provides an extreme example; its internode, pedicel and rudimentary pedicelled spikelet are fused into a cylindrical wooden box, with a lid on one side formed from the indurated lower glume of the fertile spikelet housed within.

Sterile spikelets

These occur in various guises, broadly summarized as follows:

Basal bracts of a capitate inflorescence: *Sesleria*, *Aegilops*, *Elytrophorus* and *Amphipogon*.

Reduced to a bunch of awns: *Sitanion*, *Heteranthelium* and *Pereilema*.

Mixed in racemelets and clusters: *Cynosurus*, *Lamarckia*, *Hordeum* and some Cynodonteae (Boutelouinae, Zoysiinae).

Involucral around fertile spikelet: *Chaetopoa*, *Anthephora* and *Phalaris paradoxa*.

Sterile spikelets reach a peak of development in Andropogoneae, whose raceme segments normally include a sterile pedicelled spikelet. In some genera, such as *Heteropogon*, a third spikelet morph of 'homogamous' sterile pairs occupies the lower part of the raceme. When the raceme is very short, the homogamous pairs may form an involucre around it (*Germainia*, some *Hyparrhenia*, *Exotheca* and *Themeda*). The ultimate expression of this tendency, in *Iseilema*, is a raceme composed of two homogamous pairs, a sessile spikelet and two pedicelled spikelets, all deciduous together.

Sterile branches

An involucre of bristles derived from sterile panicle branches surrounds the spikelet in *Setaria* and allies; it is deciduous with the spikelets in *Pennisetum* and allies. Sterile branches can also coalesce into a prickly cup (*Cenchrus*) or take the form of a pungent quill (*Spinifex*), apparently assuming a defensive role, as do the spinous tips of the raceme rhachis in *Cladoraphis*. More complex confections of spiny branches, sterile spikelets and foliaceous rhachis occur in *Streptolophus*, *Chlorocalymma* and *Trachys*.

Peduncle

The capitate inflorescence of *Cornucopiae* is embraced by an involucral expansion of the peduncle tip.

Spatheole

Occasionally the inflorescence acquires an extra covering by incomplete exsertion from the inflated sheath of the flag leaf. This trend is fully exploited by the compound inflorescence, where each true inflorescence is enclosed by a spatheole. The spatheole is indurated and deciduous in *Coix* and *Iseilema vaginiflorum*. Three expanded spatheoles, deciduous with the spikelet, aid wind dispersal in *Zygochloa*.

Leaf-blades

Little ovate leaf-blades form an involucre around the short racemes of *Hitchcockella* and *Perrierbambus*. A larger blade forms a canopy over the raceme of *Diandrolyra*.

The spikelet as fruit

The grass fruit is a caryopsis, defined as having a thin pericarp adherent to the seed. Sometimes the pericarp is free, but Sendulsky, Filgueiras & Burman (1987) observe that this is brought about by dissolution of the inner layers and that the fruit is still ontogenetically a caryopsis. They deprecate describing such fruits as achenes and prefer a neutral terminology for the variant forms:

Cistoid; pericarp separable, with a little persuasion, when wet. 12 genera: bamboo allies, Arundineae, Cynodonteae.

Follicoid; pericarp free, soft. 38 genera; Arundineae, Eragrostideae, Cynodonteae.

Nucoid: pericarp free, crisp. 7 genera; *Actinocladum*, *Merostachys*, *Zizaniopsis*, *Luziola*, *Pentameris*, *Pyrrhanthera*, *Dregeochloa*.

Baccoid; pericarp adherent, fleshy. 7 genera, all Bambuseae; *Olmeca*, *Decaryochloa*, *Alvimia*, *Melocalamus*, *Dinochloa*, *Melocanna*, *Ochlandra*.

The seed is composed mainly of starch-storing endosperm, accessible to the embryo through a haustorial scutellum. In *Melocalamus* and *Dinochloa* the endosperm is evanescent, starch being stored in an enlarged scutellum; the other baccoid fruits have an endosperm but store much of their starch in the pericarp. Some Pooideae, particularly Aveneae, have a high content of lipids in the endosperm, which may remain fluid in dried specimens for 50 years or more (Rosengurtt, Laguardia & Arrillaga de Maffei, 1971; Terrell, 1971). Weight for weight, lipids have about twice the energy value of carbohydrates.

The diaspore is rarely a seed; the best examples are in *Sporobolus* where the follicoid pericarp becomes mucilagious and extrudes the seed to the tip of the spikelet. It is occasionally a caryopsis (examples in *Agrostis* and *Eragrostis*), but in the bulk of the family it is a false fruit embodying various spikelet parts. False fruits, and their relative frequency, can be summarized as follows:

Floret (rarely without palea). 296 genera.

Floret with accessory sterile lemmas. 62 genera, Cynodonteae well represented.

Spikelet. 136 genera, half of them Paniceae.

Rhachis internode. 89 genera, mainly Triticeae and Andropogoneae.

Branches, racemelets and clusters. 50 genera.

Inflorescence. 13 genera with compact inflorescences that form a single fruit. In addition there is a scattering of individual species with deciduous open panicles that behave as tumbleweeds, such as *Nasella trichotoma*, *Panicum olyroides* and *Digitaria divaricatissima*.

Double abscission adds a further, but uncommon, complication. Thus, the spikelets of *Brachiaria eruciformis* and the clusters of *Pennisetum polystachion* are deciduous, but in both species the upper floret is also readily shed. Another example is *Catalepis*, whose raceme rhachis and spikelets are separately deciduous.

The usual anecdotal account of wide dispersal in the context of geographical distribution gives a very false impression. Actually, dispersal distances are small, because the primary requirement is to maintain the species in its present habitat (Harper, 1977). The optimum strategy is to move the bulk of the seed only a short distance, typically less than 10 m from the parent, but to allow occasional opportunities for travelling much greater distances. Grasses accomplish this by three partly interchangeable methods (Silberbauer-Gottsberger, 1984; Davidse, 1987):

1 *Ingestion* by large herbivores, consumption by granivorous birds or storage by rodents, using the foliage (Janzen, 1984) or grain as bait, a small proportion of the seed surviving in a viable state. Since most bamboos and 255 other genera have no obvious morphological adaptations for dispersal, it is probably the commonest method.

2 *Wind*. Hairs, often complemented by a wide variety of membranous wings and frills, are prominent in about 95 genera. They occur on any part of the inflorescence, including the preposterous plumose awns, up to 50 cm long, of *Stipa pulcherrima*. They are sometimes long enough to operate in the parachute mode, but often seem more suited to saltation along the ground. Saltation is normal for the tumbleweeds, and for the curious balls sometimes formed from conglomerations of awns shed by species of *Aristida*.

3 *External animal transport*. Many hairy spikelets are also loosely adherent, but the principal weapon is a pungent bearded callus. It occurs in many genera, becoming ferociously effective in *Heteropogon contortus* where it is unsheathed by twisting of the awns and poised malignantly outward at maturity; *Stipa*, *Aristida* and *Chrysopogon aciculatus* are almost as great a menace. Note that the term callus refers to extensions at the foot of either spikelet or floret according to context.

Most grasses rely upon the wide-spectrum vectors described above, but a few, particularly forest inhabitants, have evolved more specialized spikelet adaptations:

4 *Fruit mimics*. The glumes and lower lemma of *Lasiacis* turn black and become packed with oil-globules at maturity, attracting frugivorous birds (Davidse & Morton, 1973).

5 *Elaiosomes*. Ants may be quite important vectors, as it is becoming apparent that several puzzling little spikelet appurtenances are oil-bearing elaiosomes (Davidse, 1987). These include appendages to the rhachilla beneath the fertile floret (*Puelia*, *Ichnanthus* and allies, *Panicum cervicatum*, *Yakirra*), enlarged basal rhachilla internodes (*Echinochloa callopus*, *Brachiaria nigropedata*, *Eriochloa*) and the knob at the base of the rhachis internode in Rottboelliinae.

6 *Glue*. The best example is *Oplismenus* with viscid awns, but sticky spikelets occur elsewhere such as *Homolepis* and *Eragrostis viscosa*.

7 *Spiny burs*. In low grassland and apparently designed to catch on the legs (*Tragus*, *Buchloe*, *Cenchrus*, *Streptolophus*).
8 *Hooks*. Reflexed bristles (*Centotheca*); hooked hairs (*Pharus*, *Leptaspis*, *Pseudechinolaena*, *Ancistrachne*); retrorse barbs, the most tenacious of all (*Lophatherum*, *Setaria*, *Cenchrus*). A kind of hook is also formed from the deflexed raceme, tipped by a pungent pedicelled spikelet, of *Cymbopogon refractus*. The most elegant are the 'fish hooks' dangled on a line, the lines twining so that when one hook catches, the whole inflorescence is stripped. The homologies of hook and line are: *Streptochaeta* (cleft palea with reflexed tips on a coiled lemma awn), *Streptogyna* (springy rhachilla internode on an elongated stigma) and *Streblochaete* (barbellate callus on a coiled lemma awn).
9 *Flotation*. Spongy or aerenchymatous tissue occurs in some aquatic and seashore genera: *Rhynchoryza* (lemma tip), *Jouvea* (rhachilla), *Stenotaphrum* (inflorescence axis), *Spinifex* (raceme rhachis, this shed after the inflorescence has performed as a tumbleweed) and *Steyermarkochloa* (palea); *Thuarea* capsules (folded foliaceous rhachis) may also float. Even here, normal dispersal is over a short distance, *Stenotaphrum* remaining buoyant for only about a week.

Awns are the most conspicuous spikelet appendages, usually borne by the lemmas (on the back rather than the tip in Aveneae), but sometimes by glumes or sterile spikelets, even the palea in *Apochiton* and a few other genera. They are sometimes multiple, three being a popular number (especially Aristideae which have a three-pronged awn, and Boutelouinae) but rising to 18 on the lemma of Pappophoreae. They are occasionally supplemented (especially Cynodonteae) by terminal awned rudimentary florets deciduous with the fertile floret, and can form elaborate combinations when the diaspore is composed of several spikelets. The awn itself is sometimes deciduous from the diaspore. Awns are of two main kinds: straight or flexuous (about 160 genera) and spirally twisted (about 100 genera).

Awns may occasionally play a defensive role, threatening the eyes of predatory birds, but are usually classed as dispersal organs. It is true that they sometimes participate in wind dispersal, but their primary function seems to be more concerned with germination, i.e. finding a suitable microsite, orienting the seed for optimal moisture uptake and anchoring it against the thrust of the emerging radicle. They achieve this by hygroscopic flexion or torsion which, combined with the one-way device of the callus beard, drives the diaspore along the ground. The motion can be quite complicated when moving and static parts interact, and it has been shown (Peart, 1984; Peart & Clifford, 1987) to be associated with particular microsite preferences.

Melocalamus and *Melocanna* condense the whole process by adopting vivipary, the baccoid fruit germinating before it is shed from the parent (Vaid, 1962).

Origin of the spikelet

In the foregoing it has been tacitly assumed that the primitive spikelet contained several trimerous flowers and was borne in a panicle. This assumption is based upon the presumed homology of spikelet parts and the internal phylogeny of the family, there being no extant or fossil grasses that provide direct evidence of precursory forms. Can these forms be deduced from the inflorescence structure of neighbouring families?

The families most closely related to grasses (Dahlgren, Clifford & Yeo, 1985) have trimerous flowers – typically $P3+3A3+3\underline{G}(3)$ – with bracts subtending the flowers and branches but no prophylls. Flagellariaceae and Joinvilleaceae have terminal panicles of flowers that give no hint of spikelet formation. Restionaceae are more complex (Linder, 1984) but for the present purpose their inflorescence can be presented as a simple transformation series: the upper culm nodes each bear a panicle of flowers (*Elegia verticillaris*); the distal branchlets of the panicle condense, the bracts becoming spirally imbricate to form a spikelet (*Restio complanatus*); this contracts to a sessile capitulum of spikelets (*R. schoenoides*). Cyperaceae can be associated with this model, but not Juncaceae whose inflorescence is cymose.

Grasses differ in possessing a palea, and will fit the Restionaceae model only if the palea is derived from the union of two outer perianth members. There is no evidence that this is so, but rather the palea has every outward appearance of a prophyll. Ontogenetic evidence is ambiguous, the palea arising from two lobes but these being sited on an initial ridge (Clifford, 1987). If the palea is accepted as a prophyll, then the grass spikelet has no homologous counterpart among its immediate neighbours.

For an inflorescence whose axillary branches are furnished with both a subtending bract and a prophyll, it is necessary to look further afield, to the Marantaceae. This family has a complex synflorescence in which four types of branching can be recognized (Anderson, 1976):

1 The main axis produces an axillary branch at one or more of its distal nodes.

2 The axillary branches are multiplied by successive dichotomies, each of the ultimate branches terminating in a florescence.
3 The florescence is the visually obvious element of the synflorescence. It is spiciform and bears imbricate distichous bracts which conceal axillary florescence components.
4 The florescence component is a short cymose sequence bearing the actual flowers. It presumably represents an archetypal Commelinaceous inflorescence, now reduced to a minor role within the, otherwise monopodial, synflorescence structure.

The generalized form is exemplified by species such as *Ischnosiphon arouma*. There are many variants, one of which involves gross abbreviation of all the internodes to form a capitulum, as in *Phrynium*.

Of greater interest in the present context are species such as *Thalia geniculata*. Here the florescence component is reduced to a single flower furnished with bract and prophyll. It is thus a floret analogue, and the florescence is a spikelet analogue. Moreover, bracts and prophylls on the main axis and axillary branches are deciduous, yielding a remarkable imitation of the grass panicle. The model clearly demonstrates how a grass-like panicle can arise, and illustrates the ready mutability with which peripheral branching may be condensed and new orders of branching be assimilated below.

Marantaceae provide an instructive example of parallel evolution but cannot be considered as ancestral to grasses; they have petaloid flowers with an inferior ovary among other differences. It is, however, reasonable to postulate a common ancestor for Gramineae, Restionaceae, Flagellariaceae and Joinvilleaceae in which both bracts and prophylls were well developed.

The pseudospikelet

The incorporation of lower-order branching is evident in the spathate compound inflorescence of Andropogoneae, where the true inflorescences are invested by modified leaf-sheaths and aggregated into a false panicle. Its branches arise from a little unilateral falcate shoot with closely spaced nodes, which Stapf (1919) described as cymose. If so, it would be anomalous, but in fact the structure is monopodial with linear subtending sheaths and no prophylls. Being associated with complex racemes and C_4 photosynthesis, the compound inflorescence of Andropogoneae is clearly derived.

The spathate inflorescence of Bambuseae would seem to be a parallel development, differing in its unspecialized dichotomous branching and the presence of prophylls. This type of inflorescence must also be derived when the branches terminate in bractless racemes; by extrapolation, spatheoles subtending a single spikelet are likewise derived, the spikelet representing a reduced raceme.

The bambusoid inflorescence is more difficult to interpret when the subtending leaves are represented by little glume-like scales. It can then appear that the lowest glumes of the spikelet subtend axillary shoots; these comprise a prophyll followed by a spikelet whose lowest glumes repeat the process, eventually forming a complex of interlocking spikelets. The individual spikelets may be distant (*Glaziophyton*), fascicled (*Chimonobambusa*, *Criciuma*) or condensed into a capitulum at the culm node (*Bambusa* and other iterauctant genera). Their distinctness from ordinary spikelets with empty glumes has been emphasized by use of the special term pseudospikelet (McClure, 1966), and they have been widely regarded as primitive. The reasoning has never been clearly explained, but apparently stems from a notion that the pseudospikelet is a contracted form of the primitive bract-bearing panicle, comparable to *Phrynium* and *Restio schoenoides*. If so, it is unique among grasses, and would be expected to form an isolated clade. Instead, it appears among the bamboos, whose woody culms, elaborate branch architecture and possession of fusoid cells show them to be both monophyletic and advanced. It is therefore difficult to reconcile the retention of a bracteate capitulum with the family's internal phylogeny, or to envisage an evolutionary sequence in which it was the critical transitional form between a bracteate panicle of flowers and an ebracteate panicle of spikelets.

A more tenable proposition is that the pseudospikelet represents a contracted compound inflorescence with spatheoles reduced to scales, the extreme iterauctant condition being analogous to the nodal compaction of vegetative branches that is such a distinctive feature of bamboo morphology. There are many intermediate states between inflorescence types, leading to ambiguities in the definition of pseudospikelet and casting doubt upon its value as a descriptive term. In Andropogoneae an incipient parallel trend to contraction is displayed by species such as *Cymbopogon densiflorus*.

Anomalous spikelets

Streptochaeta, with its raceme of peculiar single-flowered spikelets, has long been a puzzle. In orthodox terms the spikelets comprise three lodicules up to 2 cm long, a palea cleft to the base, a lemma with coiled awn and four or five tiny whorled glumes rarely containing axillary buds. Actually, these unusual scales are less aberrant than they seem at first sight: the lodicules are almost as long in *Ochlandra*, the cleft palea is evidently an adaptation to the fish hook dispersal mechanism (a rudimentary divided palea also occurs in some *Micraira* species), and the basal scales could include sterile lemmas with axillary palea vestiges as in *Guaduella*. Söderstrom (1981) provided an ingenious alternative explanation by treating the scales as subtending bracts in a pseudospikelet, but the disposition of pseudospikelets in a bractless raceme is wholly anomalous. It would therefore seem that the orthodox interpretation is preferable, despite the unexplained whorled arrangement of the glumes.

Anomochloa has a spathate compound inflorescence whose flowers are invested by two spathiform bracts, the upper coriaceous with a large callus, the lower papyraceous. Since there is no precedent for either spatheole or palea to be mounted on a callus, the upper scale must therefore be a lemma, the palea missing. Moreover, large enclosing glumes occur in pairs unless the spikelet backs on to a rhachis internode; the lower scale must therefore be a spatheole and the glumes are missing.

It is concluded that the lemmaless condition postulated by Clayton & Renvoize (1986) is incorrect, and that all grass spikelets can be treated as derivatives from the trimerous bambusoid form.

References

Allred, K. W. (1982). Describing the grass inflorescence. *J. Range Manag.* **35**, 672–5.

Andersson, L. (1976). The synflorescence of Marantaceae. *Bot. Not.* **129**, 39–48.

Beetle, A. A. (1980). Vivipary, proliferation and phyllody in grasses. *J. Range Manag.* **33**, 256–61.

Blaser, H. W. (1944). Studies in the morphology of Cyperaceae: II The prophyll. *Amer. J. Bot.* **31**, 53–64.

Bogdan, A. V. (1962). Grass pollination by bees in Kenya. *Proc. Linn. Soc. London*, **173**, 57–60.

Campbell, C. S., Quinn, J. A., Cheplick, G. P. & Bell, T. J. (1983). Cleistogamy in grasses. *Ann. Rev. Ecol. Syst.* **14**, 411–41.

Clayton, W. D. & Renvoize, S. A. (1986). *Genera Graminum: grasses of the world.* Kew Bull. Addit. Ser. 13. Royal Botanic Gardens, Kew: London.

Clifford, H. T. (1961). Floral evolution in the family Gramineae. *Evolution*, **15**, 455–60.

(1987). Spikelet and floral morphology. In *Grass Systematics and Evolution*, ed. T. R. Söderstrom, K. W. Hilu, C. S. Campbell & M. E. Barkworth, pp. 21–30. Smithsonian Institution Press: Washington DC.

Connor, H. E. (1980). Breeding systems in the grasses – a survey. *New Zealand J. Bot.* **17**, 547–74.

(1981). Evolution of reproductive systems in the Gramineae. *Ann. Miss. Bot. Gard.* **68**, 48–74.

(1987). Reproductive biology in the grasses. In *Grass Systematics and Evolution*, ed. T. R. Söderstrom, K. W. Hilu, C. S. Campbell & M. E. Barkworth, pp. 117–32. Smithsonian Institution Press: Washington DC.

Dahlgen, R. M. T., Clifford, H. T. & Yeo, P. F. (1985). *The Families of the Monocotyledons*. Springer-Verlag: Berlin.

Davidse, G. (1987). Fruit dispersal in the Poaceae. In *Grass Systematics and Evolution*, ed. T. R. Söderstrom, K. W. Hilu, C. S. Campbell & M. E. Barkworth, pp. 143–55. Smithsonian Institution Press: Washington DC.

Davidse, G. & Morton, E. (1973). Bird-mediated fruit dispersal in the tropical grass genus *Lasiacis* (Gramineae; Paniceae). *Biotropica*, **5**, 162–7.

Grandbacher, F. J. (1963). The physiological function of the cereal awn. *Bot. Rev.* **29**, 366–81.

Harper, J. L. (1977). *Population Biology of Plants*. Academic Press: New York.

Janzen, D. H. (1976). Why bamboos wait so long to flower. *Ann. Rev. Ecol. Syst.* **7**, 347–91.

(1984). Dispersal of small seeds by large herbivores: foliage is the fruit. *Amer. Nat.* **123**, 338–53.

Langer, R. H. M. & Ryle, G. J. A. (1958). Vegetative proliferation in herbage grasses. *J. Br. Grassl. Soc.* **13**, 29–33.

Linder, H. P. (1984). A phylogenetic classification of the genera of the African Restionaceae. *Bothalia*, **15**, 11–76.

McClure, F. A. (1966). *The Bamboos – A Fresh Perspective*. Harvard University Press: Cambridge, Mass.

Niklas, K. J. (1985). The aerodynamics of wind pollination. *Bot. Rev.* **51**, 328–86.

Peart, M. H. (1984). The effects of morphology, orientation and position of grass diaspores on seedling survival. *J. Ecol.* **72**, 437–53.

Peart, M. H. & Clifford, H. T. (1987). The influence of diaspore morphology and soil-surface properties on the distribution of grasses. *J. Ecol.* **75**, 569–76.

Pissarek, H. P. (1971). Untersuchungen über Bau und Funktion der Gramineen-Lodiculae. *Beitr. Biol. Pfl.* **47**, 313–70.

Redman, R. E. & Reekie, E. G. (1982). Carbon balance in grasses. In *Grasses and Grasslands*, ed. J. R. Estes, R. J. Tyrl & J. N. Brunken, pp. 195–231. Norman: Oklahoma.

Rosengurtt, B., Laguardia, A. & Arrillaga de Maffei, B. R. (1971). El endosperma central lipido en la sistematica de Gramineas. *Adansonia, sér. 2*, **11**, 383–91.

Sendulsky, T., Filgueiras, T. S. & Burman, A. G. (1987). Fruits, embryos and seedlings. In *Grass Systematics and Evolution*, ed. T. R. Söderstrom, K. W. Hilu, C. S. Campbell & M. E. Barkworth, pp. 31–6. Smithsonian Institution Press: Washington DC.

Silberbauer-Gottsberger, I. (1984). Fruit dispersal and trypanocarpy in Brazilian cerrado grasses. *Plant Syst. Evol.* **147**, 1–27.

Söderstrom, T. R. (1981). Some evolutionary trends in the Bambusoideae (Poaceae). *Ann. Miss. Bot. Gard.* **68**, 15–47.

Söderstrom, T. R. & Calderon, C. E. (1971). Insect pollination of tropical rain forest grasses. *Biotropica*, **3**, 1–16.

Stapf, O. (1904). Fragmenta Phytographiae Australiae occidentalis: *Xerochloa* R.Br. *Bot. Jahrb.* **35**, 64–8.

(1919). Andropogon. In *Flora of Tropical Africa* 9, ed. D. Prain, pp. 209–10. L. Reeve & Co.: London.

Terrell, E. E. (1971). Survey of occurrences of liquid or soft endosperm in grass genera. *Bull. Torr. Bot. Cl.* **98**, 264–8.

Tran, V. N. (1973). Sur la valeur morphologique des lemmes des Graminées. *Bull. Mus. Hist. Nat. sér. 3*, **128**, Bot. 8, 33–57.

Vaid, K. M. (1962). Vivipary in bamboo, *Melocanna bambusoides* Trin. *J. Bombay Nat. Hist. Soc.* **59**, 696–7.

3

Ovule structure and diversity

J. Greenham and G. P. Chapman

Introduction

In this chapter the normal development sequence for sexual ovules of the Poaceae is described, with cytological details of the component parts. Although apomictic ovules are important in many grass species, the main focus of this chapter is the pre-fertilization sexual ovule.

A grass inflorescence matures either from its centre up and downward or from the apex to the base, and individual spikelets unfold their florets from the base to the apex. Within the florets themselves, development of the anthers and the ovule may occur at different rates and both protogyny and protandry can occur. The ovule is the central figure of this complex inflorescence. The rapid progress from a vegetative undifferentiated apex to ripening seeds, combined with the interactions between an ovule and the other parts of the developing inflorescence result in a progressive sequence of events which will vary with the species, and the particular genotype. The environment has an influence on all aspects of this development, extending in some species to a seasonal variation for sexuality or apomixis. It would perhaps be surprising therefore if, throughout the grass family, ovules were uniformly simple in structure and function. In fact, the similarity of the sporophyte can frequently mask diversity within the gametophyte, and the ovule is remarkably diverse in its various manifestations, and thus provides an important contribution to the reproductive versatility of the grasses.

Carpel and ovule ontogeny

The typical angiosperm gynoecium consists of an ovary bearing a style which bears a stigmatic surface. In the grasses there is a single unilocular gynoecium, the style is commonly bifid and the two stigmas plumose. Pistils with three styles and stigmas are common in the bamboos, and do occur sporadically in genera which are normally bifid. There is some uncertainty over the origin of the gynoecium. Philipson (1985) reviewed the evidence and came to the conclusion that the grass carpel was syncarpous and trimeric. Mehlenbacher (1970) suggested that the gynoecium was derived from three fused carpels in the bamboos, and from two carpels in other grass taxa.

The orientation of an ovule in its ovary varies during development, changing particularly as the carpel encloses the ovule and also as the embryo sac is formed. Comparisons of the structural features of ovules in the literature are complicated by the different ages of the ovules, this changing orientation, and the variations in descriptive terminology used. These points are discussed in detail by De Triquell (1986).

In the Poaceae there is one main type of mature ovule, anatropous. An anatropous ovule is more or less reflexed at the chalazal region, leaving the micropyle facing away from the style and towards the base of the gynoecium, and the ovule pressed to the adjoining placenta. The orientation of an ovule frequently changes from atropous to anatropous during its development (Maze, Bohm & Mehlenbacher, 1970).

In *Oryzopsis hendersoni*, for example, the carpel initiates beneath the floral apex which then goes on to differentiate the ovule. During development, the originally basal ovule becomes lateral, and its integuments more or less encircle the nucellus starting development first on the lemma side; concurrently, the styles are developing and the carpel is fusing around the ovule. At megasporengenesis, the single ovule is borne laterally on the lemma/ventral side of the ovary wall and towards the base of the gynoecium. Subsequent divisions in the ventral portion of the chalazal region tilt the ovule downward (Mehlenbacher, 1970).

For information on the gross morphology of a typical grass floral apex, the reader is directed to the papers by Barnard (1955, 1957), which look at the histogenesis of grass inflorescences and of the flower of wheat, and those by Bonnett (1966) in which the gross features of floret development in maize, wheat, rye, oats and barley are described and illustrated. Sharman (1960) has reported on the development of the carpel primordia in *Anthoxanthum odoratum*, and Waddington & Cartwright (1983) have developed a quantitative scale of initial spike and pistil development in barley and wheat.

Ovule morphology

While the principal events in angiosperm ovule development have been known since the late nineteenth century, there have been a number of studies over the last 20 years which have increased our information on specific aspects of grass ovules. A number of clearing techniques are now being used to complement paraffin sections in the investigation of ovule and megagametophyte development (for details, see Herr (1971, 1982) and Young, Sherwood & Bashaw (1979)). Again, the use of the electron microscope has added much detail to our knowledge of megagametophyte fine structure, and thereby generated more questions to be answered.

However, despite the world-wide importance of the grass family, there is still comparatively little information available on the development of grass ovules, particularly pre fertilization. With the increasing commercial interest in ovule culture and DNA injection techniques, it is probable that our understanding of ovule development and fertilization, particularly in the cereals, will soon improve.

The grass ovule develops from the ventral suture of the carpel, has no funicle, arising directly from the ovary wall, and consists of a nucellus and two integuments. The ovule is usually regarded as an integumented megasporangium and originates from hypodermal cells under the epidermis of the placenta. A primordium develops from the ovary placenta by periclinal divisions, forming a conical projection with a rounded tip. Further divisions in the protoderm produce a ring of tissue. These annular meristems usually produce an inner and then an outer integument, which subsequently more or less enclose the nucellus, leaving a small passage, the micropyle. Meanwhile, the megagametophyte is developing from a hypodermal cell in the nucellus, which consists of both vegetative and sporogenous cells.

Ovule vascularization

The venation and traces to the ovule are branches of the carpel vascular system, and in the ovary wall there are one, two or three vascular strands. In many grasses there are two lateral bundles, the median bundles passing on to the styles, plus an ovule supply formed from the marginal veins of the adjacent strands. These bundles are usually visible in cleared tissue. The vascular supply to most grass ovules is a single vascular strand traversing the placenta and ending in the chalazal region. The primary vascular tissue does not usually extend to the nucellus, and only occasionally to the integuments. In *Elymus virginicus* Beaudry (1951) found the vascular bundle entered the ovary at the base, and ran medially in the ventral ovarian wall to a position opposite the antipodals. Weir & Dale (1960) also found a large placental vascular strand in *Zizania aquatica* which traversed the posterior or lemma wall of the ovary and terminated in the chalazal region, again quite close to the antipodal region of the embryo sac.

The integuments

In the grasses the ovary is normally bitegmic. Generally, in the panicoid grasses the outer integument on the upper side of the ovule covers only about half of the ovule, while in the festucoid grasses the outer integument is more fully developed (Chandra, 1963). Even in the festucoid grasses the outer integument usually fails to cover the inner integument completely, which thus forms the micropyle. Fusion between the basal region of the ovule and the integuments at the attachment to the placenta produces the chalaza. At the chalaza there is less obvious differentiation between the integuments and the base of the ovule, and the chalaza depending on the orientation of the ovule can be short or long, and more or less contiguous with the ovule. Each of the integuments is normally originally composed of about two cell layers, but there can be subsequent division during development. The outer integument frequently becomes thicker at its base and occasionally integumentary cells in other regions enlarge and divide differentially.

The two ovular envelopes, the integuments, enclose and protect the central tissue, the nucellus. The integuments and the nucellar epidermis can be cutinized and prior to fertilization these tissues may control access to the megasporophyte, thus influencing virus invasion, desiccation, pollen tube entry and the impact of maternal tissue on the gametophyte. Usually following fertilization, the outer integument is lost, and the inner integument sclerifies and forms the testa of the grass caryopsis. The nucellar epidermis is also lost or modified during embryo development.

Nucellus

The nucellus provides the mass of the ovule, is usually several cells thick, and can be more or less persistent. It is the nucellar cells that differentiate and become the megaspore mother cells. Although Mehlenbacher (1970) reported multiple putative archesporial cells (i.e. with dense cytoplasm and large nuclei) in *Oryzopsis hendersoni*, only one subsequently functioned as a megaspore mother cell. In those grasses with apomixis, the distinction between archesporial cells and sporophytic cells becomes blurred.

The two main types of nucellus in the Poaceae are distinguished by the location of the megaspore mother cell. In a crassinucellate nucellus a large megaspore mother cell is located deep within the nucellus, and this is regarded as a primitive condition. In a tenuinucellate nucellus the megaspore mother cell differentiates just under the epidermis and is small. The differences in the position of the megaspore mother cell are linked to the number and extent of the divisions in the nucellar epidermis as the archesporial cell develops. In a tenuinucellate nucellus the thin layer of tissue left surrounding the megaspore can be disrupted by the subsequent expansion of the embryo sac, leading to the inside layer of the inner integument and the embryo sac wall becoming adjacent. An example of a crassinucellate nucellus is *Paspalum* and of a teniunucellate, *Eleusine*.

Differential growth of the ovule

Outgrowths of the ovule tissue are quite common in the grasses and can be divided into developments of the integuments, the nucellus or the embryo sac. This asymmetrical growth may have a role either in influencing transport to strategic portions of the developing ovule; or the orientation of pollen tubes. These developments usually originate in the integuments and the nucellar cells; they are produced only occasionally by the embryo sac.

In an early report True (1983) reported an outgrowth of the integument in *Zea mays* which entered the style base. A similar development of the outer integument has also been reported in *Zizania* (Weir & Dale, 1960). Maze *et al.* (1970), comparing the development of the outer integument in two related species, found in *Oryzopsis miliacea* the outgrowths to be transitory, occasionally absent and obliterated as the ovule developed, whereas in *Stipa tortilis*, the outer integument developed a large, persistent bump in the chalazal region. Maze *et al.* (1970) postulated that these bumps had some unknown role in fertilization. Mehlenbacher (1970) described a similar development of the outer integument in *Oryzopsis hendersoni* which reached 50 μm in size, and persisted until after fertilization, when the outgrowth cells died.

The epidermal cells of the micropylar nucellus adjacent to the embryo sac wall and the enclosed egg cell may increase in size or divide in some grasses (Maze & Bohm, 1974). In *Oryzopsis miliacea* Maze *et al.* (1970) found that the cells of the nucellar epidermis at the micropylar region were larger and that the pollen tube subsequently grew between these enlarged cells. In *Stipa tortilis*, which had the persistent integumentary bump, the nucellar epidermis at the micropyle did not enlarge conspicuously.

It is not uncommon to find that the cells around the micropyle are enlarged compared with cells near by, and frequently both the integuments and the nucellus are involved in this hypertrophy. The extent of this enlargement may be only slight, as illustrated but not discussed by Artschwager & McGuire in *Sorghum* (1949), while in a series of papers by Narayanaswami (1953, 1955, 1956) a whole range of hypertrophy at the micropyle is illustrated and discussed. In *Echinochloa frumentacea* the cells of the nucellar epidermis at the micropyle are only slightly enlarged. In *Setaria italica* the pair of nucellus cells closest to the micropyle enlarge, elongate and may protrude into the micropylar canal. Concurrently, one portion of the outer integument is forming a hood for the micropyle and another is developing a hump in the stylar region. In *Pennisetum typhoideum* one portion of the outer integument develops up into the stylar region. In this species the nucellar epidermis is never covered by the inner integument, and the nucellar epidermis over the embryo sac is four to six layers thick. These cells in the micropylar region enlarge and become densely protoplasmic. Narayanaswami suggested that these cells in *Pennisetum* were glandular and involved in pollen tube nutrition. Maze & Bohm (1974), in a summary of their work on ovule development, also reported the projection of the nucellar cells into the micropyle in *Agrostis interrupta*. They referred to enlarged cells occurring in *Avena* and in *Hordeum* (Cass & Jensen 1970).

At present there is no clear indication of the function of these enlarged cells. The development of both enlarged nucellar cells and integumentary projections in the same species may suggest that they play different roles. It is possible to speculate that the integumentary bumps could be simply a response to a localized release of pressure on the integumentary tissue. The developments around the micropyle of cells which give the appearance of being active secretory cells are more likely to be involved in dynamics of pollen tube development. It is interesting that Wagner *et al.* (1989) noted that, following enzymatic digestion of *Zea mays* ovules, the isolated embryo sac retained some attached nucellus cells, although the technique had removed the antipodals.

Philipson (1977) reported in *Cortaderia* the development of the synergids as haustorium through the cells lining the micropyle. The tips of the synergids grew beyond the micropyle, between the ovule and the ovary wall, exposing the megagametophyte to subsequent fertilization by the pollen tube. Occasionally, the haustoria grew through the integuments to reach the area between the ovary wall and the ovule. She also described the development of transfer walls and associated endoplasmic reticulum at the tip of the haustoria.

Fate of the ovule tissues

In the dicotyledons epidermal cells of the inner integument replace the nucellar tissue and form a nutritive layer around the archesporium and megagametophyte. Swamy & Krishnamurthy (1970) suggest that the presence of such an endothelium does not occur outside the dicotyledons, and that its absence is a characteristic of the monocotyledons. In grasses the outer integument usually disintegrates post fertilization, and the inner one survives to become the testa. There is also a varying degree of adhesion of the integuments, both to each other and to the developing fruit.

Morrison (1975) described the ultrastructure of developing wheat grains. At anthesis a mature wheat ovule had the inner epidermis of the pericarp and the outer integument separated by two thin osmiophilic, *cuticle-like* lamella, which were also located between the nucellus and the inner integument and between the two integuments. He did not test for cutin, but simply described this different layer. In wheat the nucellar epidermis became the hyaline layer, whereas in *Zea mays* Russell (1979) reported that at the two-celled megagametophyte stage the nucellar cells start to degenerate and that the eight-celled megagametophyte was surrounded by several layers of crushed nucellar cells.

Archesporial tissue

In angiosperm ovules the archesporial tissue is reduced to a single cell, the megaspore mother cell, and there are no parietal cells. In apomictic grasses there are a range of variants on this basic theme, since nucellar tissue can elaborate an embryo sac without meiosis and this diploid embryo sac together with the one of archesporial origin can coexist inside the same ovule. Moreover, since the unreduced embryo sac can sometimes complete embryogenesis from its egg cell with or without fertilization, its status within the framework of classical morphology is ambiguous. Biologically, the separation between gametophyte and sporophyte is blurred here and one is reminded of pollen grains (nominally gametophytic) which can, in culture, elaborate haploid or polyhaploid sporophytes. In grasses there can thus exist a reproductive nucellus where nominally sporophytic cells mimic features of the gametophyte.

Apospory and diplospory

Conventionally, the archesporium is both described and defined as producing megaspores by meiosis. If putative archesporial tissue generates megaspores and an embryo sac without meiosis, what distinguishes this sac from an unreduced embryo sac arising from the nucellus? Genetically, the two situations would seem identical. The archesporium is an ill-defined region in the nucellus, where the megaspore mother cell is usually described as differentiated by denser cytoplasm and a larger nuclei. The distinction between unreduced embryo sacs which arise from an archesporial cell (diplospory) and those which arise from supposedly non-archesporial cells (apospory) therefore merits reconsideration.

Polygonum **and** ***Oenothera*** **embryo sacs**

The familiar eight-nucleate *Polygonum* embryo sac derived from a single post-meiotic megaspore, which is characteristic of most angiosperms, also typifies the sexual embryo sac of the grasses. By contrast, apomictic embryo sacs are often four-nucleate with an egg apparatus, one polar nucleus and no antipodals. Although the arrangement specifically recalls *Oenothera*, in the latter genus the four nuclei derive from a post-meiotic megaspore. Perhaps the situation in grasses should be called 'pseudo-*Oenothera*'.

Pseudogamy

In pseudogamous apomicts this unreduced 'pseudo-*Oenothera*'-type embryo sac receives two haploid male gametes at the embryo sac: one fuses with the single diploid polar nucleus resulting in triploidy, the other distintegrates and the diploid egg cell develops parthenogenetically. It is noteworthy that this mechanism recreates the chromosome numbers from sexual fusion ($2n:3n$), but from a 'non-antipodal' embryo sac (see Chapter 5).

Megasporogenesis

There are two phases in the normal developmental transition from female sporophyte to female gametophyte: megasporogenesis, the meiotic event that turns the sporophyte 'nucellar' cell into the megaspore, followed by megagametogenesis, the mitotic event that turns the megaspore into the gametophyte containing the female gamete.

The nucellus and its surrounding integuments are female sporophyte tissue. One of the nucellar cells becomes the megaspore mother cell which develops into the megagametophyte. The megagametophyte can be meiotic and gametophytic in origin or mitotic and (thus sporophytic), both types usually originating in the nucellus. In the former there is a transition from sporophyte to the gametophyte stage, which is enclosed by maternal tissue.

Meiosis in the megaspore mother cell

The meiotic division of the megaspore mother cell may be linear, producing a line of megaspores, or asymmetric, producing a T or an inverted T. The second division in one of the dyad pairs may also be omitted producing a triad of megaspores. A T-shaped tetrad or triad is found occasionally in many species including: *Bambusa* and *Coleanthus* (Schnarf, 1929), *Bromus* (Beck & Horton, 1932), *Eragrostis cilianensis* (Stover, 1937), *Echinochloa frumentacea* (Narayanaswami, 1955) and *Poa* (Nygren, 1950) quoted in Narayanaswami (1955). In some grasses a T-shaped arrangement of megaspores is the norm as in *Festuca* (Maze & Bohm, 1977).

On division, the megaspore mother cell may or may not produce cell walls, and the cells particularly at the micropylar end may be plus or minus cell walls after division. Division may also produce either a dyad, a triad or a tetrad of megaspores. In *Polygonum*-type development it is the basal or chalazal spore which develops to form the megagametophyte, the three micropylar megaspores usually degenerating quickly. In the apomictic *Poa alpina* the tetrad is very variable, and any megaspore may develop into the embryo sac (Hakansson, 1943).

Muniyamma (1976) described megaspore development in *Agrostis pilosula*. The archesporial cell is hypodermal and develops into the megaspore mother cell. After the first meiotic division the inner cell of the dyad is larger, and the second division produced a T-shaped triad in 70% of the ovules observed; in the remaining 30% the smaller outer cell of the dyad did not divide and only a linear triad was produced. A linear tetrad was found by Maze & Bohm (1974) in *Agrostis interrupta*.

The polarity of the megaspores with the chalazal dyad cell dividing before the micropylar found in *Agrostis pilosula* is also found in *Poa pratensis* (Tinney, 1940; Nielsen, 1946) and in *Panicum miliaceum* and *P. miliare* (Narayanaswami, 1955).

Usually, the megaspore mother cell (MMC) is differentiating and dividing early on in the ovule's development as the integuments start to envelop the nucellus. It is possible to sample accurately the various stages in the process by correlating meiosis in the MMC with the external ovary devlopment. Bennett, Chapman & Riley (1971) looked at meiosis in the ovules of Triticale, *Triticum aestivum* and *Secale* and found it took 20–45 h and was coupled to anther development, a correlation later confirmed in *Hordeum vulgare* (Bennett, 1977). However, in other species male meiosis is earlier, e.g. *Brachiaria ruziziensis* (Gobbe, Longly & Louant, 1982), while in others the male gametogenesis is later than that of the female. In *Eragrostis tef* the two meiotic events coincide, and Longly, Rabau & Louant (1985) suggested that this parallel development is integral to autogamy. However, further studies of cross-pollinated rye by Xiang-Yuan & DeMason (1984) found that meiosis in both the megasporocytes and the microsporocytes also began concurrently and that the megaspores and the microspores reached maturity together.

Callose in megaspore walls

It may be significant that callose production and its deposition in the megaspore cell walls is associated with the formation of the megaspores. Rodkiewicz (1970) has described callose deposition in 43 species from 14 families of the angiosperms. All of the 32 dicot species and all but 3 of the 11 monocots had callose. The three monocot species which lacked callose were not grasses, and all three had tetrasporic embryo sacs.

In Rodkiewicz's (1970) general description of *Polygonum*-type megasporogenesis callose appeared first in the chalazal portion of the megasporocyte, i.e. around the functional spore, extended to the micropylar end of the tetrad and then started to decrease first at the chalazal end as the functional megaspore developed into a megagametophyte. Rodkiewicz suggested that callose plays a part in isolating the megaspores, particularly the functional one, and helps to explain the subsequent degeneration of three out of four of the tetrad cells.

An early paper by Cooper (1937) describes wall thickening of the micropylar end of dividing megaspores in *Zea*, *Euchlaena* and their hybrid. The presence of callose around grass MMC was subsequently reported by Schwab (1971), detected in *Diarrhena* using aniline blue fluorescence. The fluorescence appeared as a thin layer around the megaspore mother cell at prophase I of meiosis, while at the dyad stage the two cells were surrounded and divided by callose with more fluorescence occurring in the micropylar cell. At the tetrad stage the micropylar megaspore was almost obscured by the strong fluorescence, attributed to callose. Russell (1979) also described an aniline blue fluorescence in the megasporocyte wall in *Zea*, with the fluorescence slightly more intense in the wall at the micropylar end. After division the cross-walls fluoresced and the functional megaspore decreased in fluorescence at the chalazal wall first (where plasmodesmata occur), whereas the non-functional megaspores increased in fluorescence. He suggested that these events were related to the imposition of an internal polarity on to the functional megaspore.

Megaspore mother cell: cytology and biochemistry

Mogensen (1977) reported double membrane structures in the megaspore mother cell of *Zea*, which he presumed were dedifferentiated plastids and mitochondria. He also observed that the MMC had numerous ribosomes but that the nucellar cells had more.

The megaspore mother cell organelles disappear during meiosis. In *Capsella* autophagic vacuoles remove ribosomes, mitochondria, plastids and endoplasmic reticulum from the MMC. After meiosis, the functional megaspore goes on to produce new ribosomes from its nucleolus.

Megagametogenesis

In the *Polygonum*-type embryo sac there are three mitoses following meiosis, producing an eight-nucleate embryo sac. This *Polygonum*-type development has been widely reported in the sexual grasses, although there is one report (Mohamed & Gould, 1966) of an *Adoxa*-type sexual megasporocyte in the *Bouteloua curtipendula* complex in which one mitotic division of the four megaspore nuclei produces an eight-nucleate embryo sac. In apomicts the development is still monosporic, but may be either pseudo-*Oenothera* with only two mitoses leaving the embryo sac four-nucleate or pseudo-*Polygonum*.

The functional megaspore nucleus (usually that furthest from the micropyle in sexual grasses) undergoes three mitotic divisions to form an eight-nucleate megagametophyte containing three antipodals, an egg apparatus consisting of an egg cell and two synergids, and two polar nuclei. The egg nucleus and one polar nucleus are sisters, as are the two synergids. The two polar nuclei may fuse pre or post fertilization to form a central cell or secondary nucleus.

The appearance of the sexual megagametophyte and the subsequent development of its constituent cells varies between genera and families, and apomixis (see Chapter 5) introduces a whole range of new megagametophyte types.

Alternation of generations

The interrelations between the sporophyte nucellus and the gametophyte embryo sac are of great interest. The first indication of the separation of the generations may be this deposition of callose around the megaspore mother cell and its meiotic tetrad. Plasmodesmatal continuity appears to be transiently lost or greatly reduced at meiosis, so that prior to megaspore mitosis, the developing megagametophyte is cut off from direct contact with the sporophyte nucellus.

Carroll & Mayhew (1976) studied barley stripe mosaic virus and found no plasmodesmata between the megagametophyte and the nucellus, which led them to presume that virus transmission is via early infection of the archesporial cell. Diboll (1968) also found that plasmodesmata occurred between all the cells of the *Zea* megagametophyte, but not in its outer wall.

Diboll & Larson (1966) described the mature megagametophyte wall of *Zea* as a series of lamellae – the innermost is the wall of the functional megaspore, the outer layers are remnants of nucellus cells – and they did not see plasmodesmata from the megaspore into the nucellus. However, Russell (1979), in a later study of maize megagametophyte development, found a few plasmodesmata in the megaspore mother cell at its chalazal end and, after meiosis, the functional megaspore also had a few plasmodesmata at the chalazal end. Seen too, but not explained, was an electron-dense layer on the inner surface of the megaspore wall enclosing the megagametophyte. This layer persisted, as did the plasmodesmata, to the binucleate stage as the degeneration of the nucellus continued. Plasmodesmata were still visible at the four-celled stage.

Cass, Peteya & Robertson (1985) found that, at the binucleate stage of megagametophyte development, there was a loss of part of the wall area in irregular patches, which they suggested was due to loss of the remnants of callose from the megaspore wall.

The surrounding nucellus starts to degenerate at the megaspore stage, as the embryo sac enlarges. The megagametophyte is enlarging to occupy a larger proportion of the nucellus, and the first growth phase of the megagametophyte is as the functional megaspore divides to form the embryo sac. Within the embryo sac the antipodals start to divide, the polar nuclei may be migrating and internal cell walls are laid down.

Internal divisions of the megagametophyte

The mitotic divisions of the megaspore give rise to synergids, egg cells, polar nuclei and antipodals. The number, presence and position of these cells vary. In monosporic megagametophytes vacuolation occurs at the two- to four-nucleate stage and the two pairs of nuclei are separated by the vacuole. At the four-nucleate stage, the division of the micropylar pair of nuclei gives the synergids (sister cells) and the egg cell and one polar nucleus (sister cells). The division of the chalazal nuclei gives three antipodals and one polar nucleus. The synergids and antipodals are ephemeral; the egg cell and polar nucleus contribute post fertilization to the embryo and the endosperm, respectively.

The shape of the embryo sac can vary. Occasionally, the basal region (i.e. the region away from the micropyle) is more or less enlarged, as in *Poa* (Anderson, 1927) and *Agrostis* (Muniyamma, 1976) or it can change with time. In *Oryzopsis miliacea* Maze *et al.* (1970) found that the shape of the megagametophyte was wider at the micropyle pre fertilization,

Fig. 3.1. Longitudinal view of the mature maize megagametophyte reconstructed from light micrographs of 1 μm thick sections. Antipodals (AN), apical pocket (AP), central cell (CC), egg (E), filiform apparatus (FA), nucellus (NC), polar nuclei (PN), synergid (SY). Magnification ×1000. (Diboll & Larson, 1966.)

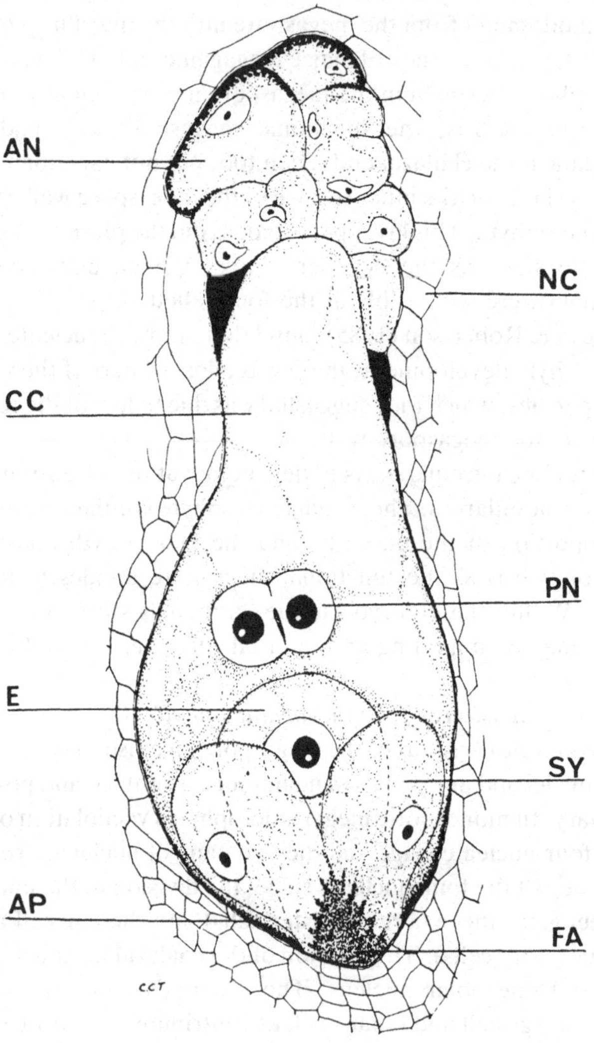

uniform at fertilization and wider at the chalazal end post fertilization, whereas in the related *Stipa tortilis* the diameter was constant throughout.

The development of the megagametophyte is different in the pooid and the panicoid grasses: in the panicoid grasses the antipodals are at the basal or chalazal end of the megasporocyte; in the pooid grasses they come to lie laterally along the placental side of the megasporocyte. Chandra (1963) summarized the differences between ovule development in the festucoid and the panicoid grasses as follows. In the panicoid grasses the outer integument develops first on the upper side of the ovule and grows to cover only about half the ovule, there are periclinal divisions in the nucellar epidermis which produce parietal-like cells, and the antipodals remain at the chalazal end of the megagametophyte. In the festucoid grasses the outer integument grows to cover most of the ovule, there are no periclinal divisions in the nucellus epidermis, and the antipodals come to occupy a lateral position in the megagametophyte. Weir & Dale (1960) have reported periclinal divisions in the nucellar protoderm of *Zizania aquatica*, but these differences are not absolute, and Maze *et al*. (1970) have questioned the value of periclinal division as a taxonomic character. They described periclinal division in *Stipa*, which did not, however, produce parietal cells. In the sexual grasses there is usually one embryo sac per ovule.

Megagametophyte: orientation

The megagametophyte is orientated in one of two ways to the long axis of the ovary: it may be parallel, as in *Bambusa*, *Oryza* and *Zea*, or at right angles/perpendicular, as in the Festuceae, the Hordeae and the Aveneae. There are two long axes: the micropyle to the chalaza, i.e. the ovule axis, and the antipodals to the egg cell, i.e. the embryo sac axis. Chandra (1963) found that the two ovule axes are not parallel in festucoid grasses and as a result the antipodals tend to be lateral, while in the panicoid grasses the antipodals are terminal and the two long axes are parallel.

Megagametophyte: polarity

As the megagametophyte is developing, new cell organelles are being formed prior to and during the transition from sporocyte to gametophyte, and polarization is occurring as the megagametophyte divides and increases in size. The initial enlargement is along the ovule axis, and a vacuole forms and surrounds the central nucleus. Subsequent mitosis produces two nuclei, then four nuclei separated by a vacuole. New internal walls start to be laid down first around the antipodals and the egg apparatus, and the whole embryo sac is contained in a wall derived from the megaspore. There is a pause in the growth of the megagametophyte following mitosis and prior to fertilization. Within the constituent cells of the megagametophyte, new organelles form and polarity begins to be apparent.

The egg cell has a vacuole at the micropyle end and its nucleus at the chalazal/basal end. The synergids have a filiform apparatus at the micropyle end, a median constriction or pointed end, and a concentration of their cell organelles and vacuoles at the chalazal end. The central egg cell and the synergids have their cell walls modified at the chalazal end.

Morphology of the megagametophyte wall

The megagametophyte wall is adjacent to and may include remnants of sporophytic tissue. The cell layers around the megagametophyte are produced by the nucellus, and the wall of the megaspore mother cell becomes part of the wall of the megasporocyte. The expansion of the megagametophyte disrupts the vegetative portion of the nucellus, and the surrounding tissue may be lost or subsequently reabsorbed. The major expansions in size are on vacuolation at the two- to four-nucleate stage and then post fertilization. Norstog (1974) found that post-fertilization megagametophytic growth in barley was accommodated by a breakdown of the nucellus and the production of a nucellar lysate by actual lysis rather than physical disruption, and Russell (1979) described nucellar cell degeneration *preceding* megagametophyte expansion in maize.

On p. 58 reference was made to the blurred distinction between gametophyte and sporophyte and on p. 62 to the alternation of generations. How far the boundaries of sexual and apomictic embryo sacs genuinely differ ultrastructurally seems at present only conjectural.

Central cell

In 1898 Navashin recognized the fusion of the two polar nuclei in *Lilium* with the secondary male nucleus to give the primary endosperm nucleus in the central cell. The polar nuclei which form the central cell originate from opposite ends of the embryo sac. The exception described by Stover (1937) in *Eragrostis ciliansis* is apparently based on a misinterpretation (De Triquell, 1986).

Cass, Peteya & Robertson (1986) described the development of the horizontal wall separating the egg apparatus from the micropylar polar nuclei in barley. This wall formation is followed by the expansion of the megagametophyte and the migration of the polar nuclei to the central cell region. Cass & Jensen (1970) described a bridge of cytoplasm linking the egg apparatus to the antipodals. The bridge contained the partially fused polar nuclei. In both *Zea* and *Hordeum* the cytoplasm of the central cell surrounds and invaginates the egg apparatus. The vacuolation of the central cell makes it the largest cell in the megagametophyte.

Fusion of the polar nuclei occurs post fertilization in *Hordeum* (Pope,

1937), and precedes zygote formation in *Elymus virginicus* (Beaudry, 1951). Triple fusion also precedes syngamy in *Zea mays* (Diboll & Larson, 1966). The coupled polar nuclei usually lie adjacent to the egg apparatus in *Hordeum*, but are occasionally located antipodally (Cass & Jensen, 1970). Mogensen (1984) describes the post-fertilization migration of the primary endosperm nucleus towards the antipodals in barley.

The central cell has contact with both the former megaspore wall at the nucellus, and also the new cell walls laid down during mitotic division at the antipodals and the egg apparatus. In *Zea* Diboll & Larson (1966) observed plasmodesmata between all the cells of the embryo sac. There was a thickening of the central cell wall at the antipodals, as well as an apical pocket around the egg apparatus, which contained dictyosomes and associated lipid vesicles. Cass *et al.* (1986) reported membrane contact between the central cell and the egg apparatus following expansion of the megagametophyte. Wagner *et al.* (1989) found that the central cell and the egg apparatus detached themselves from the nucellus and the antipodals following enzymatic digestion in *Zea*.

The central cell itself is largely vacuole, with its cytoplasm lining the walls of the megagametophyte, the adjacent embryo sac cells and particularly the polar nuclei. In maize and barley cytoplasmic starch grains are common particularly around the nuclei, and the starch content of the central cell falls rapidly after one or two endosperm divisions. The nucleolus of the central cell nuclei occupies about half the nuclear volume, and the presence of nucleolar vacuoles suggests that there is protein synthesis, but the endoplasmic reticulum is not well developed, suggesting that the role of the central cell is to store rather than to produce.

Tilton & Lersten (1981), in a review article on angiosperm reproduction, suggested that the central cell generally plays a passive role in unfertilized ovules. Nishiyama & Yabuno (1978) proposed that the central cell had a role in sexual isolation, based on the results of a diallel cross of 10 *Avena* species, where the endosperm failed to develop concurrently with the embryo in certain unbalanced crosses. Normally, the fusion and division of the endosperm precedes fusion and division in the zygote (Beaudry, 1951).

In some apomictic species there may be only one polar nucleus, and in non-pseudogamous apomicts this also develops without fertilization.

The antipodals and their role

These are very interesting cells, with potential roles in early endosperm nutrition and the transfer of nutrients from the ovary to the megagametophyte. They have all the cell organelles associated with intense metabolic activity, and may increase in size markedly post fertilization. This and their location led Brink & Cooper (1944) to suggest that they linked the vascular tissue to the developing endosperm, albeit for a brief period, and provided a growth factor for the endosperm.

The antipodals are located in the chalazal region/basal region or slightly laterally of the chalazal pole. They are usually compact and located chalazally at least initially or laterally as the endosperm develops. The displacement of the antipodals to a lateral position is caused by the growth of the megagametophyte past them. They may proliferate before fertilization and only persist for a short time post fertilization.

In the Poaceae there is great variation in the subsequent development of the initial three antipodal cells into antipodal tissue. The original three antipodal cells may divide to give a large number of cells. An extreme case is the 300 antipodals described in a bamboo, *Sasa paniculata* (Yamaura, 1933). Cass & Jensen (1970) reported up to 100 antipodals per embryo sac at fertilization in *Hordeum*, with no further division post fertilization. Diboll & Larson (1966) found about 20 antipodal cells in *Zea mays* with between one and four nuclei each. Individual antipodal cells may be multinucleate or polyploid; endoreplication can occur or a multinucleate syncytium can be formed. Walls are absent in parts of the antipodal mass in *Zea* forming a syncytium. In *Hordeum jubatum* Brink & Cooper (1944) found that they continued to enlarge after fertilization. In inviable *Elymus virginicus* × *Agropyron repens* seed Beaudry (1951) reported endomitosis and more than 100 chromosomes per cell, compared with $2n = 14$. He regarded this post-syngamic activity as the abnormal condition and linked it to a subsequent failure of endosperm development. In *Chloris gayana*, prior to the fusion of the polar nuclei, the three antipodal cells become very large and may occupy more than half the megagametophyte volume (Chikkannaiah & Mahalingappa, 1975). In *Pennisetum* the antipodal cells are multinucleate (Narayanaswami, 1953) and in *Agrostis* Maze & Bohm (1974) reported three to four nuclei per cell. Muniyamma (1976) thought that the multiple nuclei he found in *Agrostis pilosula* were due to a hormone imbalance, as did Maze & Bohm (1974) in *A. interrupta*, whereas Diboll & Larson (1966) thought the multinucleate antipodals they found in *Zea* were due to incomplete wall formation during division. In *Poa alpina* non-functional megagametophytes can have antipodals at both ends (Hakansson, 1943).

The antipodals are vacuolate and show no polarization in *Zea*, they are papillate at both their junction with the nucellus and the central cell

(Diboll & Larson, 1966). Cass & Jensen (1970) described wall formation following mitosis in barley, and found that it was similar to the wall at the egg apparatus end, with a horizontal wall separating the central cell from the antipodals and then vertical walls separating the antipodal cells themselves. Diboll & Larson (1966) found that cell wall formation within the antipodal mass in *Zea* was incomplete, with perforations giving cytoplasmic continuity. The inner face of the antipodals adjacent to the nucellus was papillate and the nucellar cells at the antipodal region were vacuolate. The high rate of respiration associated with the tissue made them suggest that transport was occurring. They also reported plasmodesmata between all the cells of the megagametophyte, but none with the nucellus. Maze & Lin (1975) also reported that no plasmodesmata occurred between the antipodals and the nucellus in *Stipa*. In recent work on *Zea* Wagner *et al.* (1989) found that the antipodals are separated from the central cell and egg apparatus by enzymatic digestion; Morrison (1955) also found that they separated easily from the rest of the embryo sac in wheat.

At fertilization, the antipodals increase in activity and organelle number. Diboll (1968) also found an increase in the wall thickening in the papillate area adjacent to the nucellus.

Diboll & Larson (1966), Diboll (1968) and Maze & Lin (1975) have all reported that the antipodals are apparently very active cells in *Zea* and *Stipa*, with large nucleoli, a high RNA content (suggesting that they are involved in protein synthesis) and numerous mitochondria and plastids. Antipodals are also frequently multinucleate, with a well-developed endoplasmic reticulum, numerous mitochondria and dictyosomes. They inferred that antipodals have a role in transferring materials from the nucellus to the megagametophyte and possibly the endosperm. Jones & Rost (1989) studied the megagametophyte in rice and found that the antipodals stained for RNA, and they suggested that they were synthesizing enzymes. The role of the antipodals appears to be replaced two or three days after fertilization by the endosperm tissue.

The antipodals have a high protein content in *Hordeum* (Cass & Jensen, 1970), a high lipid content with vesicles in *Zea* (Diboll & Larson, 1966) and in *Stipa elmeri* (Maze & Lin, 1975) and they usually have some starch, though this is reportedly absent in *Hordeum* (Cass & Jensen, 1970).

Two views of the antipodals are that they are either secretory or sacrificial. These views might be merged to suggest that, leading up to their demise a few days after fertilization, the first function is increasingly replaced by the second. Two other points of view are possible, one that they are versatile and, since the male nuclei entering both the egg and the central cell can vary, that they might develop more than one response.

There is no clear evidence in favour of any one view, although abundant DNA replication so typical of actively secretory tissue hardly suggests no function. Grass seeds are endospermic, indeed conspicuously so, and it may be that the need to establish nutrient support for the zygote is of such importance that the endosperm itself is, from the moment of its formation, at the service of a tissue already primed. To extend speculation further, if a consistent feature of grass evolution is to integrate genetic material from beyond the species or even the genus, endosperm establishment might be precarious unless subject to antipodal support. To secure such advantages, a self-destroying surge of antipodal activity for a few days after fertilization is a negligible cost.

Brink & Cooper (1944) and Beaudry (1951) reported that the antipodals hypertrophied in grass hybrids, postulating that such a lack of development had a role in hybrid seed failure, a topic worth re-examination with so many intergeneric hybrids now available.

Egg apparatus

The egg apparatus is usually at the micropylar end of the megagametophyte. The egg is larger than the synergids in *Stipa* (Maze & Lin, 1975) and in *Festuca* (Maze & Bohm, 1977) and has numerous small vacuoles at the micropyle end in *Zea* (Diboll & Larson, 1966) and *Festuca* (Maze & Bohm, 1977).

The egg is frequently pear-shaped and its walls form common surfaces with the synergids. Diboll & Larson (1966) and Diboll (1968) reported an attenuated cell wall at the chalazal end, reduced to plasma membrane only along the junction of the egg apparatus with the polar nuclei in *Zea*. The DNA staining of the egg nucleus reduces as fertilization approaches and the nucleolus enlarges, although the cytoplasmic DNA is high. Again, although the endoplasmic reticulum is reduced, the RNA levels are high. These facts, combined with the presence of abundant starch, suggest a state of metabolic preparedness.

Egg synergids

The synergids of the angiosperms degenerate rapidly post fertilization and probably have either a nutritive, a transfer or a secretory role in the development of the egg cell and orientation of the pollen tube (see Chapter 4). They are located close to the egg cell and the subsequent zygote/embryo; they usually have abundant cytoplasmic organelles, including dense mitochondria, and storage products; and they have a polysaccharide invagination of the cell surface, known as a filiform apparatus, particularly in the region adjacent to the micropyle/ovary wall.

The synergids are strongly polarized in *Zea*, and have their nuclei and organelles at the upper/micropylar end, the vacuole at the lower/chalazal end and a filiform apparatus or wall invagination at the micropylar end. The filiform apparatus was first described by Schacht in 1857, in *Zea* synergids. Some Poaceae also have hooked-shaped projections at the micropylar end of the synergid (Chikkannaiah & Mahalingappa, 1975). This constriction in the cell divides the synergid, and the upper portion is in contact with the embryo sac wall. It is here that the filiform apparatus, a secretory or transfer area, develops. The cellulose wall extends microtubules into the cytoplasm, which the cytoplasm envelops, thus increasing the surface area, and mitochondria are associated with the apparatus. Filiform apparatus is also found in transfer cells. It may secrete chemotropic substances that help in the transport of substances from the nucellus to the embryo and the endosperm, although Diboll (1968), unusually, reported that the endoplasmic reticulum and ribosomes were sparse in *Zea mays* synergids and the filiform apparatus.

Diboll & Larson (1966) described an attenuated wall at the chalazal end of the synergids of *Zea*, which was also found by Maze & Lin (1975) in *Stipa*. They are separated from the central nucleus by two plasmalemmas only, and plasmodesmata have also been reported between the synergids and the egg.

The synergids have an important role at fertilization, opening the tip of the pollen tube, and placing the released male gametes between the egg and the secondary nucleus. Cass & Jensen (1970) found that the synergid degenerated after pollination but before the pollen tube entered the megagametophyte. In *Eragrostis tef* they degenerated pre fertilization, but left the filiform apparatus intact (Longly *et al.*, 1985). Both synergids degenerate in wild rice (Weir & Dale, 1960); in *Oryzopsis miliacea* only one synergid degnerates and fertilization is assumed to occur via the intact synergid (Maze *et al.*, 1970). Hakansson (1943) claimed that the synergids themselves were fertilized in *Poa alpina*.

The ovule is thus differentiated and primed for development by either sexual or apomictic means.

References

Anderson, A. M. (1927). Development of the female gametophyte and caryopsis of *Poa pratensis* and *Poa compressa*. *Journal of Agricultural Research*, **34**, 1001–18.

Artschwager, E. & McGuire, R. C. (1949). Cytology of reproduction in *Sorghum vulgare*. *Journal of Agricultural Research*, **78**(12), 667.

Barnard, C. (1955). Histogenesis of the inflorescence and flower of *Triticum aestivum* L. *Australian Journal of Botany*, **3**, 1–20.

(1957). Floral histogenesis in the monocotyledons. I. The Gramineae. *Australian Journal of Botany*, **5**, 1–20.

Beaudry, J. R. (1951). Seed development following the mating *Elymus virginicus* L. × *Agropyron repens* (L.) Beauv. *Genetics*, **36**, 109–33.

Beck, R. & Horton, J. S. (1932). Microsporogenesis and embryology in certain species of *Bromus*. *Botanical Gazette*, **93**, 42–54.

Bennett, M. D. (1977). The time and duration of meiosis. *Philosophical Transactions of the Royal Society, London B*, **266**, 39–81.

Bennett, M. D., Chapman, V. & Riley, R. (1971). The duration of meiosis in pollen mother cells of wheat, rye and triticale. *Proceedings of the Royal Society, series B*, **178**, 259–75.

Bonnett, O. T. (1966). *Inflorescences of Maize, Wheat, Rye, Barley, and Oats: their initiation and development.* University of Illinois College of Agriculture, Agriculture Experiment Station Bulletin 721.

Brink, R. A. & Cooper, D. C. (1944). The antipodals in relation to abnormal endosperm behaviour in *Hordeum jubatum* × *Secale cereale* hybrids. *Genetics*, **29**, 391–406.

Carroll, T. W. & Mayhew, D. E. (1976). Occurrence of virions in developing ovules and embryo sacs of barley in relation to the seed transmissibility of barley stripe mosaic virus. *Canadian Journal of Botany*, **54**, 2497–512.

Cass, D. D. & Jensen, W. A. (1970). Fertilization in barley. *American Journal of Botany*, **57**, 62–70.

Cass, D. D., Peteya, D. J. & Robertson, B. L. (1985). Megagametophyte development in *Hordeum vulgare*. 1. Early megagametogenesis and the nature of cell wall formation. *Canadian Journal of Botany*, **63**, 2164–71.

(1986). Megagametophyte development in *Hordeum vulgare*. 2. Later stages of wall development and morphological aspects of megagametophyte cell differentiation. *Canadian Journal of Botany*, **64**, 2327–36.

Chandra, N. (1963). Some ovule characters in the systematics of the Gramineae. *Current Science*, **32**, 277–9.

Chikkannaiah, P. S. & Mahalingappa, M. S. (1975). Antipodal cells in some members of Gramineae. *Current Science*, **44**, 22–3.

Cooper, D. C. (1937). Macrosporengenesis and embryo sac development in *Euchleana mexicana* and *Zea mays*. *Journal of Agricultural Research*, **55**, 539–51.

De Triquell, A. A. (1986). Grass gametophytes: their origin, structure and relation to the sporophyte. International Grassland Symposium, 1986. Ed. T. R. Söderstrom, K. W. Hilu, C. S. Campbell & M. E. Barkworth.

Diboll, A. G. (1968). Fine structural development of the megagametophyte of *Zea mays* following fertilization. *American Journal of Botany*, **55**, 797–806.

Diboll, A. G. & Larson, D. A. (1966). An electron microscope study of the mature megagametophyte in *Zea mays*. *American Journal of Botany*, **53**, 391–402.

Gobbe, J., Longly, B. & Louant, B. P. (1982). Calendrier des sporogenèses et gametogenèses femelles chez le diploide et le tetraploide induit de *Brachiaria ruziziensis* (Gramineae). *Canadian Journal of Botany*, **60**, 2032–6.

Hakansson, A. (1943). Die Entwicklung des Embryosacks und die Befruchtung bei *Poa alpina*. *Hereditas*, **29**, 25–61.

Herr, J. M. (1971). A new clearing-squash technique for the study of ovule development in angiosperms. *American Journal of Botany*, **58**, 785–90.

(1982). An analysis of methods for permanently mounting ovules cleared in four-and-a-half type clearing fluids. *Stain Technology*, **57**(3), 161–9.

Jones, T. J. & Rost, T. L. (1989). Histochemistry and ultrastructure of rice (*Oryza sativa*) zygotic embryogenesis. *American Journal of Botany*, **76**(4), 504–20.

Longly, B., Rabau, T. & Louant, B. P. (1985). Developpement floral chez *Eragrostis tef*: dynamique des gametophytogenèses. *Canadian Journal of Botany*, **63**, 1900–6.

Maze, J. & Bohm, L. R. (1974). Embryology of *Agrostis interrupta* (Gramineae). *Canadian Journal of Botany*, **52**, 365–79.

(1977). Embryology of *Festuca microstachys* (Gramineae). *Canadian Journal of Botany*, **55**, 1768–82.

Maze, J. & Lin, Shu-Chang (1975). A study of the mature megagametophyte of *Stipa elmeri*. *Canadian Journal of Botany*, **53**, 2958–77.

Maze, J., Bohm, L. R. & Mehlenbacher, L. E. (1970). Embryo sac and early ovule development in *Oryzopsis miliacea* and *Stipa tortilis*. *Canadian Journal of Botany*, **48**, 27–41.

Mehlenbacher, L. E. (1970). Floret development, embryology, and systematic position of *Oryzopsis hendersoni* (Gramineae). *Canadian Journal of Botany*, **48**, 1741–8.

Mogensen, H. L. (1977). Ultrastructural analysis of female pachynema and the relationship between synaptonemal complex length and crossing over in *Zea mays*. *Carlsberg Research Communications*, **42**, 475–97.

(1984). Quantitative observations on the pattern of synergids degeneration in barley. *American Journal of Botany*, **71**, 1448–51.

Mohamed, A. H. & Gould, F. W. (1966). Biosystematic studies in the *Bouteloua curtipendula* complex. II. Megasporogenesis and embryo sac development. *American Journal of Botany*, **53**, 166–9.

Morrison, I. N. (1975). Ultrastructure of the cuticular membranes of the developing wheat grain. *Canadian Journal of Botany*, **53**, 2077–87.

Morrison, J. W. (1955). Fertilization and postfertilization development in wheat. *Canadian Journal of Botany*, **33**, 168–76.

Muniyamma, M. (1976). A cytoembryological study of *Agrostis pilosula*. *Canadian Journal of Botany*, **54**, 2490–6.

Narayanaswami, S. (1953). The structure and development of the caryopsis in some Indian millets. 1. *Pennisetum typhoideum* Rich. *Phytomorphology*, **3**, 98–111.

(1955). The structure and development of the caryopsis in some Indian millets. IV. *Echinochloa frumentacea* Link. *Phytomorphology*, **5**, 161–71.

(1956). Structure and development of the caryopsis in some Indian millets. VI. *Setaria italica*. *Botanical Gazette*, **117**, 112–22.

Nielsen, E. L. (1946). The origin of multiple macrogametophytes in *Poa pratensis*. *Botanical Gazette*, **108**, 41–50.

Nishiyama, I. & Yabuno, T. (1978). Causal relationships between the polar nuclei in double fertilization and interspecific cross-incompatibility in *Avena*. *Cytologia*, **43**, 453–66.

Norstog, K. (1974). Nucellus during early embryogeny in barley: fine structure. *Botanical Gazette*, **135**, 97–103.

Nygren, A. (1950). Cytological and embryological studies in arctic *Poae*. *Symbolae Botanicae Upsaliensis*, **10**, 1–64.

Philipson, M. N. (1977). Haustorial synergids in *Cortaderia* (Gramineae). *New Zealand Journal of Botany*, **15**, 777–8.

Philipson, W. R. (1985). Is the grass gynoecium monocarpellary? *American Journal of Botany*, **72**(12), 1954–61.

Pope, M. (1937). The time factor in pollen-tube growth and fertilization in barley. *Journal of Agricultural Research*, **54**, 525–9.

Randolph, L. F. (1936). Developmental morphology of the caryopsis of maize. *Journal of Agricultural Research*, **53**, 881–916.

Rodkiewicz, B. (1970). Callose in cell walls during megasporogenesis in angiosperms. *Planta (Berlin)*, **93**, 39–47.

Russell, S. D. (1979). Fine structure of megagametophyte development in *Zea mays*. *Canadian Journal of Botany*, **57**, 1093–110.

Schnarf, K. (1929). Embryologie der Angiospermen. *Handb. Pflanzenanat. Berlin*, **11**, 1–690.

Schwab, C. A. (1971). Callose in megasporogenesis of *Diarrhena* (Gramineae). *Canadian Journal of Botany*, **49**, 1523–4.

Sharman, B. C. (1960). Developmental anatomy of the stamen and carpel primordia in *Anthoxanthum odoratum* L. *Botanical Gazette*, **121**, 192–200.

Stover, E. L. (1937). The embryo sac of *Eragrostis cilianensis* (All.) Link. A new type embryo sac and a summary of grass embryo sac investigations. *Ohio Journal of Science*, **37**, 172–81.

Swamy, B. G. L. & Krishnamurthy, K. V. (1970). On the so-called endothelium in the monocotyledons. *Phytomorphology*, **20**, 262–9.

Tilton, V. R. & Lersten, N. R. (1981). Ovule development in *Ornithogalum caudatum* (Liliaceae) with a review of selected papers on angiosperm reproduction. III Nucellus and megagametophyte. *New Phytologist*, **88**, 477–504.

Tinney, F. W. (1940). Cytology of parthenogenesis in *Poa pratensis*. *Journal of Agricultural Research*, **60**, 351–60.

True, R. H. (1983). On the development of the caryopsis. *Botanical Gazette*, **18**, 212–26.

Waddington, S. R. & Cartwright, P. M. (1983). A quantitative scale of spike initial and pistil development in barley and wheat. *Annals of Botany*, **51**, 119–30.

Wagner, V. T., Song, Y. C., Matthys-Rochon, E. & Dumas, C. (1989). Observations on the isolated embryo sac of *Zea mays* L. *Plant Science*, **59**, 127–32.

Weir, C. E. & Dale, H. M. (1960). A developmental study of wild rice, *Zizania aquatica* L. *Canadian Journal of Botany*, **38**, 719–39.

Xiang-Yuan, Xi & DeMason, D. A. (1984). Relationship between male and female gametophyte development in rye. *American Journal of Botany*, **71**(8), 1067–79.

Yamura, A. (1933). Karyologische und embryologische studien über einige *Bambusa. Arten Botanical Magazine, Tokyo*, **47**, 551–5.
Young, B. A., Sherwood, R. T. & Bashaw, E. C. (1979). Cleared-pistil and thick-sectioning techniques for detecting aposporous apomixis in grasses. *Canadian Journal of Botany*, **57**, 1668–72.

4

Fertilization and early embryogenesis

H. Lloyd Mogensen

Pollen tube growth and male gamete delivery

Once the pollen grain comes in contact with the stigma surface (Fig. 4.1*a*), several complex physical and biochemical events occur including adhesion, hydration, secretion and enzyme activation (Knox & Heslop-Harrison, 1970; Heslop-Harrison, 1979, 1982; Heslop-Harrison & Heslop-Harrison, 1980; Reger, 1989). In self-incompatible grass species a two-gene (*S*,*Z*) polyallelic gametophytic recognition system is operative which arrests normal pollen tube growth in the case of an incompatible combination (i.e. pollen from the same plant), thereby preventing self-fertilization and promoting outbreeding within the species (Lundquist, 1956; Nettencourt, 1977; Heslop-Harrison, 1982; Heslop-Harrison & Heslop-Harrison, 1985). However, regardless of the compatibility combination, contrary to the situation in non-grass systems (Gaude & Dumas, 1987), pollen attachment, hydration and initial tube growth occur equally well in the grasses (Heslop-Harrison & Heslop-Harrison, 1982*a*). Following incipient pollen hydration, tube emergence through the single germination aperture begins within 1.5 min in rye and within 7 min in corn (Heslop-Harrison, 1979; Heslop-Harrison, Reger & Heslop-Harrison, 1984).

It is when the pollen tube tip comes in contact with the proteinaceous coating of the stigma cells that the self-incompatibility reaction may begin, resulting in cessation of tube growth at or near the stigma surface (Heslop-Harrison, 1982; Shivanna, Heslop-Harrison & Heslop-Harrison, 1982). According to a hypothesis based primarily upon work with grasses (Heslop-Harrison & Heslop-Harrison, 1982*a*; Heslop-Harrison, 1982, 1983), the incompatibility factors (*S*,*Z* gene products) on the female side are located within the stigma surface secretions and in the secretions within the intercellular spaces of the pollen tube transmitting

This work was supported by NSF grant DCB-8501995, USDA grant 83-CRCR-1-1270, and by a grant from the Northern Arizona University Organized Research Fund.

Fig. 4.1. (*a*) Pollinated silk of corn. Scale bar = 0.5 mm. (Photograph courtesy of M. L. Rusche.) (*b*) Pollen tube (pt) within the secondary stigma branch (sb) of wheat 15 min after pollination. Stained with decolourized aniline blue and photographed in ultraviolet light to show callose of pollen grain (pg) and tube wall. Scale bar = 50 μm. (Mogensen & Ladyman, 1989.)

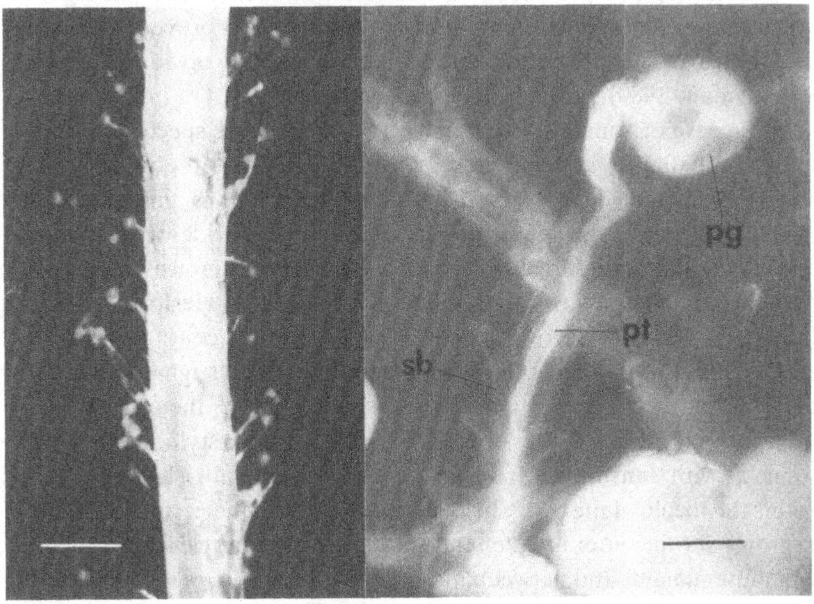

(*a*) (*b*)

tracts in the form of glycoproteins having lectin-like properties. The incompatibility genes on the male side are transcribed and translated some time after meiosis of the microsporocyte, resulting eventually in the synthesis of specific sugar sequences or arrays in the polysaccharide (primarily pectin) composing the major portion of pollen tube wall precursor vesicles, of which there are large numbers already formed in mature grass pollen (over 1 million/pollen grain in rye; Fig. 4.6*a*; Heslop-Harrison & Heslop-Harrison, 1982*b*). Since pollen tube growth occurs at the tip as the result of the deposition and incorporation of the polysaccharide vesicle contents into the growing wall (Cass & Peteya, 1979; Heslop-Harrison & Heslop-Harrison, 1982*b*), the specific sugar sequences or arrays become positioned toward the outside of the pollen tube wall such that they can be challenged by the female-side factors. In the case of an incompatible combination, it is envisioned that the binding specificities of the pistil-side glycoproteins are complementary to the sugar sequences or arrays of the carbohydrate in the pollen tube tip wall.

The resultant binding prevents normal incorporation of precursor vesicle contents into the wall and, consequently, tube growth ceases (Heslop-Harrison & Heslop-Harrison, 1982*a*; Heslop-Harrison, 1983). It is interesting that the mode of action of a recently developed chemical hybridizing agent for wheat also appears to be based upon a specific targeting of the pollen tube wall precursor vesicles such that their normal incorporation into the tube wall malfunctions and tube growth is prevented. In this case, however, no interaction with female tissues is involved (Mogensen & Ladyman, 1989).

In the case of compatible combinations, or in those species lacking a self-incompatibility system, such as corn (*Zea mays*), the pollen tube, having interacted with the superficial stigmatic secretions, grows through the cuticle layer of the stigma papilla. Such entry is evidently facilitated by pollen wall intine-derived acid hydrolases and transferases released as the tube tip emerges from the pollen grain (Knox & Heslop-Harrison, 1970, 1971; Heslop-Harrison, 1982). The pollen tube continues to grow into the secretion-filled intercellular spaces of the stigma cells, first through the secondary stigma branch (Fig. 4.1*b*), then within the transmitting tissue of the primary stigma branch (stylodium) which connects with transmitting tissue in the upper ovary wall. The pollen tube enters the locule of the ovary by penetrating a thin cuticle, then continues to grow over the outer integument, through the micropyle formed by the inner integument, and between nucellar cells before entering into one of the two synergids flanking the egg cell (Figs 4.2 and 4.8; Heslop-Harrison, 1982; Mogensen, 1982; Van Lammeren, 1986).

The above description of the pollen tube pathway is based on the organization of the usual grass pistil which consists of two primary branches from the basal ovary, each having numerous secondary branches with receptive trichomes. However, the organization of the 'silk' of corn is viewed as being basically the same, only greatly elongated, with the two primary branches adnate along most of their lengths (Fig. 4.1*a*; Weatherwax, 1916; Heslop-Harrison *et al.*, 1984). Various time periods have been reported, but in small grain cereals it typically takes less than 1 h after pollination for the pollen tube to reach the embryo sac, whereas in corn it may require as long as 20–25 h (Weatherwax, 1919; Pope, 1937; Diboll, 1968; Mogensen, 1982; You & Jensen, 1985; Engell, 1988; Van Lammeren, 1986).

Before the pollen tube reaches the ovule, one or both of the synergids has begun to degenerate, and the pollen tube consistently enters a degenerated synergid before terminating growth and discharging the male gametes through a terminal pore (Fig. 4.8; Cass & Jensen, 1970;

Fig. 4.2. Phase-contrast image of the micropylar region of the ovule of barley showing the final pathway of the pollen tube (pt) as it reaches the degenerated synergid (ds). ii = inner integument; oi = outer integument; nu = nucellus; z = zygote. Scale bar = 50 μm. (Mogensen, 1982.)

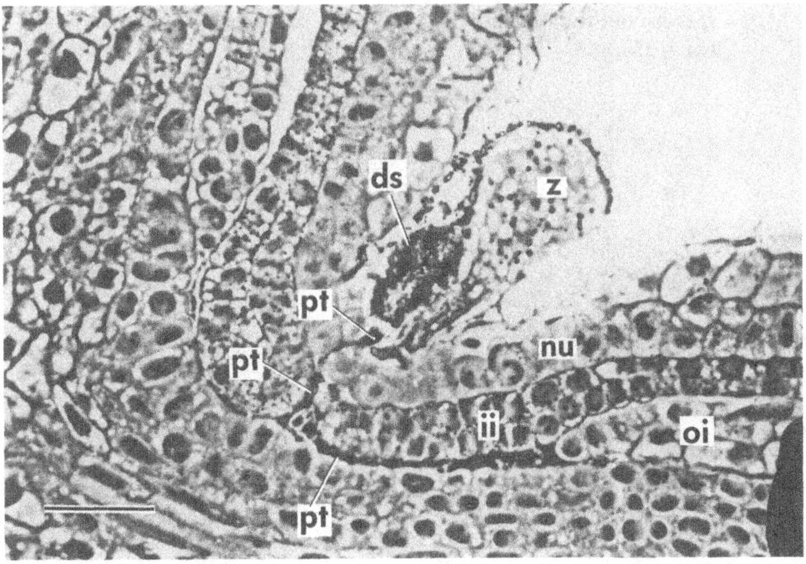

Mogensen, 1982, 1984; You & Jensen, 1985). It is of note that the pollen tube never enters a living cell throughout its entire pathway, and that the degenerated synergid plasma membrane is lost by the time the sperms are discharged. Since the inner vegetative cell plasma membrane of the pollen grain, which previously enclosed each sperm cell (Fig. 4.3*b*), is also lost at the time of pollen tube discharge, the plasma membranes of the sperm cells are free to interact with those of the female gametes to effect the unique process of double fertilization as described in the next section.

The male sex cells and fertilization

The sperm cells within the pollen grain

As is the case in about 30% of flowering plants investigated, the sperm cells of grasses are formed before anther dehiscence, and the pollen is therefore traditionally termed 'trinucleate' (Fig. 4.3*a*; Brewbaker, 1967). However, since the sperms are true cells and not naked nuclei, a more appropriate terminology is 'tricellular' pollen, referring to the vegetative cell (tube cell) and the two sperm cells (Fig. 4.3*b*).

Fig. 4.3. (*a*) Mature pollen grain of wheat stained with acetocarmine to show the two sperm nuclei (sn) and the vegetative nucleus (vn). Scale bar = 25 μm. (Mogensen & Ladyman, 1989.) (*b*) Drawing based on Fig. 4.3*a* demonstrating the cellular nature of the sperm cells, the inner vegetative cell plasma membrane surrounding each sperm, and the lack of physical connections between the sperms or between the sperms and the vegetative nucleus (vn). sc = sperm cytoplasm. Scale bar = 25 μm.

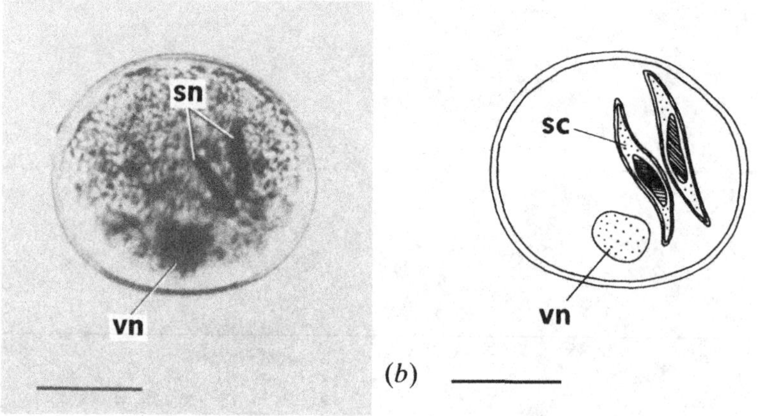

(*a*) (*b*)

Following their formation from division of the generative cell, the sperms undergo considerable modification before pollen maturity. In barley the change from nearly spherical to long, crescent-shaped cells is accompanied by reductions in sperm cell volume and surface area of 51% and 30% respectively, and by a 50% reduction in the number of mitochondria per cell (Cass & Karas, 1975; Mogensen & Rusche, 1985). The formation and subsequent loss of cytoplasmic extensions containing mitochrondria is thought to be a primary mechanism of the sperm cell transformation. Since the mean volume and surface area per mitochondrion are significantly less in mature sperms than in younger ones, it is not likely that the reduction in mitochondrial number results from their fusion, and no examples of mitochondrial degeneration have been observed (Mogensen & Rusche, 1985). Such morphological alterations of sperm cell size and shape may facilitate their movement through the pollen tube (Tatintseva, 1983), and cytoplasmic organelle reduction may play a role in effecting the strict maternal cytoplasmic inheritance that occurs in the grasses (Robertson, 1936; Conde, Pring & Levings, 1979; Hagemann & Schröder, 1984). Although plastids are often reported to be absent from grass sperm cells, a condition characteristic of most angiosperms (Sears, 1980; Whatley, 1982), they are clearly present in the generative cells and sperms of wheat, as well as many other grasses (Hagemann & Schröder, 1984; Wagner & Mogensen, unpublished). The ultimate fate of the male cytoplasm is discussed later in this chapter.

Recent studies applying the techniques of serial ultrathin sectioning, computer-assisted quantitation and three-dimensional reconstruction have demonstrated that, in certain species, the two sperms of a pair may be quite different from each other with respect to size, shape, cytoplasmic organelle content and association with the vegetative nucleus. This phenomenon of sperm dimorphism occurs most markedly in *Plumbago zeylanica* (a dicot with tricellular pollen) in which one sperm contains nearly all the plastids and relatively few mitochrondria, whereas the other sperm contains an average of over six times more mitochondria and has a long extension which forms an intimate and persistent association with the vegetative nucleus (Russell & Cass, 1981; Russell, 1984*b*). Other examples of sperm dimorphism have been shown in *Brassica* (McConchie *et al.*, 1987*b*) and spinach (Wilms, 1986), which are also dicots with tricellular pollen. Interestingly, in *Plumbago*, the plastid-rich sperm preferentially fertilizes the egg, suggesting the existence of a gamete-level recognition system (Russell, 1985). Such distinctions between sperms of a pair have important implications for plant biotechnology since it may be possible to isolate morphotypes and, following genetic transformation, utilize them for specific *in vitro* fertilizations (Keijzer, Wilms & Mogensen, 1988). Another significant result of recent work in sperm cell biology is the recognition of what has been termed the 'male germ unit', referring to the linkage between not only one (or both) sperm cell and the vegetative nucleus, but also between the two sperm cells (Dumas *et al.*, 1984). This close association between the units of male heredity has been demonstrated in several plants to date (Barnes & Blackmore, 1987; Hu & Yu, 1988; Wagner & Mogensen, 1988), and is considered important in male gamete delivery to the female target cells (Dumas *et al.*, 1984).

The grasses, however, appear to be exceptional in that no contacts occur between the two sperms or between either of the sperms and the vegetative nucleus in the mature pollen grain (Fig. 4.3*b*; Cass & Karas, 1975; Mogensen & Rusche, 1985; McConchie, Hough & Knox, 1987*a*; Rusche, 1988; Rusche & Mogensen, 1988). Sperm dimorphism too seems to be slight or lacking, at least at the time of pollen maturity (Fig. 4.3*b*; McConchie *et al.*, 1987*a*; Rusche & Mogensen, 1988). Yet, it has long been known that 'directed' or preferential fertilization occurs in corn: the sperm cell carrying B-chromosomes most often fusing with the egg cell (Roman, 1948). It may be that, in the grasses, differences between sperms of a pair occur only at the biochemical level (plasma membrane

Fig. 4.4. (*a*) Transmission electron micrograph of sperm cells within the pollen grain of barley 15 min after pollination. Note the close association between the two sperm cells (unlabelled arrowheads). sc = sperm cytoplasm, vn = vegetative nucleus. Scale bar = 1 μm. (*b*) Different plane of section within pollen grain of Fig. 4.4*a* showing the vegetative nucleus (vn) completely surrounding one end of one sperm cell. sn = sperm nucleus. Scale bar = 1 μm. (Mogensen & Wagner, 1987.)

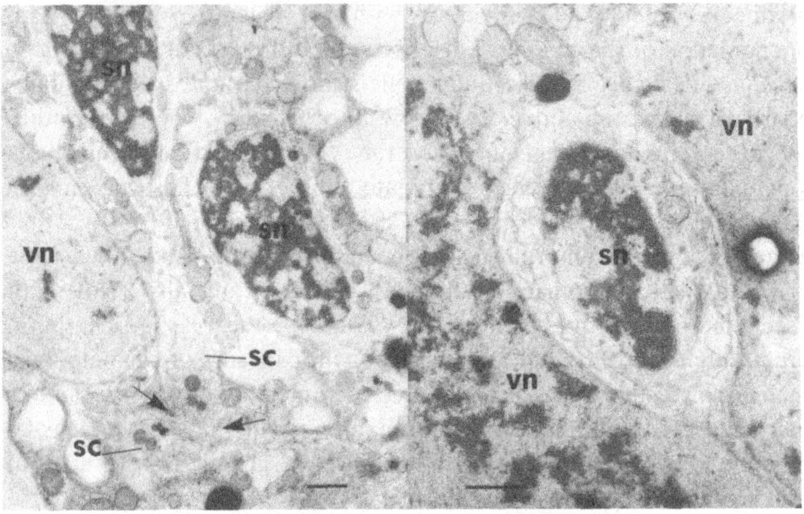

(*a*) (*b*)

recognition molecules), or that post-pollination morphogenic events occur which result in more pronounced sperm dimorphism. Certainly, several changes, including the formation of a distinct male germ unit, do take place after pollination, as discussed in the next section.

The sperm cells within the activated pollen grain and pollen tube

Male gamete modifications occurring within the developing pollen grain are equalled, if not exceeded, after pollination. Within 5 min after pollination in barley the two sperms can be found in close association with the vegetative nucleus (Mogensen & Wagner, 1987). Within 15 min after pollination, but before the sperm pair has exited the pollen grain, the sperms become closely associated at several places along their surfaces (Fig. 4.4*a*). The associations do not involve a physical connection between the vegetative cell plasma membranes surrounding each sperm, but the membranes do come to within 60 nm of each other with no intervening background cytoplasm (Fig. 4.4*a*). Unlike the condition

Fig. 4.5. Manual reconstruction of complete male germ unit of barley from a pollen grain fixed approximately 15 min after pollination (same pollen grain as in Fig. 4.4), showing close association of sperm cells (s) and vegetative nucleus (vn). Scale bar = 1 μm. (Mogensen & Wagner, 1987.)

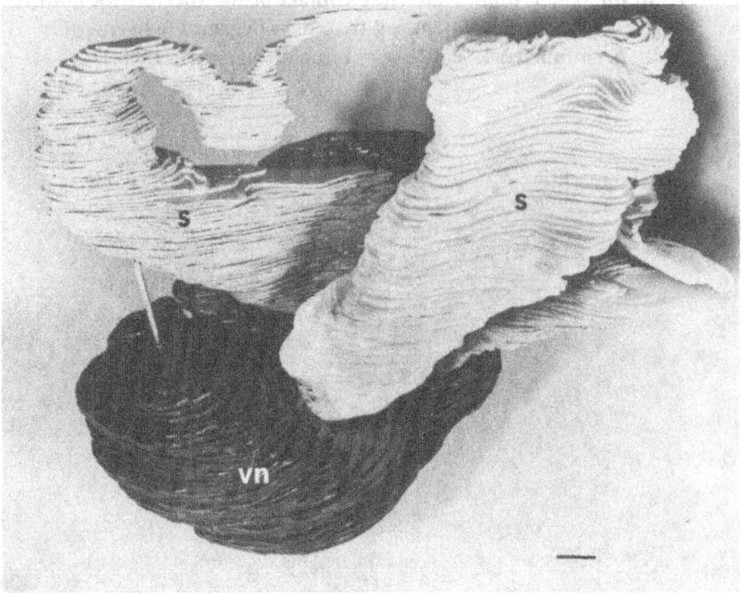

within the mature barley pollen grain at anthesis (Cass & Karas, 1975; Mogensen & Rusche, 1985), the sperms now have rather extensive cytoplasmic projections at one or both ends (Figs 4.4*a* and 4.5), similar to the sperm cells in mature pollen of wheat and corn (Fig. 4.3*b*; Zhu *et al.*, 1980; Mogensen, 1986; McConchie *et al.* 1987*a*; Rusche, 1988; Rusche & Mogensen, 1988). The sperms are now more closely associated with the vegetative nucleus, which has been observed to completely surround a sperm cell extension along a distance of approximately 3 μm (Figs 4.4 and 4.5).

The sperm cells remain together after leaving the pollen grain, and within the early pollen tube they are seen to be connected by a common, fibrillar cell wall (Fig. 4.6*a*). The vegetative nucleus, however, does not remain associated with the sperms and is typically found within the pollen grain after the sperms have passed well into the pollen tube (Chandra & Bhatnager, 1974; Mogensen & Wagner, 1987). In later stages, when the pollen tube has reached nearly to the level of the ovule, the sperm cells are still connected, but their walls are reduced or absent (Mogensen & Wagner, 1987). At this stage the sperms are connected at both ends over

Fig. 4.6. **(*a*) Barley sperm pair within a pollen tube (pt) shortly after exiting the pollen grain (pg). Note close association and apparent fusion of fibrillar cell walls of the sperm cells. sh = stigma hair; sn = sperm nucleus; wp = wall precursor vesicles. Scale bar = 1 μm. (Mogensen & Wagner, 1987.) (*b*) Sperm cell of barley after discharge from the pollen tube. Note close contact of the sperm with both the egg cell (e) and the central cell (cc). ds = degenerated synergid; sn = sperm nucleus. Scale bar = 1 μm. (Mogensen, 1988.)**

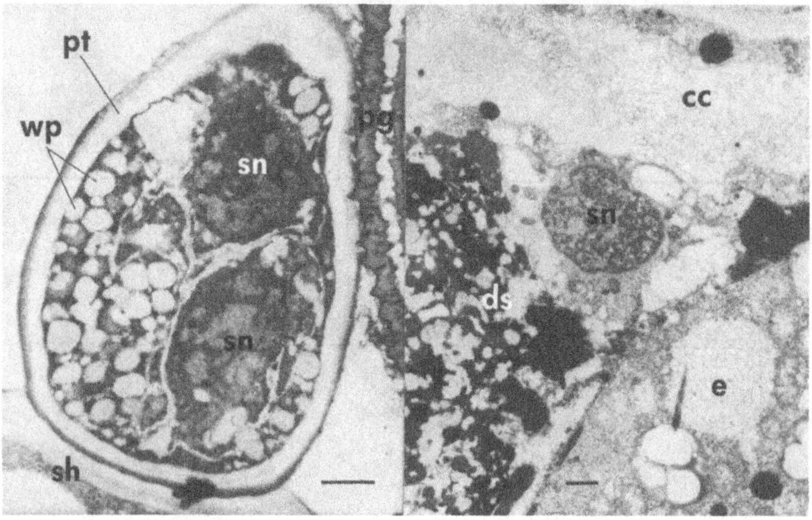

(*a*) (*b*)

rather extensive areas, and each cell has a prominent nucleus and flattened cytoplasmic projections (Fig. 4.10*a*).

Similar post-pollination morphogenesis has been observed in corn, where the male germ unit, including the vegetative nucleus, is formed after pollen activation and is maintained at least during early passage within the pollen tube (Maxine Rusche, unpublished). Formation of intimate associations between the sperms and the vegetative nucleus as a response to pollen activation and tube growth seems to underscore the significance of the so-called male germ unit, even though its precise function or functions are not yet clear. In addition to facilitating the transmission, timing and order of arrival of the male gametes to the embryo sac (Dumas *et al.*, 1984), physical connections between sperm cells of a pair may help reduce the occurrence of heterofertilization, the condition where the sperm fertilizing the egg comes from a different pollen tube from the sperm fusing with the polar nuclei of the same embryo sac (Sprague, 1932). The intimate association between the vegetative nucleus and sperm cells may facilitate the transfer of materials between these entities; i.e. vegetative nucleus mRNA or its translation products, for instance (Mogensen & Wagner, 1987). Experimental approaches to the question of male germ unit function are clearly needed.

The sperm cells within the embryo sac

Although early reports on fertilization in grasses described male nuclei within the embryo sac (Pope, 1937; Morrison, 1954), the first clear examples demonstrating that the sperms are still cellular at this stage were provided in 1970 for barley (Cass & Jensen, 1970). Sperm cells having a nucleus and a distinct area of surrounding cytoplasm were seen at the light microscope level within the synergid, as well as in contact with the egg cell (Cass & Jensen, 1970). More recent work on barley, using the techniques of serial thick sectioning, serial ultrathin sectioning of re-embedded thick sections and computer-generated three-dimensional reconstructions has shown that the sperms just before gamete fusion are nearly spherical cells with an intact plasma membrane enclosing a curved (horseshoe-shaped) nucleus and cytoplasmic components including mitochondria, plastids, dictyosomes, endoplasmic reticulum, ribosomes, lipid bodies and vacuoles (Figs 4.6*b* and 4.10*b*; Mogensen, 1982; Mogensen, unpublished). There is no longer a vegetative cell plasma membrane surrounding the sperm cells, and since the degenerated synergid membrane is no longer present, the plasma membranes of the male and female gametes are able to come into intimate contact (Fig. 4.6*b*). In fact, a given sperm plasma membrane may be in contact with both the egg and central cell plasma membranes simultaneously before fusion (Mogensen, 1982), thereby setting the stage for gamete-level recognition systems which may subsequently determine the fate of each sperm cell. Mitochrondrial counts from complete serial ultrathin sections of two pre-fusion sperm cells within the same embryo sac (Mogensen, unpublished) show that the sperm apparently destined to fuse with the egg (as judged from its location against the egg cell) had 38 mitochondria, whereas the other sperm, presumably destined to fuse with the central cell, had a total of 61 mitochondria. Whether the sperms at this stage are significantly different from those at the time of pollination, and whether the difference in mitochondrial number between these two sperms may represent true sperm dimorphism cannot be determined until a larger sample is obtained.

Another feature of sperm cells within the embryo sac of barley is the lack of any connection or contact between them. The sperms apparently become separated from each other at the time of discharge from the pollen tube. This also appears to be the case for the sperms of cotton (Jensen & Fisher, 1968) and *Plumbago* (Russell, 1984*a*).

In two embryo sacs to date (Mogensen, unpublished) more than two sperm cells have been seen within the same embryo sac. In each case a sperm nucleus was in contact with the egg nucleus, yet two sperm cells were also present within the degenerated synergid. It was not clear whether the central cell had yet been fertilized. These observations are significant because they provide cytological evidence that supports the occurrence and mechanism of heterofertilization, a phenomenon which has long been recognized from genetic studies in both corn (Sprague, 1932) and barley (R. T. Ramage, personal communication).

The sperm nuclei within the female gametes

In barley the male nucleus retains its condensed chromatin while traversing the egg cytoplasm to the female nucleus. Inside the egg, no sperm plasma membrane is present and no cytoplasmic organelles can be recognized specifically as being of male origin (Fig. 4.7*a*). Although the sperm mitochondria are smaller on average than those of the female, the egg does contain, before fertilization, some mitochondria which are in the same size range as those of the sperm (Mogensen, 1982). Upon reaching the egg nucleus, the male chromatin becomes progressively more diffuse as karyogamy takes place. Nuclear fusion occurs by the formation of several bridges between the male and female nuclear membranes as has been described for other species (Fig. 4.7*b*; Jensen, 1974; Hause & Schroder, 1987). As nuclear fusion becomes complete, the male chromatin is still identifiable at one side of the zygote nucleus up to at least 6 h after pollination (Pope, 1937). Fragments of nuclear membranes are often seen within the zygote nucleus in the vicinity of the male chromatin, and by this time a new nucleolus is beginning to form in association with the male chromatin (Mogensen, 1982).

As in the case of the egg, after entrance of the male nucleus into the central cell of barley, it no longer has its own plasma membrane or any identifiable male cytoplasmic components (Mogensen, 1982). Upon reaching the already partially fused polar nuclei, the male nucleus fuses with one polar nucleus in a manner similar to that described for the egg and sperm nuclei. During the fusion process, the three nuclei migrate chalazally to become located near the antipodal cells. After fusion with the polar nucleus, the male chromatin quickly becomes diffuse and one or more new nucleoli begin to appear. Soon the male chromatin is no longer distinguishable from that of the polar nucleus and the only evidences of

Fig. 4.7. (*a*) Barley sperm nucleus (sn) within egg cytoplasm near egg nucleus (en). Scale bar = 1 μm. (*b*) Sperm nucleus (sn) fusing with egg nucleus (en). Scale bar = 1 μm. (Mogensen, 1982.)

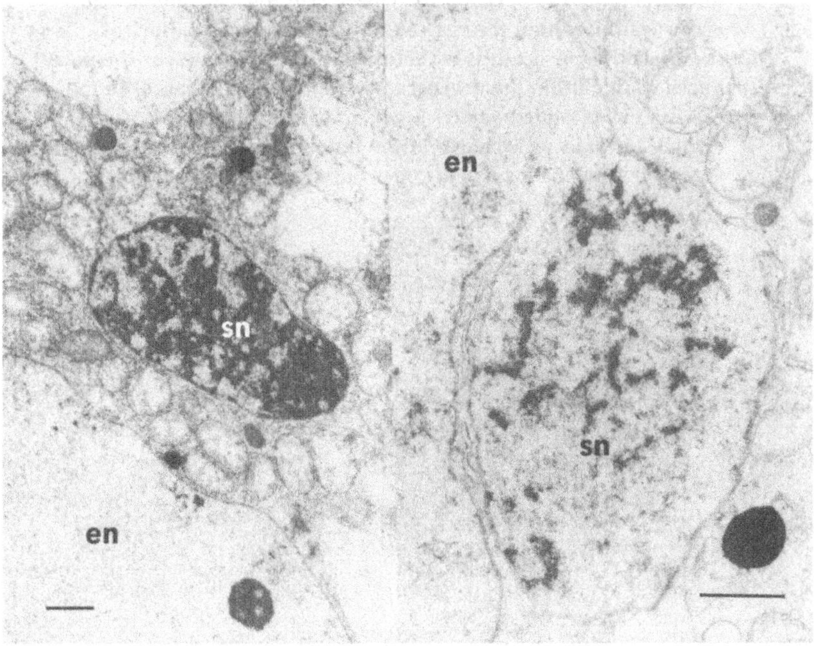

(*a*) (*b*)

fusion are the presence of extra nucleoli within the triple fusion nucleus, and the relocation of the nucleus from adjacent to the egg apparatus to the chalazal end of the embryo sac (Mogensen, 1982).

The fate of the male cytoplasm

Despite the presence of mitochondria and sometimes plastids in grass sperm cells, it is known from genetic evidence that these DNA-containing organelles are inherited strictly maternally (Anderson, 1923; Robertson, 1936; Conde *et al.*, 1979). The obvious question is, what happens to the male cytoplasm? Even though mechanisms exist which reduce the number of these organelles during microgametogenesis (i.e. exclusion of organelles from the generative cell during the first microspore mitosis, and discarding of organelle-containing cytoplasmic vesicles during sperm maturation), ultrastructural studies show that sperm cells, at least in barley, still contain relatively large numbers of mitochondria (up to 61) and some plastids by the time they reach the female gametes (Mogensen, 1982; Mogensen, unpublished). Clearly, the

Fig. 4.8. Composite drawing based upon four serial thick sections from the same embryo sac of barley showing early post-fusion stage. The pollen tube (pt) has entered the degenerated synergid (ds) and discharged the two sperm cells. One sperm nucleus (unlabelled arrowhead) is within the egg cell (e), the other sperm nucleus (unlabelled arrowhead) is within the central cell (cc) near the polar nuclei (pn). Within the degenerated synergid and adjacent to the egg cell is a cytoplasmic body (cb), which is interpreted to be sperm cytoplasm that was excluded at the time of fusion between egg and sperm. ii = inner integument; nu = nucellus. Dashed lines outline vacuoles. (Mogensen, 1988.)

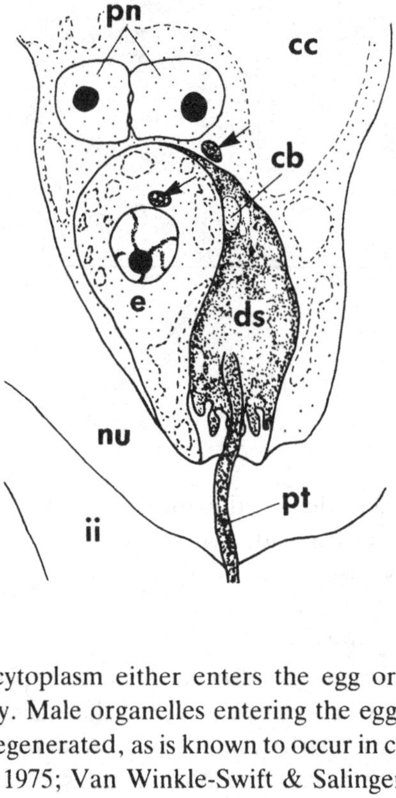

sperm cytoplasm either enters the egg or is excluded at the time of syngamy. Male organelles entering the egg could conceivably be selectively degenerated, as is known to occur in certain species of lower plants (Sager, 1975; Van Winkle-Swift & Salinger, 1988), or their replication could be suppressed (Tilney-Basset, 1976). If the male cytoplasm is excluded, as is contended to be the case in the grasses (Hagemann & Schröder, 1984), it should be possible to detect it using modern morphological techniques at the right stage. With such a goal in mind, I examined large numbers of barley embryo sacs that had been fixed from 30 to 60 min after hand-pollination. Observations of serial thick sections (about 3 μm) with phase-contrast microscopy revealed in some of the embryo sacs what appeared to be an enucleated cytoplasmic body adjacent to a recently

Fig. 4.9. (*a*) Electron micrograph of a cytoplasmic body (cb) within the degenerated synergid (ds) and next to the egg cell (e) of barley showing numerous mitochondria (m). Scale bar = 1 μm. (*b*) Same cytoplasmic body (cb) as in Fig. 4.9*b*, sectioned about 1.4 μm from the plane of Fig. 4.9*b*, showing one of the three plastids (p) within this structure. Scale bar = 1 μm. (Mogensen, 1988.)

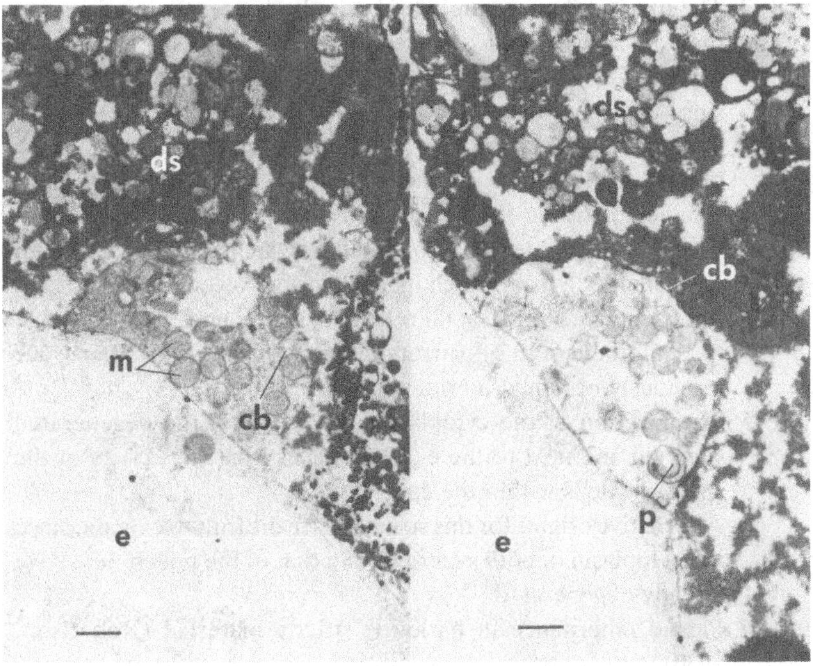

(*a*) (*b*)

fertilized egg cell (Fig. 4.8). Ultrastructural examination of ultrathin sections (68 sections) from re-embedded thick sections (3 sections) containing such a cytoplasmic body showed that it was indeed enucleate, that it was enclosed by a limiting membrane, that it was of the same general shape (mostly spherical to oval, but flat on the side appressed to the egg cell) as pre-fusion sperm cells, and that it contained 59 mitochondria, 3 plastids, 7 dictyosomes and a rather large vacuolate area (Figs 4.9 and 4.10*c*; Mogensen, 1988). Computer-assisted measurements estimate the cytoplasmic body to have a volume of 93 μm^3 and a surface area of 90 μm^2. This compares with a mean volume estimate of 105 ± 4 μm^3 and a mean surface area estimate of 96 ± 9μm^2 for the two pre-fusion sperm cells observed within the degenerated synergid of barley described earlier.

Several factors support the proposition that this cytoplasmic body represents the entire male cytoplasm left outside the egg at the time of gametic fusion:

1 Many of these structures have been observed at the light microscope level within recently fertilized embryo sacs; none has been seen prior to fertilization.
2 The size and shape of the cytoplasmic body are roughly similar to that of intact sperm cells within the embryo sac. Strict size comparisons at this point may not be particularly meaningful since the quantitative data are so limited.
3 The cytoplasmic structure contains organelles known to be present within barley sperm cells.
4 The cytoplasmic body contains a non-membrane-bound, vacuolated area devoid of organelles, which would be expected in such a body from which its nucleus had recently exited. That this vacuolate area is smaller than the sperm nucleus may be due to some invasion of surrounding cytoplasm into the space previously occupied by the nucleus.
5 The location of the cytoplasmic body within the degenerated synergid and next to the egg cell places it in the vicinity of the sperm nucleus within the egg cell.
6 Alternative origins for this structure are difficult to explain, since the cytoplasm of both synergids and that of the pollen tube have already degenerated.
7 Plastid inheritance in barley is strictly maternal (Robertson, 1936).

The mechanism resulting in the transfer of only the male nucleus into the egg is not entirely clear, but it is envisioned that initial cellular contact between egg and sperm results in plasma membrane fusion and the formation of cytoplasmic bridges, as described for other species where there is apparently complete transfer of the sperm contents (Russell, 1983). In the case of barley, however, it may be that the opening created between the two gametes is restricted to one end of the sperm cell. If the nucleus were positioned near that end, with the other organelles polarized at the opposite end of the cell, transfer of the sperm nucleus

Fig. 4.10 (*facing page*). Computer-reconstructed stereoscopic pairs. (*a*) Sperm pair within a pollen tube near the ovule. Scale bar = 1 μm. (Mogensen & Wagner, 1987.) (*b*) Sperm cell within degenerated synergid. Scale bar = 1 μm. (*c*) Cytoplasmic body within degenerated synergid and against the egg cell (yellow) showing mitochondria (red) and plastids (green). m = mitochondria; sc = sperm cytoplasm; sn = sperm nucleus. All material from barley. Scale bar = 1 μm.

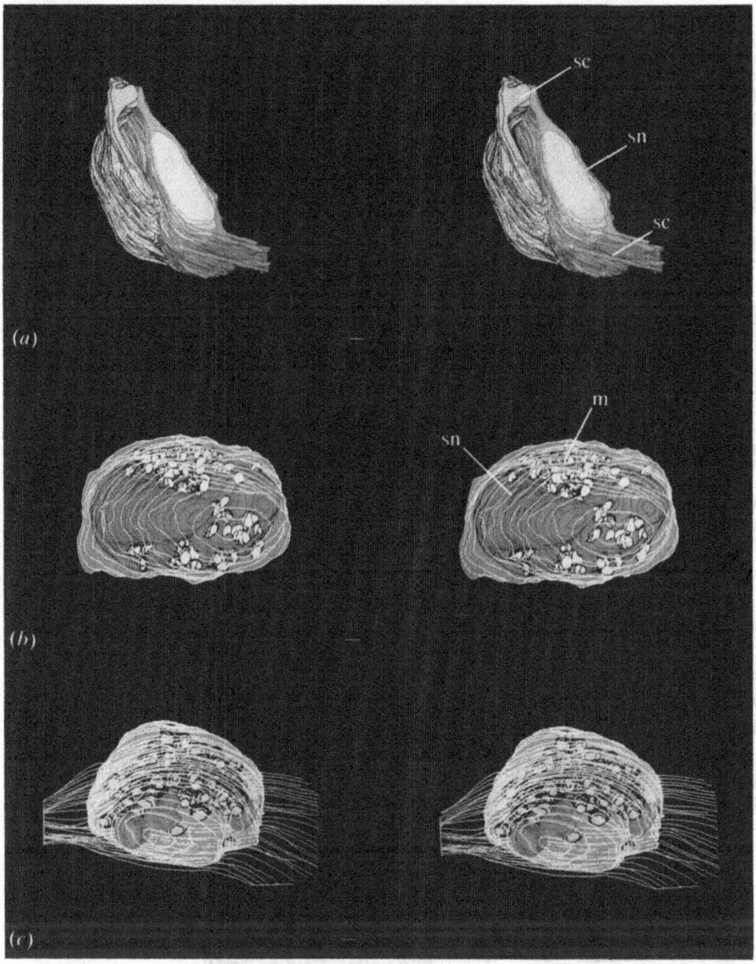

This image is available in colour for download from www.cambridge.org/9780521188913

facing page 90

This image is available in colour for download from
www.cambridge.org/9780521188913

facing page 91

only would be facilitated. Subsequent reformation of the fertilized egg plasma membrane would prevent any further transfer of sperm cell contents. Numerous dictyosomes and vesicles are present within both the egg and sperm cells which could serve as the source of new plasma membrane (Fig. 4.11*a–d*; Mogensen, 1988). Three-dimensional reconstruction of the cytoplasmic body discussed above does show a distinct polarization of the mitochondria toward one end of the body and/ or away from the egg cell side. The vacuolate area is located away from most of the mitochondria and is positioned against the egg cell side of the cytoplasmic body. The plastids are not far away from the vacuolate area, however (Fig. 4.10*c*). Ultrastructural observations of critical fusion stages are still needed to further elucidate the process of syngamy in grasses. A mechanism not unlike that proposed here for sperm–egg fusion in barley has been proposed for other plants (Jensen & Fisher, 1968; Jensen, 1974; Went & Willemse, 1984).

Although the cytoplasmic bodies are difficult to find and they apparently degenerate very quickly, it may be significant that, so far, only one cytoplasmic body has been found within a given embryo sac. This

Colour plate (facing page)
***Top row.* Left: *Panicum stapfianum* occurs in Transvaal, Orange Free State and E. and S. Cape Province. Its caespitose growth habit distinguishes it from *P. coloratum*, a much more widely distributed species with which it integrades. Right: *Phragmites australis* is an aquatic/semi-aquatic species now widely distributed. Exceptionally long-lived clones of this species are referred to by Richards (Chapter 6).**
***Middle row.* Left: *Brachiaria decumbens* (Surinam grass) is a highly successful pasture grass of the tropics. An apomictic grass, its genus has affinities with *Panicum*, *Urochloa* and *Erichloa*. Note the large outer glume enclosing the florets. Right: *Aristida ramosa* (purple wire grass) grows on semi-arid sites of low fertility in the open forests of Australia. *Aristida* is a large genus with upwards of 300 species whose prominent awns can make it unsuitable for stock. Note the anthers and stigmas exserted laterally from the cylindrical lemma.**
***Bottom row.* Left: *Oplismenus aemulus*. *Oplismenus* is a small tropical genus of about five species unusual in distributing its fruits by means of sticky brown awns, visible in the photograph behind the dehiscing anther and discussed by Clayton in Chapter 2. The generic name is an allusion to 'hoplite', the heavily armed foot soldiers of ancient Greece. Right: *Poa bulbosa*. Bulbous *Poa* grass is important for maintaining ecological stability in over-grazed steppes and is discussed by Kernick in Chapter 7. As an agamic complex it shows wide ecological amplitude.**
(Photograph of *Poa bulbosa* courtesy of the Royal Botanic Gardens, Kew; other photographs courtesy of Professor L. Watson.)

Fig. 4.11. Proposed mechanism of male cytoplasmic exclusion during sexual fusion in barley. (*a*) The plasma membranes of the sperm (sp) and egg are in contact. The sperm nucleus is located near the egg cell. (*b*) The plasma membranes have fused and cytoplasmic bridges have formed between sperm and egg near the sperm nucleus. (*c*) Only the sperm nucleus (sn) passes into the egg cell. (*d*) Membrane has reformed, leaving the male cytoplasm (sc) outside the zygote. ds = degenerated synergid; en = egg nucleus.

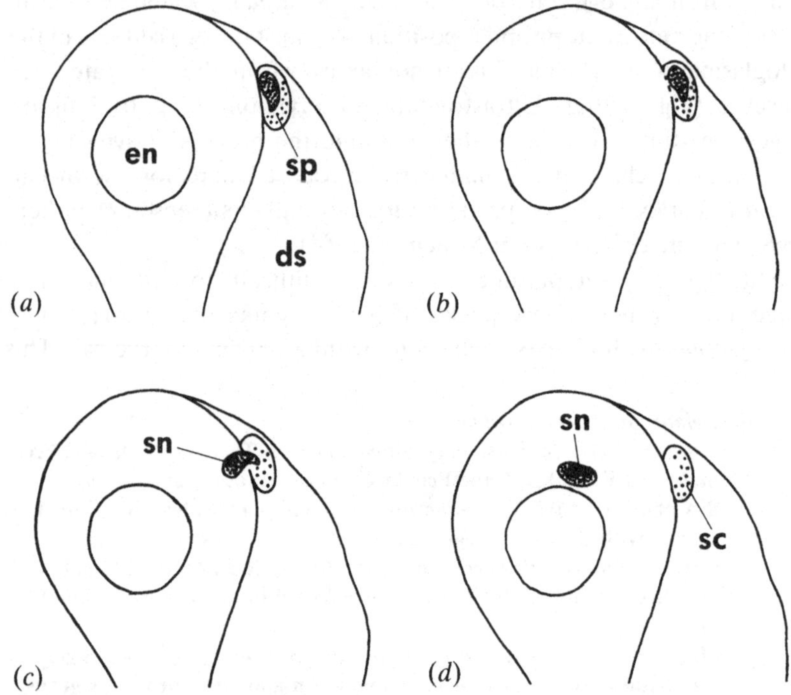

suggests that the fusion process between the sperm and the central cell may be fundamentally different from that occurring between sperm and egg and that paternal cytoplasmic organelles may be transferred and become functional in the developing endosperm (Mogensen, 1988).

Recent studies indicating that the entire cytoplasmic component of mature pollen may be devoid of DNA in those species demonstrating uniparental–maternal inheritance is intriguing (Miyamura, Kuraoiwa & Nagata, 1987; Corriveau & Coleman, 1988). However, these studies were based primarily on direct staining techniques with DAPI, and await corroboration from investigations employing techniques of molecular biology.

Fig. 4.12. Barley embryogeny. (*a*) Zygote. Perinuclear cytoplasm contains plastids (p) with starch grains, mitochondria, dictyosomes and ribosomes. Endoplasmic reticulum is sparse. Note peripheral vacuoles (v). cc = central. Scale bar = 10 μm. (*b*) Five-celled embryo. The two cells of the suspensor (su) are divided by a longitudinal wall. ds = degenerated synergid; ps = persistent synergid; es = endosperm. Scale bar = 10 μm. (Norstog, 1972.)

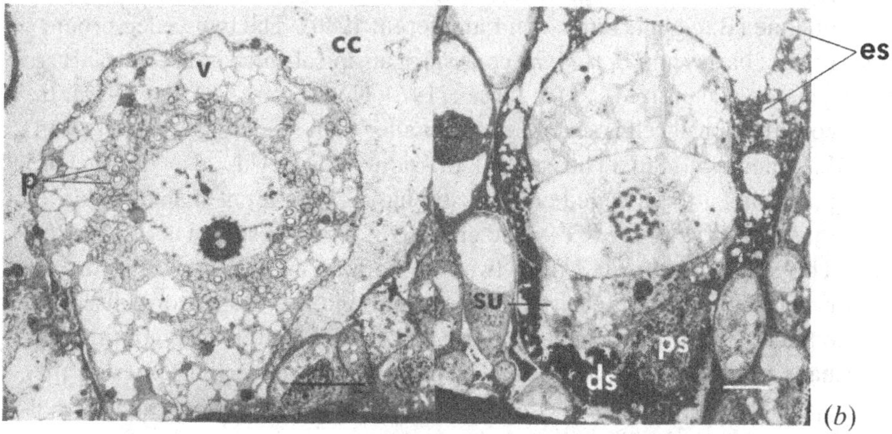

(*a*) (*b*)

Embryogeny

The zygote

There is apparently little structural change occurring in the zygote of barley when compared to the egg (Norstog, 1972; Mogensen, 1982); however, Diboll (1968) and Van Lammeren (1986) report several changes in the zygote of corn. An increase in the amount of rough endoplasmic reticulum, polysome formation, mitochondria and plastid enlargement, and more dictyosomes producing secretion vesicles all point to elevated synthetic activities as the result of fertilization. Histochemical studies also indicate an increase in enzyme activity within corn ovules about the time of fertilization (Singh & Malik, 1986, cited in Van Lammeren, 1986). Figure 4.12*a* shows that the zygote of barley is pyriform with a large nucleus centrally located within the bulbous portion of the cell. The perinuclear cytoplasm, containing numerous plastids with starch grains, is surrounded by peripheral vacuoles (Cass & Jensen, 1970; Norstog, 1972; Engell, 1988).

Early embryo

Although the zygote of barley appears to be in early mitotic prophase at the time of its formation (Figs 4.8 and 4.12*a*; Norstog, 1972;

Engell, 1988), the first division is not completed until about 21–25 h after fertilization (Pope, 1937; Engell, 1988). In corn the zygote divides about 8–12 h after fertilization (Diboll, 1968; Van Lammeren, 1986). There is little uniformity of division patterns during early grass embryogeny (Norstog, 1972; Esau, 1977; Raghavan, 1986; Van Lammeren, 1986; Engell, 1988). In barley the division of the zygote is transverse (Cass & Jensen, 1970; Norstog, 1972), whereas in corn and wheat it is more oblique (Batygina, 1969; Van Lammeren, 1986). The two-celled proembryo of barley and *Poa annua* consists of an apical and basal cell which are nearly equal in size (Cass & Jensen, 1970; Norstog, 1972; Esau, 1977). In corn the apical cell is several times smaller than the large basal cell (Van Lammeren, 1986). The second division in barley is also transverse, producing a three-celled embryo still having an overall shape similar to that of the zygote. A five-celled embryo of barley is shown in Fig. 4.12*b*. This stage has resulted from two longitudinal divisions within a three-celled embryo, one each in the terminal and the suspensor cells. The embryo is now essentially cylindrical in outline and its cells continue to have peripheral vacuoles; however, there are more ribosomes in the apical and suspensor cells than in the median cell. Plasmodesmata occur within internal cell walls of the embryo, but, as in the zygote, no plasmodesmata are present in the walls adjacent to the nucellus or endosperm (Norstog, 1972). Cell divisions occur at least up to the 20-celled embryo stage without cell enlargement, thus the cells become increasingly smaller, particularly those of the embryo proper (Norstog, 1972; Engell, 1988). The 20-celled embryo of barley shown in Fig. 4.13 is somewhat club-shaped with a several-celled suspensor containing larger vacuoles than the cells of the embryo proper. The cytoplasm of the embryonic cells, with the exception of the vacuoles which are no longer peripherally located, is of uniform density and contains smaller nuclei and smaller, more numerous mitochondria than in the zygote. At this stage the embryo is no longer attached to the nucellus and is completely surrounded by free-nucleate endosperm (Norstog, 1972).

Early endosperm

The endosperm of grasses is initially free-nucleate. In corn the primary endosperm nucleus, resulting from the triple fusion of the two polar nuclei and the sperm nucleus, divides about 4 h after fertilization (Diboll, 1968), forming approximately eight endosperm nuclei by the time the zygote divides (Van Lammeren, 1986). The endosperm of barley has a much higher concentration of ribosomes than the embryo cells, as well as more rough endoplasmic reticulum and larger dictyosome vesicles

Fig. 4.13. Twenty-celled embryo. es = endosperm; nu = nucellus; su = suspensor. Scale bar = 10 μm. (Norstog, 1972.)

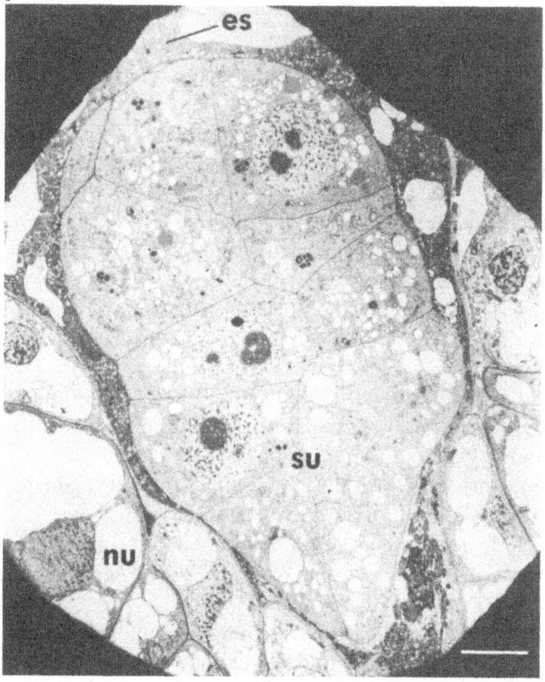

(Fig. 4.13; Norstog, 1972). In wheat, the primary endosperm nucleus divides as early as 6 h after pollination, division of zygote occurs about 22 h after pollination and endosperm cell wall formation begins about 2 days after anthesis and is completed within 4–5 days after anthesis (Bennett *et al.*, 1973; Mares *et al.*, 1977). For further details, later stages of embryo and seed development, and physiological aspects see Raghavan's (1986) excellent review.

References

Anderson, E. G. (1923). Maternal inheritance of chlorophyll in maize. *Bot. Gaz.* **76**, 411–18.

Barnes, S. H. & Blackmore, S. (1987). Preliminary observations on the formation of the male germ unit in *Catananche caerulea* L. (Compositae: Lactuceae). *Protoplasma*, **138**, 187–9.

Batygina, T. B. (1969). On the possibility of separation of a new type of embryogenesis in angiospermeae. *Rev. Cytol. Biol. Beg.* **32**, 335–41.

Bennett, M. D., Rao, M. K., Smith, J. B. & Bayliss, M. W. (1973). Cell development in the anther, the ovule, and the young seed of *Triticum aestivum* L. var. Chinese Spring. *Phil. Trans. Roy. Soc. Lond.* **266B**, 39–91.

Brewbaker, J.L. (1967). The distribution and phylogenetic significance of binucleate and trinucleate pollen grains in the angiosperms. *Amer. J. Bot.* **54**, 1069–83.

Cass, D. D. & Jensen, W. A. (1970). Fertilization in barley. *Amer. J. Bot.* **57**, 62–70.

Cass, D. D. & Karas, I. (1975). Development of sperm cells in barley. *Can. J. Bot.* **53**, 1051–62.

Cass, D. D. & Peteya, D. J. (1979). Growth of barley pollen tubes *in vivo*. I. Ultrastructural aspects of early tube growth in the stigmatic hair. *Can. J. Bot.* **57**, 386–96.

Chandra, S. & Bhatnagar, S. P. (1974). Reproductive biology of *Triticum*. II. Pollen germination, pollen tube growth, and its entry into the ovule. *Phytomorphology*, **24**, 211–17.

Conde, M. F., Pring, D. R. & Levings, C. S., III (1979). Maternal inheritance of organelle DNA's in *Zea mays–Zea perennis* reciprocal crosses. *J. Heredity*, **70**, 2–4.

Corriveau, J. L. & Coleman, A. W. (1988). Rapid screen method to detect potential biparental inheritance of plastid DNA and results for over 200 angiosperm species. *Amer. J. Bot.* **75**, 1443–58.

Diboll, A. G. (1968). Fine structural development of the megagametophyte of *Zea mays* following fertilization. *Amer. J. Bot.* **55**, 797–806.

Dumas, C., Knox, R. B., McConchie, C. A. & Russell, S. D. (1984). Emerging physiological concepts in fertilization. *What's New in Plant Physiology*, **15**, 17–20.

Engell, K. (1988). Embryology of barley II: synergids and egg cell, zygote and embryo development. In *Sexual Reproduction in Higher Plants*, ed. M. Cresti, P. Gori & E. Pacini, pp. 383–8. Springer-Verlag: Berlin.

Esau, K. (1977). *Anatomy of Seed Plants.* John Wiley & Sons: New York.

Gaude, T. & Dumas, C. (1987). Molecular and cellular events of self-incompatibility. *Int. Rev. Cytol.* **107**, 333–66.

Hagemann, R. & Schröder, M. B. (1984). New results about the presence of plastids in generative and sperm cells of Gramineae. In *Proceedings of the Eighth International Symposium on Sexual Reproduction in Seed Plants, Ferns and Mosses*, ed. M. T. M. Willemse & J. L. van Went, pp. 53–5. Pudoc: Wageningen, The Netherlands.

Hause, G. & Schröder, M. B. (1987). Reproduction in *Triticale* 2. Karyogamy. *Protoplasma*, **139**, 100–4.

Heslop-Harrison, J. (1979). Aspects of the structure, cytochemistry, and germination of the pollen of rye (*Secale cereale* L.). *Ann. Bot. Suppl.* **1**, 1–47.

(1982). Pollen–stigma interaction and cross-incompatibility in the grasses. *Science*, **215**, 1358–64.

(1983). Self-incompatibility: phenomenology and physiology. *Proc. R. Soc. Lond. B*, **218**, 371–95.

Heslop-Harrison, J. & Heslop-Harrison, Y. (1980). Cytochemistry and function of the Zwischenkorper in grass pollens. *Pollen Spores*, **22**, 5–10.

(1982*a*). The pollen–stigma interaction in the grasses. 4. An interpretation of the self-incompatibility response. *Acta Bot. Neerl.* **31**, 429–39.

(1982*b*). The growth of the grass pollen tube: 1. Characteristics of the

polysaccharide particles ('P-particles') associated with apical growth. *Protoplasma*, **112**, 71–80.
(1985). Surfaces and secretions in the pollen–stigma interaction: a brief review. *J. Cell Biol. Suppl.* **2**, 287–300.
Heslop-Harrison, Y., Reger, B. J. & Heslop-Harrison, J. (1984). The pollen–stigma interaction in the grasses. 6. The stigma ('silk') of *Zea mays* L. as host to the pollens of *Sorghum bicolor* (L.) Moench and *Pennisetum americanum* (L.) Leeke. *Acta Bot. Neerl.* **33**, 205–27.
Hu, S.-Y. & Yu, H.-S. (1988). Preliminary observations on the formation of the male germ unit in pollen tubes of *Cyphomandra betacea* Sendt. *Protoplasma*, **147**, 55–63.
Jensen, W. A. (1974). Reproduction in flowering plants. In *Dynamic Aspects of Plant Ultrastructure*, ed. A. G. Robards, pp. 481–503. McGraw Hill: New York.
Jensen, W. A. & Fisher, D. B. (1968). Cotton embryogenesis: the entrance and discharge of the pollen tube in the embryo sac. *Planta*, **78**, 158–93.
Keijzer, C. J., Wilms, H. J. & Mogensen, H. L. (1988). Sperm cell research: the current status and applications for plant breeding. In *Plant Sperm Cells as Tools for Biotechnology*, ed. C. J. Keijzer & H. J. Wilms, pp. 3–8. Pudoc: Wageningen, The Netherlands.
Knox, R. B. & Heslop-Harrison, J. (1970). Pollen-wall proteins: localization and enzymatic activity. *J. Cell Sci.* **6**, 1–27.
(1971). Pollen-wall proteins: fate of intine-held antigens in compatible and incompatible pollination of *Phalaris tuberosa* L. *J. Cell Sci.* **9**, 239–51.
Lundquist, A. (1956). Self-incompatibility in rye I. Genetic control in the diploid. *Hereditas*, **42**, 293–348.
Mares, D. J., Stone, B. A., Jeffery, C. & Norstog, K. (1977). Early stages in the development of wheat endosperm. II. Ultrastructural observations on cell wall formation. *Aust. J. Bot.* **25**, 599–613.
McConchie, C. A., Hough, T. & Knox, R. B. (1987*a*). Ultrastructural analysis of the sperm cells of mature pollen of maize, *Zea mays*. *Protoplasma*, **139**, 9–19.
McConchie, C. A., Russell, S. D., Dumas, C., Tuchy, M. & Knox, R. B. (1987*b*). Quantitative cytology of the sperms of *Brassica compestris* and *B. oleracea*. *Plants*, **170**, 446–52.
Miyamura, S., Kuraoiwa, T. & Nagata, T. (1987). Disappearance of plastid and mitochondrial nucleoids during the formation of generative cells of higher plants revealed by fluorescence microscopy. *Protoplasma*, **141**, 149–59.
Mogensen, H. L. (1982). Double fertilization in barley and the cytological explanation for haploid embryo formation, embryoless caryopses and ovule abortion. *Carlsberg Res. Commun.* **47**, 313–45.
(1984). Quantitative observations on the pattern of synergid degeneration in barley. *Amer. J. Bot.* **71**, 1448–51.
(1986). Preliminary observations on the three-dimensional organization of the wheat sperm cell. In *Pollination '86*, ed. E. G. Williams, R. B. Knox & D. Irvine, pp. 116–71. University of Melbourne Press.
(1988). Exclusion of male mitochondria and plastids during syngamy in barley as a basis for maternal inheritance. *Proc. Natl Acad. Sci. USA*, **85**, 2594–7.
Mogensen, H. L. & Ladyman, J. A. R. (1989). A structural study on the mode of

action of CHA chemical hybridizing agent in wheat. *Sex. Pl. Reprod.* **2**, 173–83.

Mogensen, H. L. & Rusche, M. (1985). Quantitative ultrastructural analysis of barley sperm. I. Occurrence and mechanism of cytoplasm and organelle reduction and the question of sperm dimorphism. *Protoplasma*, **128**, 1–13.

Mogensen, H. L. & Wagner, V. T. (1987). Associations among components of the male germ unit following *in vivo* pollination in barley. *Protoplasma*, **138**, 161–72.

Morrison, J. W. (1954). Fertilization and post fertilization development in wheat. *Can. J. Bot.* **33**, 168–76.

Nettencourt, D. de (1977). *Incompatibility in Angiosperms.* Springer-Verlag: Berlin, Heidelberg & New York.

Norstog, K. (1972). Early development of the barley embryo: fine structure. *Amer. J. Bot.* **59**, 123–32.

Pope, M. (1937). The time factor in pollen-tube growth and fertilization in barley. *J. Agr. Res.* **54**, 525–9.

Raghavan, V. (1986). *Embryogenesis in Angiosperms, a Developmental and Experimental Study.* Cambridge University Press.

Reger, B. J. (1989). Stigma surface secretions of *Pennisetum americanum. Amer. J. Bot.* **76**, 1–5.

Robertson, D. W. (1936). Maternal inheritance in barley. *Genetics*, **22**, 104–13.

Roman, H. (1948). Directed fertilization in maize. *Proc. Natl Acad. Sci. USA*, **34**, 46–52.

Rusche, M. L. (1988). The male germ unit of *Zea mays* in the mature pollen grain. In *Plant Sperm Cells as Tools for Biotechnology*, ed. H. J. Wilms & C. J. Keijzer, pp. 61–7. Pudoc: Wageningen, The Netherlands.

Rusche, M. L. & Mogensen, H. L. (1988). The male germ unit of *Zea mays*: quantitative ultrastructure and three dimensional analysis. In *Sexual Reproduction in Higher Plants*, ed. M. Cresti, P. Gori & E. Pacini, pp. 221–6. Springer-Verlag: Berlin.

Russell, S. D. (1983). Fertilization in *Plumbago zeylanica*: gametic fusion and fate of the male cytoplasm. *Amer. J. Bot.* **70**, 416–34.

(1984*a*). Timetable of fertilization. In *Pollination '84*, ed. E. G. Williams & R. B. Knox, pp. 69–77. University of Melbourne Press.

(1984*b*). Ultrastructure of the sperm of *Plumbago zeylanica.* II. Quantitative cytology and three-dimensional organization. *Plants*, **162**, 385–91.

(1985). Preferential fertilization in *Plumbago*: ultrastructural evidence for gamete-level recognition in an angiosperm. *Proc. Natl Acad. Sci. USA*, **82**, 6129–32.

Russell, S. D. & Cass, D. D. (1981). Ultrastructure of the sperms of *Plumbago zeylanica.* 1. Cytology and association with the vegetative nucleus. *Protoplasma*, **107**, 85–107.

Sager, R. (1975). Patterns of inheritance of organelle genomes: molecular basis and evolutionary significance. In *Genetics and Biogenesis of Mitochondria and Chloroplasts*, ed. C. W. Birkey, Jr., P. S. Perlman & T. J. Byers, pp. 252–67. The Ohio State University Press: Columbus, Ohio.

Sears, B. B. (1980). Elimination of plastids during spermatogenesis and fertilization in the plant kingdom. *Plasmid*, **4**, 233–55.

Shivanna, K. R., Heslop-Harrison, Y. & Heslop-Harrison, J. (1982). The pollen–stigma interaction in the grasses. 3. Features of the self-incompatibility response. *Acta Bot. Neerl.* **31**, 307–19.

Singh, M. B. & Malik, C. P. (1976). Histochemical studies on reserve substances and enzymes in female gametophyte of *Zea mays. Acta Biol. Acad. Sci.* **27**, 231–4. (As cited in Van Lammeren, 1986.)

Sprague, G. F. (1932). The nature and extent of hetero-fertilization in maize. *Genetics*, **17**, 358–68.

Tatintseva, S. S. (1983). Male gamete form of angiosperms. In *Fertilization and Embryogenesis in Ovulated Plants*, ed. O. Erdelska, pp. 113–15. Centre of Biological and Ecological Science, Slovak Academy of Science: Bratislava, Czechoslovakia.

Tilney-Bassett, R. A. E. (1976). The control of plastid inheritance in *Pelargonium*. IV. *Heredity*, **37**, 95–107.

Van Lammeren, A. A. M. (1986). *Embryogenesis in* Zea mays *L. A structural approach to maize caryopsis development* in vivo *and* in vitro. Agricultural University: Wageningen, The Netherlands.

Van Winkle-Swift, K. P. & Salinger, A. P. (1988). Loss of mt^{+}-derived zygotic chloroplast DNA is associated with a lethal allele in *Chlamydomonas monioca. Curr. Genet.* **13**, 331–7.

Wagner, V. T. & Mogensen, H. L. (1988). The male germ unit in the pollen and pollen tubes of *Petunia hybrida*: ultrastructural, quantitative and three-dimensional features. *Protoplasma*, **143**, 101–10.

Weatherwax, P. (1916). Morphology of the flowers of *Zea mays. Bull. Torrey Bot. Club*, **43**, 127–44.

(1919). Gametogenesis and fecundation in *Zea mays* as a basis of xenia and heredity in the endosperm. *Bull. Torrey Bot. Club*, **46**, 73–90.

Went, van J. L. & Willemse, M. T. M. (1984). Fertilization. In *Embryology of Angiosperms*, ed. B. M. Johri, pp. 273–317. Springer-Verlag: New York.

Whatley, J. J. (1982). Ultrastructure of plastid inheritance: green algae to angiosperms. *Biol. Rev.* **57**, 527–69.

Wilms, H. J. (1986). Dimorphic sperm cell in the pollen grain of *Spinacia*. In *Biology of Reproduction and Cell Motility in Plants and Animals*, ed. M. Cresti & R. Dallai, pp. 193–8. University of Siena: Siena.

You, R. & Jensen, W. A. (1985). Ultrastructural observations of the mature megagametophyte and fertilization in wheat (*Triticum aestivum*). *Can. J. Bot.* **63**, 163–78.

Zhu, C., Hu, S. V., Xu, L. V., Li, H. R. & Sen, J. H. (1980). Ultrastructure of sperm cell in mature pollen grain of wheat. *Sci. Sinica*, **23**, 371–9.

5

Apomictic reproduction

E. C. Bashaw and Wayne W. Hanna

A significant number of plant species form their seed by a vegetative (asexual) method of reproduction known as apomixis. This is a natural method for cloning plants through seed propagation and is one of the most interesting and dynamic methods of reproduction in the plant kingdom. Apomixis mimics the normal method of pollination and fertilization that takes place in the flower of sexual plants and, except in special cases discussed later, the male and female sex cells (sperm and egg) do not unite to form a zygote. Instead, the embryo of the seed develops from an unreduced vegetative (somatic) cell in the ovule of the female and receives no genetic material from the male. Unless a mutation occurs, or there is some degree of sexual reproduction, the offspring of an apomictic plant are identical. These plants, called obligate apomicts, produce progeny which are exact replicas of the maternal parent. Obligate apomictic reproduction is analogous to growing clones from the buds of a potato but with the convenience of seed propagation. Apomixis is not limited to a single cytological mechanism nor is it restricted to a few related taxa. As noted by Harlan & de Wet (1963) 'apomixis is widespread and recurs over and over in unrelated groups and has evolved independently on hundreds of occasions throughout the plant and animal kingdom'.

Apomixis is known in over 300 species of at least 35 different plant families (Hanna & Bashaw, 1987) including numerous important genera of Rosaceae and Gramineae. For general information on apomixis in various species the reader is referred to comprehensive reviews on apomixis by Stebbins (1941), Gustafsson (1947), Nygren (1954) and Nogler (1984). Apomixis is especially prevalent among perennial forage grasses and has been reported in more than 125 species representing most of the tribes. Some apomictic grasses have been the subjects of considerable cytogenetic research and are well recognized for their peculiar method of reproduction. Research on apomixis in the popular turf and forage grass *Poa pratensis* began in the early 1930s (Muntzing, 1933) and current research continues to reveal exciting new information on apomixis in this species and its relatives. Tinney (1940) published a remarkably accurate description of the cytological basis for apomixis in *P. pratensis* which set the stage for modern research on apomictic mechanisms in the grasses. Other well-known apomictic grasses include *Paspalum dilatatum*, *P. notatum*, *Panicum maximum*, *Eragrostis curvula* and *Cenchrus ciliaris*. These include some of the most important forage grasses in the world. Discussions on the nature, significance and breeding of apomictic forage grasses have been published by several investigators and will be included in subsequent sections.

Apomixis has not been reported in the grain crops with the exception of grain sorghum where partial apomixis (facultative) has been observed in some lines (Hanna, Schertz & Bashaw, 1970). Hanna *et al* (1970) reported apospory in a polygynaceous line of grain sorghum. Subsequently, they confirmed facultative apomixis through progeny tests using a genetic marker for disease resistance. When the apomictic line, which is self-sterile and recessive, was crossed with a pollen parent having a dominant marker gene for susceptibility to *Periconia circinata*, up to 25% of the progeny were maternal and resistant to the disease. This polygynaceous line and four additional lines with low levels of apomixis and their F_1 progenies were investigated by Tang, Schertz & Bashaw (1980). Only very low levels of apomixis were transmitted to progenies indicating limited prospects for enhancing apomixis through conventional breeding techniques by using this germplasm. At present, geneticists have not been able to increase the level of apomixis in sorghum through breeding and little effort has been devoted to the search for new sources of genes. Apomictic cross-fertile wild 'grassy' relatives of several crops have been reported. *Tripsacum dactyloides* is well known for its facultative apomictic tetraploid strains and can be crossed with *Zea mays*. For many years Soviet scientists have been involved in efforts to transfer apomixis from *T. dactyloides* to corn (Petrov *et al.*, 1973). At least one species (*Elymus rectisetus*) related to wheat reproduces by apomixis (Hair, 1956; Crane & Carman, 1987). It seems possible that apomixis may be present among wild relatives of other cultivated plants but has not been discovered because of limited investigations.

Apomixis is known to be transmitted from parent to offspring and therefore has important implications for breeding when cross-compatible sexual strains and species are available for hybridization. Consider for a moment the economic and scientific impact of a world centre for hybridization where elite lines are crossed with native and exotic germplasm and the offspring are always true-breeding, obligate apomictic hybrids. Developing nations would need only to supply appropriate germplasm and could confine their efforts to testing the vast array of apomictic hybrids that might be produced at a central location. Superior obligate apomictic genotypes could be increased and distributed immediately as new cultivars. With a little imagination and a great deal of basic research, biotechnology and genetic engineering should make it possible in the future to transmit genes for obligate apomixis freely among many species of the Gramineae. This chapter is devoted to discussion of the mechanisms of apomixis in the grasses and review of the current state of our knowledge in this area.

The cytological basis for apomixis

Apomixis in a species is usually suspected when the progeny of individual plants lack appreciable variation or are completely uniform and all or part of the offspring are maternal in appearance. The majority of the apomictic grasses are facultative apomicts such as *Poa pratensis* and individual plants are capable of both sexual and apomictic reproduction. Thus, the progenies from these plants would be expected to comprise some identical types exactly like the female parent and others that differ in one or more characteristics. It should be noted, however, that within facultative apomictic species some plants and some strains may be completely sexual while others are obligate apomicts. While it is technically correct to call these species facultative apomicts, most plant breeders reserve the term 'facultative' for dual modes of reproduction in the same plant.

The potential for a particular method of reproduction can be determined accurately only by a thorough study of megasporogenesis and embryo sac development. When this information is combined with the data from progeny tests or test crosses using dominant markers to assess variation and observe for maternal phenotypes, one can usually be confident of the method of reproduction. Most apomictic grasses are polyploid, highly heterozygous, generally of hybrid origin and frequently behave like derivatives of wide crosses. Consequently, many of the apomictic grasses are meiotically irregular and sexual embryo sacs of facultative apomicts are often subject to a high frequency of abortion. Therefore, some plants which are cytologically facultative may produce aborted sexual embryo sacs and rarely produce off-type progeny. Thus, both a progeny test and cytological evaluation are essential for accurate investigations.

In general the origin of the nucleus which gives rise to the embryo determines the type of mechanism of apomixis. Three apomictic mechanisms are recognized in higher plants: adventitious embryony, apospory and diplospory. Adventitious embryony involves development of an embryo directly from the nucleus of a somatic cell without formation of an embryo sac. The cell nucleus functions as an embryonic initial and the endosperm must be derived from a sexual embryo sac in the same ovule. Adventitious embryony is common in citrus but has never been reported as a regular occurrence in grasses.

Apospory

Apospory is by far the most common mechanism of apomixis in grasses and accounts for more than 95% of known apomictic species. Cytologically, apospory involves development of the embryo from an unreduced nucleus in an embryo sac derived from a somatic cell of the ovary. Generally, aposporous embryo sacs arise in the nucellus of the ovule but also may develop in the integuments and ovary wall in some species. The origin of aposporous cells in the grass ovule is quite different from the normal pattern of sexual megasporogenesis illustrated in Fig. 5.1. Early development of the megaspore mother cell (MMC) is usually identical in aposporous and sexual ovules. In both cases the MMC differentiates in the hypodermal layer of the nucellus in the micropylar region during the enlargement stage of the young ovule (Fig. 5.1*a*). Meiosis in both aposporous and sexual ovules generally results in a linear arrangement (tetrad) of megaspores (Fig. 5.1*b*). At this point the similarity between apospory and sexual embryo sac development ends. In the sexual ovule the chalazal megaspore becomes the embryo sac initial and the other three megaspores abort. This functional megaspore divides mitotically to produce a two-nucleate embryo sac which elongates rapidly and the nuclei migrate to opposite ends of the sac (Fig. 5.1*c*). There is no further development in the somatic cells of the ovule which are typically senescent. Two additional mitotic divisions produce eight nuclei, four at each end of the embryo sac. As the embryo sac enlarges to maturity, differentiation of the nuclei occurs producing two synergids, an egg cell, two polar nuclei and three antipodal cells (Fig. 5.1*d*). In many grass species the antipodal cells undergo additional divisions producing a cluster of cells which may become quite extensive as shown in Fig. 5.1*e*. The fully differentiated embryo sac is considered to be of the *Polygonum* type. The antipodals usually persist for some time after pollination and

Fig. 5.1. Megasporogenesis and embryo sac development in ovules of a sexual forage grass. (*a*) Differentiated megaspore mother cell. (*b*) Functional megaspore (m) and remnants of three degenerated megaspores (dm). (*c*) Early embryo sac with two nuclei (arrows). (*d*) Mature sexual embryo sac with egg (e), polar nuclei (p) and antipodals (a). (*e*) Sexual embryo sac with antipodal cluster (a). Synergids have disintegrated at mature embryo sac stage. (*f*) Early embryo (e) and endosperm (arrows) with remnants of antipodals (a). (Bashaw, 1980*a*.)

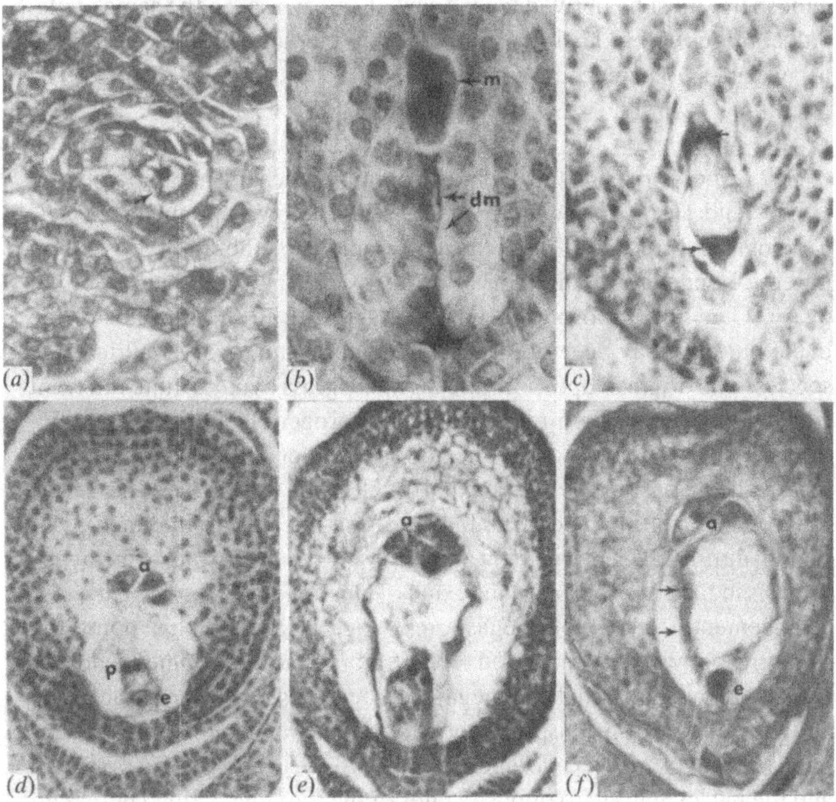

are visible after the embryo and endosperm begin to develop (Fig. 5.1*f*). Typically, the sporogenous tissue of the aposporous ovule persists through meiosis and aborts shortly thereafter.

Aposporic development usually begins at about the time of meiosis of megasporogenesis and is first evident as unusual enlargement of one or more somatic cells (Figs. 5.2*a*,*b*) which are normally approaching senescence at that stage in sexual ovules. Early aposporic initials usually

Fig. 5.2. Origin and development of embryo sacs in ovules of an aposporous forage grass. (*a*) Aposporous cells (arrows) developing adjacent to degenerated megaspore (m). (*b*) and (*c*) Aposporous cells in various stages of development (note resemblance to sexual egg and polar nuclei (*c* arrow)). (*d*) Multiple aposporous embryo sacs in mature ovule (note absence of antipodal cells). (Bashaw, 1980*a*.)

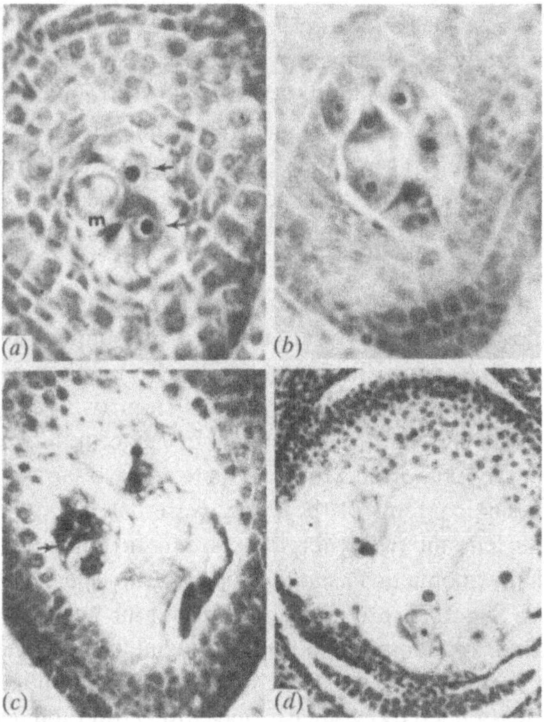

have prominent nuclei and dense cytoplasms, whereas normal nucellar cells cease growth soon after the ovule is fully developed and their nuclei are inconspicuous. The nucleus of the aposporous cell initially undergoes one to several mitotic divisions, and at anthesis the number of nuclei in each sac and the total number of aposporic embryo sacs may be quite variable, even in the same species or plant (Fig. 5.2*c*,*d*).

The degree of differentiation of aposporous embryo sacs is usually a characteristic of the particular species. In some species one or more of the aposporous sacs may develop to the extent that they closely resemble the typical sexual embryo sac. For example, in *Poa pratensis* it is common to find one fully differentiated aposporous sac that cannot readily be distinguished from the normal *Polygonum*-type sexual female gametophyte. This sac even mimics the proliferation of antipodal cells which occurs in the sexual gametophyte. Generally, one differentiated sac will be formed along with one or more less developed sacs. Occasionally in *Poa pratensis* and some other facultative apomicts one sees two identical fully differentiated sacs, presumably one sexual and one apomictic. In these cases identification of the aposporous member can be made only on the basis of its location in the ovule, the sexual member being located centrally in the micropylar region. Twin seedlings are common in aposporous apomicts and triplets are fairly frequent. Often in *Poa pratensis* one seedling will be maternal while the other differs distinctly from the female parent and is obviously of sexual origin. The potential for simultaneous sexual and apomictic reproduction in the same ovule poses a difficult problem in studies of the inheritance of apomixis in these facultative apomicts.

In *Paspalum dilatatum*, *Cenchrus ciliaris* and numerous other apomictic species the aposporous embryo sacs undergo very limited differentiation (Bashaw & Holt, 1958; Bashaw, 1962). The number of sacs of nucellar origin in these species usually ranges from two to eight and sometimes involves a cluster of sacs that cannot be accurately counted. Brown & Emery (1958) reported a four-nucleate embryo sac as typical of the Panicoideae. The contents of these sacs were described as a polar nucleus, an egg nucleus and two synergid cells. This arrangement has not been observed in the apomictic *Paspalum* and *Panicum* species which we have investigated. Synergids are never present in four-nucleate embryo sacs of these grasses and most aposporous sacs contain only one or two nuclei. Often there are one or two large nuclei resembling polars and a smaller nucleus surrounded by cytoplasm and closely resembling a sexual egg. Typically, three is the maximum number of nuclei observed and these appear as two polars and an egg nucleus (Fig. 5.2*c*).

A significant number of ovules of some strains of apomictic *C. ciliaris* have a single aposporous embryo sac with no additional development in the nucellus. These sacs are centrally located in the micropylar region of the ovule and occupy about the same location and total area as the sexual gametophyte in sexual plants of this species. Typically, this single aposporous sac has an egg cell and one polar nucleus and, except for the absence of the second polar nucleus and antipodals, it closely resembles the sexual embryo sac. Observation of early megasporogenesis has shown that these sacs orginate from the nucellus at about the time the sporogenous tissue aborts. Apparently, only one cell of the nucellus becomes meristematic and there is no activity of other somatic cells.

Fig. 5.3. Origin and development of the embryo sacs in ovules of a diplosporous forage grass. (*a*) Megaspore mother cell. (*b*) Unreduced embryo sac with two nuclei. (*c*) Fully differentiated diplosporous embryo sac with egg (e), polar nuclei (p) and antipodals (a). (Voigt & Bashaw, 1976.)

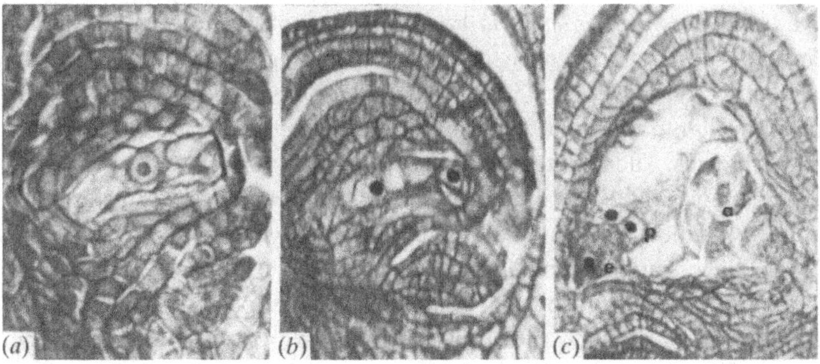

Diplospory

In diplospory the embryo develops in an unreduced embryo sac derived from the megaspore mother cell through mitotic division. The most direct and most valuable form of diplospory for clonal propagation is called the '*Antennaria*'-type named for the genus in which it was first observed. Meiosis is completely eliminated in this mechanism of apomixis. The megaspore mother cell begins to enlarge and elongate at about the time meiosis would normally occur (Fig. 5.3*a*). In most prominent diplosporous grasses there are three successive mitotic divisions leading to an eight-nucleate embryo sac. Differentiation generally occurs as in sexual embryo sacs and a typical *Polygonum*-type female gametophyte is formed. We know of no report of undifferentiated embryo sacs in diplosporous grasses. In most cases the mature diplosporous embryo sac (Fig. 5.3*c*) cannot be distinguished from that of a sexual plant in the species. Compared to apospory there are relatively few diplosporous grasses. The more prominent ones include some species of *Poa* and *Eragrostis*, *Calamagrostis* and *Tripsacum dactyloides*. Nogler (1984) describes two additional forms of diplospory in other families, the *Taraxacum* and *Ixeris* types in which the unreduced gametophyte originates indirectly by a modified meiosis. Crane & Carman (1987) reported both *Antennaria* and *Taraxacum* forms of diplospory in *Elymus rectisetus*.

Cytological identification of diplospory is extremely difficult and usually impossible in mature ovules where the embryo sac may appear completely normal. Evidence of diplospory requires critical study of early megasporogenesis at the time meiosis would be expected to occur (Voigt & Bashaw, 1972). Observation of mitotic metaphase instead of meiosis in the megaspore mother cell and complete absence of the normal linear tetrad of megaspores denotes diplospory. Absence of the pre-megaspore dyad indicates the *Antennaria* type of diplospory. The position of the embryo sac at the two- to four-nucleate stage provides additional evidence of diplospory. The megaspore mother cell elongates extensively, well into the chalazal region of the ovule, before the first division of the nucleus. The resultant two-nucleate embryo sac extends downward to the hypodermal cell layer of the ovule, whereas at the corresponding stage (two-nucleate) of a sexual ovule the embryo sac will be situated a few cell layers up in the ovule. Obviously, only one embryo sac develops in the ovule of a diplosporous apomict. As with apospory the extent of apomixis (obligate or facultative) is determined by observing the phenotypic variation of the progeny.

Haploid parthenogenesis

Haploid parthenogenesis is an asexual method of seed formation and technically may be considered a form of apomixis. As in most species, it is a relatively rare phenomenon in the forage grasses and a non-recurrent method of reproduction. Haploid parthenogenesis differs significantly from other types of apomixis in both cytological basis and fate of the resultant progeny. The mechanism of non-reduction, essential for development of clonal progeny through seed, is missing in haploidy and normal meiosis occurs. Thus, the only similarity with other forms of apomixis is spontaneous (parthenogenic) development of the egg nucleus into an embryo.

Pseudogamy

While embryogenesis is initiated without union of the sperm and egg, fertilization of the polar nuclei is believed to be essential for seed development in all known apomictic grasses. In some cases precocious embryo development may begin before pollination, but we know of no grass species in which mature seed will develop without pollination. The requirement for pollination in order to promote development of a viable seed is known as pseudogamy. Interestingly, there are reported cases in which pseudogamy in an apomictic grass can be stimulated by pollen from an entirely different species. This has been reported in some species that are difficult or impossible to cross (Simpson & Bashaw, 1969). We recently observed that seed set in apomictic *Pennisetum flaccidum* could be increased from normally less than 10% to 80% by pollinating at anthesis with *P. mezianum* pollen. This could be very important as a method for increasing seed production in some apomicts where pollen quality is too low for effective stimulation.

The genetic basis for apomixis

Biologists have long recognized that apomixis is readily transmitted from parent to progeny, but lack of inheritance data needed for critical genetic analysis has posed a serious problem in understanding apomixis. Early theories on the causal aspects of apomixis were based primarily on generalizations gathered from observing the nature and behaviour of apomicts. It was recognized that most apomictic plants are polyploid and cytologically resemble species hybrids. Thus, it is not surprising that hybridity and polyploidy dominated these early concepts of the factors that cause apomixis. Physiological factors and hormones were also postulated as direct causes of apomixis. For example, one popular theory proposed that apomixis was induced by wound hormones produced when plants were injured. Stebbins (1941) ruled out polyploidy as a cause of apomixis because sexual polyploids greatly outnumber the apomicts. Gustafsson (1947) analysed existing evidence favouring each theory and concluded that no single phenomenon could be individually responsible for apomixis. He noted that results of a few crosses between sexual and apomictic plants indicated a genetic basis for apomixis. Both hybridity and polyploidy accompany apomixis and may facilitate its expression by causing meiotic irregularity and sterility. Numerous authors have characterized apomixis as an 'escape from sterility', thus comprising a mechanism which certainly would be most obvious in species which have serious cytological abnormalities resulting in low fertility. Modern molecular genetics will undoubtedly reveal an important role of physiological factors including hormones in the initiation of meristematic activity in somatic cells, but one would not expect these factors to be independent causes of apomixis. No doubt gene-mediated physiological or biochemical pathways are somehow responsible for changing the role of specific cells and altering the normal senescent destiny of cells which give rise to apomictic embryos.

Several factors make apomicts poor subjects for genetic study. First, it is necessary to identify cross-compatible sexual or partially sexual plants to allow for hybridization. Discrete segregation ratios require crossing plants which are completely sexual with obligate apomicts so that only the two parental reproductive phenotypes are produced. Most apomictic species are facultative and, because alternative modes of reproduction may occur in any individual ovule, it is not possible to know the origin (sexual or apomictic) of all progeny. Furthermore, there is evidence that environment affects method of reproduction in these plants. Most earlier genetic studies were conducted with facultative apomicts and the results were confusing and difficult to interpret. Results of studies involving crosses between completely sexual and obligate apomictic plants are much more reliable, but the lack of segregation of the progeny of apomictic hybrids seriously limits the amount of data normally available from progeny tests and necessitates growing large progeny populations. We have gradually accumulated data that allows new insight into the genetics of apomixis and, in a few species, we are able to ascribe specific gene action.

Discovery of sexual plants in species previously considered obligate apomicts provided the parents needed for critical genetic studies. Many of these crosses produce progeny which are either obligate apomicts or completely sexual. For the most reliable genetic studies, facultative apomixis must not appear in the hybrids or advanced generation progeny. Accurate analysis requires discrete reproduction ratios from the F_1 population and succeeding generations. One must still contend with the fact that polyploidy and the hybrid nature of apomicts present difficult problems. For example, release of sexuality in an apomict also releases the chromosomal aberrations and other abnormalities inherent to the apomict but not expressed because of the asexual method of reproduction. Inbreeding depression, including the failure of some plants to flower, often limits the data. In spite of these difficulties we have been able in recent years to establish that apomixis is genetically controlled and to develop a reasonable concept of the inheritance of mode of reproduction in some species.

The most conclusive information on inheritance of apomixis in forage grasses has come from research on species in three genera, *Paspalum*, *Panicum* and *Cenchrus*. Burton & Forbes (1960) were able to demonstrate genetic control of apospory in *Paspalum notatum* using a novel approach for hybridization. The tetraploid types of this species are obligate aposporous apomicts. No natural sexual tetraploids have been identified but there are sexual diploids in the species. By doubling the chromosome number of a diploid strain (Pensacola bahiagrass) the investigators produced a sexual autotetraploid type. Crossing this sexual tetraploid with apomictic tetraploids produced hybrids and progenies which were phenotypically either obligate apomicts or completely sexual. Analysis of the data from hybrids and their progenies gave ratios which indicated that apomixis was conditioned by a few recessive genes. From these studies they postulated a genotype of *aaaa* for the apomictic parent.

Research with *Panicum maximum* using sexual plants of different origins produced different results regarding gene action, but data from these studies also confirm a genetic basis for apomixis. This species is an autotetraploid long considered to be an obligate apomict. Hanna *et al.* (1973) used naturally occurring sexual tetraploids discovered among plant introductions as female parents in crosses with natural tetraploid apomictic accessions. Their results indicated that sexuality was dominant to apomixis and that method of reproduction was probably controlled by at least two loci. Savidan (1983), working with the same species, used the approach employed by Burton and Forbes with bahiagrass, induction of sexual autotetraploids from natural diploids to use as parents. When these plants were crossed with apomictic tetraploids, he obtained data which fit a single-gene model for inheritance of apomixis. Savidan's results indicated that apomixis was dominant and led him to postulate that genotypes of the sexual and apomictic parents were *aaaa* and *Aaaa*, respectively. It should be noted at this time that the expression of obligate apospory involves the simultaneous occurrence of three separate events: activation of somatic nucellar cells to meristematic status, suppression (destruction) of the sporogenous tissue (sexual megaspores) and autonomous development of an unreduced egg into an embryo (parthenogenesis). In view of the potential for interference with this sequence of events, which could vary with genotype and environment, one might not expect agreement of data from hybridization of diverse parents.

The discovery of a sexual plant, presumed to be a mutant, enabled Taliaferro & Bashaw (1966) to investigate the inheritance of method of reproduction in aposporous *Cenchrus ciliaris* (buffelgrass). This plant, TAM-CRD B-1s (Bashaw, 1969) proved to be phenotypically obligate sexual but heterozygous for method of reproduction. Selfed progeny from B-1s segregated for method of reproduction at a ratio of 13 sexuals to 3 obligate apomicts. As with the bahiagrass hybrids described earlier, progeny of B-1s were either obligate sexual or obligate apomicts. The behaviour of B-1s provides a good illustration of the heterozygous nature of apomictic species; identical plants were not found among 1000 selfed progeny.

When B-1s was crossed with two different obligate apomictic cultivars, the F_1 hybrids comprised a ratio of 5 : 3 obligate sexual plants to obligate apomicts. The data from selfing and crossing with both cultivars fit the ratios expected if mode of reproduction is controlled by two different genes with epistasis favouring dominant expression of the gene for sexuality. On the basis of these data we postulated that the genotype of sexual B-1s was *AaBb* and that of the two apomictic cultivars was *Aabb*. This hypothesis assumes that dominant gene *B* conditions sexual reproduction and is epistatic to dominant gene *A* which conditions apospory. The double recessive *aabb* would be expected to reproduce sexually because of the absence of dominant gene *A*. Based on this hypothesis one would expect the ratio that was obtained in these crosses. Further confirmation regarding genetic control of apomixis came from hybridization of B-1s with obligate apomictic *C. setigerus* (birdwood-grass). The F_1 hybrids from this cross comprised a ratio of 1:1 sexual to obligate apomictic plants, indicating that *C. setigerus* represents the only other genotype for apomixis (*AAbb*) possible under this hypothesis.

Consistent with the above hypothesis is the assumption that gene *A* controls all of the processes that result in development of unreduced nucellar embryo sacs and abortion of the normal sexual sporogenous tissue. Autonomous development of an embryo from an unreduced nucleus is not accounted for and probably will never be explained until we understand the biochemical basis for apomixis. Nevertheless, the hypothesis seems valid if certain assumptions are accepted. A theoretical explanation for the cytological–biochemical basis for apospory in *C. ciliaris* is shown diagrammatically in Fig. 5.4. If we assume that gene *A* conditions production of a growth factor which induces apospory by 'activating' nucellar cells to become meristematic and form unreduced gametophytes (Fig. 5.4*a*), it seems reasonable that the same factor at a given concentration could suppress (abort) the sexual mechanism. The action of the dominant *B* gene (epistatic) might be to nullify all action of gene *A* and restore sexuality (Fig. 5.4*b*). The presence of dominant gene *B* or the double recessive *aabb* (Fig. 5.4*c*) would result in sexual reproduction. This theory allows for facultative apomixis if the level of growth factor is not high enough to suppress the sexual mechanism (Fig. 5.4*d*). In that case both sexual and aposporous embryo sacs would be formed. It is obvious that critical biochemical investigations are needed before it is possible to understand fully the genetics of apomixis responsible for these ratios.

Hybridization studies with facultative apomictic species in several genera document a gene basis for apomixis, but in most cases the data are less conclusive than those involving obligate apomicts. The most reliable data have come from crosses in which good genetic markers are available for use in identifying offspring of sexual origin. In these cases extreme heterozygosity of the facultative apomictic female parent allows one to be reasonably confident that maternal progeny are from apomictic gametophytes. Harlan *et al.* (1964) conducted extensive hybridization

Fig. 5.4. Hypothetical model of gene action responsible for apospory in apomictic *Cenchrus ciliaris*. Ovules shown at linear tetrad stage of normal megasporogenesis with chalazal members degenerating. Fate of the functional megaspore and activation or suppression of nucellar cells controlled by genes *A* and *B*.

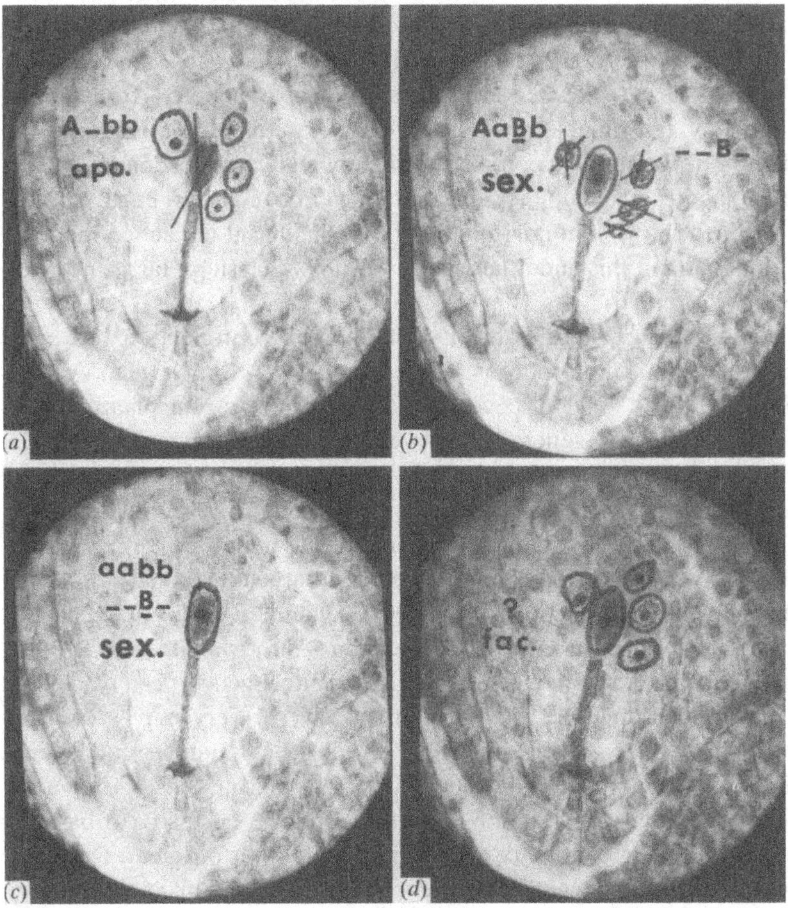

studies with apomictic species (primarily facultative) in the *Bothriochloa–Dichanthium* complex. Their research clearly demonstrated genetic control of apomixis, probably by no more than one dominant gene per genome. When predominantly apomictic facultative plants were crossed with obligate apomicts, sexual hybrids were sometimes recovered but most progeny were obligate apomicts. The composite F_2 ratio was 15:1 apomictic to sexual progeny. Crosses of obligate sexual to obligate apomictic parents produced three apomictic plants to one sexual. When both parents were obligately sexual, only sexual hybrids were obtained. They theorized that apomixis was dominant to, but independent of, sexuality. Tetraploid apomictic plants were assigned the genotype *AAaa* while the sexual genotype was assumed to be *aaaa*.

Numerous hybridization studies have been conducted with *Poa pratensis* and its relatives but we still lack a good concept of the genetics of mode of reproduction in this species. As with *Bothriochloa* and *Dichanthium*, the results clearly show patterns of transmission of apomixis from parent to progeny, but reliable ratios have been very elusive in this species. In addition to facultative apomixis, extremely high chromosome numbers, both euploid and aneuploid, are common in *Poa pratensis*. Furthermore, it has been shown that environment has a significant impact on mode of reproduction in this and other species (Knox & Heslop-Harrison, 1963; Hovin *et al.*, 1976).

Evidence for genetic control of method of reproduction has been reported for one diplosporous forage grass, *Eragrostis curvula* (Voigt & Burson, 1983). Voigt discovered a few partially sexual plants in this species which was generally believed to be an obligate apomict. In subsequent hybridization studies when highly sexual plants were crossed with highly apomictic selections, both methods of reproduction were represented in the F_1 generation. Some of the hybrids were facultative but most were either highly apomictic or completely sexual. A ratio of one highly apomictic to one completely sexual hybrid was obtained. The presence of facultative hybrids that were intermediate in sexuality prevented establishing conclusively the probable genotypes involved in mode of reproduction. As noted previously the very nature of most apomictic species prevents development of genetic data essential for a clear understanding of gene action. Different mechanisms of apomixis, meiotic irregularity, variation in fertility and the necessity to base all conclusions on phenotypic evaluation pose a serious problem. Development of a good concept of the genetic basis for apomixis will require critical information on the biochemical action involved in the various cytological phenomena associated with each mechanism of apomixis. Available information suggests that control of apomixis in most species is relatively simple, probably involving no more than two genes per genome. More research is needed to identify additional apomictic species and broaden our concept of the causal mechanisms involved. As will be seen in the following section, the limited data available has had a major impact on modern concepts of the potential for breeding apomictic species.

Progress in breeding apomictic grasses

Successful breeding of apomictic species is a relatively recent achievement and thus far only a few artificially bred cultivars of any apomictic crop have been released. For many years the only source of cultivars of apomictic species was selection and increase of the best natural apomictic ecotypes. Natural ecotypes continue to be an important source of forage grass cultivars of some species. Research on breeding apomicts is still in the pioneering stages but results obtained so far are promising. Cultivars have been produced in some apomictic forage grasses through controlled hybridization of sexual and apomictic strains and promising breeding programmes have been developed in several others. Results of basic research, especially in interspecific hybridization, indicate good prospects for extensive use of apomixis as a breeding tool.

Early research on the mechanisms of apomixis and subsequent inheritance studies set the stage for breeding experiments with apomicts. Eventually it was possible to demonstrate that method of reproduction can be controlled and manipulated in a plant breeding programme just as any other genetic character. Success in the breeding of apomicts ultimately depended on finding female parents that could be used effectively in gene transfer. This generally involves discovering or synthesizing cross-fertile sexual plants for use in hybridization with apomictic accessions. In most apomictic species the pollen is viable enough to assure fertilization, although quality is often variable and in some apomicts the pollen is sterile. Many apomictic species are polymorphic and the abundance of apomictic ecotypes provides widely diverse male parents for use as germplasm. Once suitable parents have been acquired, the breeding of apomictics is relatively simple. Hybridization of sexual and apomictic plants effectively breaks the apomictic barrier and both methods of reproduction appear in the F_1 or subsequent generations. Obligate apomictic progeny breed true and there is no need for further selection to achieve uniformity. Heterosis, if present in an apomictic hybrid, is permanent and all characteristics of the hybrid are maintained without change. The breeder only needs to select and test superior combinations, any of which may be released as a new cultivar without further effort.

Breeding of apomictic species is most efficient when only sexual or obligate apomictic phenotypes are possible among the progeny. Once obligate apomictic offspring are identified, there is no further concern for variation. Most rapid progress is possible when the female parent is heterozygous for method of reproduction and obligate apomictic progeny are relatively prevalent in the F_1 generation. Heterozygous sexual hybrids provide an additional source of obligate apomictic lines but careful selection is necessary to assure success. We have found that progeny of many sexual hybrids are subject to extreme inbreeding depression and one needs to be very cautious in choosing which sexual hybrid to use as germplasm.

Development of apomictic cultivars has been reported in at least three species of forage grasses, *Cenchrus ciliaris*, *Poa pratensis* and *Panicum maximum*. Soon after the discovery of *C. ciliaris* sexual clone B-1s, we initiated an extensive hybridization programme which led to the genetic information reported earlier. Concurrently with the genetic studies we began a breeding programme based first upon identification and testing of obligate apomictic self progeny of sexual clone B-1s, which are heterozygous for method of reproduction. Promising obligate apomictic progeny were selected and increased and tested in replicated field trials. Subsequently, a superior apomictic strain was identified and released as 'Higgins' buffelgrass (Bashaw, 1968). Hybridization of B-1s with two obligate apomictic ecotypes produced a wide range of phenotypes, many of which we had never seen among natural accessions (Taliaferro & Bashaw, 1966). As noted earlier, a predictable proportion (3/8) of the F_1 hybrids were obligate apomicts and bred true. Based on these observations, a simple scheme (Fig. 5.5) was proposed for breeding apomictic buffelgrass. Primary emphasis was placed on selection and testing of hybrids recovered in the F_1 generation to take advantage of maximum hybrid vigour. Apomictic progeny may be selected in succeeding generations but the proportion of superior plants expected is much smaller than among the F_1 population and serious sterility problems begin to appear among sexual progeny after the F_1 generation. The apomictic breeding scheme developed for buffelgrass proved to be very successful and two hybrid cultivars 'Nueces' and 'Llano' have been released (Bashaw, 1980*b*). Both cultivars are obligate apomictic true-breeding F_1 hybrids. Development of these *C. ciliaris* cultivars clearly demonstrated the potential for use of apomixis as a 'tool' in breeding and stimulated new interest in improving apomictic grasses.

Because of the difficulty encountered in recovering obligate apomictic progeny, the breeding of facultative apomicts is much less efficient than with predominantly obligate apomictic species in which some completely sexual plants or strains are available. Nevertheless, most facultative apomicts do produce some progeny which are obligate or highly apomictic. Extensive progeny testing is usually essential to identify these offspring. Fortunately, absolute uniformity is not essential in a forage grass cultivar and a considerable proportion of variant plants can be

Fig. 5.5. Diagram of methods used to produce apomictic buffelgrass cultivars. Sexual parent is heterozygous for method of reproduction. (Taliaferro & Bashaw, 1966.)

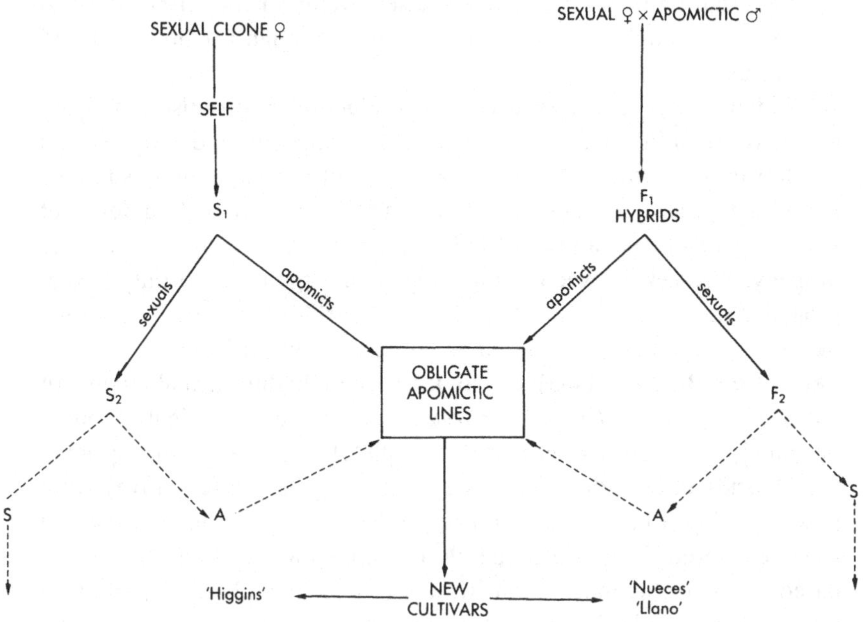

tolerated. Furthermore, we have observed that sexual progeny tend to be at a disadvantage in a pasture situation and are soon eliminated.

Most research on breeding facultative apomicts has been conducted with *Poa pratensis*. Results of early research with this species demonstrated that most accessions are capable of both sexual and apomictic reproduction in the same ovule and that mode of reproduction can be highly variable among florets on the same inflorescence. Sexual florets tend to be cross-pollinated under natural conditions but selfing occurs readily when pollen from other sources is not available. Thus, natural populations afford an abundance of genotypes of both hybrid and selfed origin. With this range of natural variation it is not surprising that most cultivars have been derived from selection among natural ecotypes. High chromosome numbers and aneuploidy, which prevent accurate genetic analysis, cause few sterility problems since a portion of the offspring are always apomictic and fertile.

In recent years successful breeding programmes with *P. pratensis* using hybridization have been initiated in several countries. In the USA, Pepin & Funk (1971) have developed four cultivars through intraspecific hybridization. In Germany, Nitzsche (1983) has been successful with both intra- and inter-specific hybridization. These investigators have shown that highly apomictic strains as well as some nearly obligate apomicts can be produced in this species. Many of the strains derived from the breeding programmes are uniform enough for both turf and pasture use.

Savidan (1983) reported successful production of promising obligate apomictic hybrids of *Panicum maximum* using either derived sexual autotetraploids, noted earlier, or natural facultative apomicts as female parents. Crosses between facultative apomicts with a high level of sexuality (40%) gave a ratio of 1:3 sexual plants to apomicts among the progeny. Crosses between completely sexual plants gave only sexual progeny. Savidan noted that 10 promising hybrids from sexual × apomictic crosses were being evaluated as potential new cultivars.

Voigt & Burson (1983) found that when highly sexual plants of facultative apomictic *Eragrostis curvula* were crossed with highly apomictic plants, most hybrids were either completely sexual or highly apomictic. Hybrids intermediate in mode of reproduction (facultative) were produced, but a high proportion of predominantly apomictic offspring were recovered. They subsequently developed a breeding programme based on use of the most apomictic and most sexual plants as parents in crosses. By careful selection of the parents they were able to reduce the proportion of facultative apomictic plants among both F_1 and backcross progeny.

Apomixis offers some novel and fascinating approaches to plant breeding that are rarely, if ever, possible with purely sexual seed-propagated species. Interspecific and intergeneric hybridization, discussed in a subsequent section, have tremendous possibilities in apomictic genera. Because apomixis can assure at least some fertility in the wide cross, valuable germplasm is preserved for use in the breeding programme. Apomictic species possess a unique mechanism for gene transfer that is just beginning to be exploited. We have long recognized that most apomicts occasionally produce offspring (called B_{III} hybrids by Rutishauser, 1948) through fertilization of the unreduced egg. Harlan & de Wet (1963) reported that this phenomenon afforded an opportunity to make wide crosses in the *Bothriochloa–Dichanthium* complex that normally were impossible. Most reports indicate that the progeny derived in this manner are fertile, true-breeding obligate apomicts. Fertilization of the unreduced egg has dynamic potential for incorporation of genes and even whole alien genomes into the somatic chromosome complement of the apomictic female parent. Meiotic disturbances common in wide crosses are of no consequence in obligate apomicts if sufficient viable pollen is available to stimulate pseudogamy.

Fertilization of the unreduced egg has generally been regarded as too infrequent to be useful as a breeding tool. However, we have found from recent studies that the frequency of fertilization of the unreduced egg can be increased substantially by thorough emasculation of the apomictic female parent and use of abundant pollen from male parents. Using an experimental gametocide for emasculation, we were able to obtain 1.5% B_{III} hybrids in crosses between an obligate apomictic *Cenchrus ciliaris* pentaploid and tetraploid apomictic *C. setigerus* (Bashaw & Hignight, 1990). Thirteen hybrids obtained in this manner were morphologically diverse, highly fertile obligate apomicts with 63 chromosomes. Using this method we have been able to combine valuable characteristics in true-breeding hybrids in an agamic complex where desirable sexual plants are extremely rare.

At present there is no explanation why fertilization of the unreduced egg is so infrequent under natural conditions. This is a matter of serious concern because the ability to increase the incidence of B_{III} hybrids would significantly enhance a breeding programme. We suggest two possibilities that might be researched in the future. Because cross-pollination, and in many cases foreign pollen, can stimulate pseudogamy in apomicts, it seems possible that the timing of pollination following thorough emasculation might influence fertilization of the unreduced egg. For example, if the pollen tube were to enter the embryo sac just prior to the first precocious division of the unreduced egg nucleus, fertilization might be more likely. A second possibility envisions repulsion of the pollen tube by the egg cell. During the study of embryo sac development in ovules of obligate apomictic *C. ciliaris* which were cleared shortly after anthesis, we often observed the pollen tube with a sperm near the polars, but rarely saw a sperm anywhere near the egg nucleus. We noted that the pollen tube rarely penetrated the cytoplasm of the egg cell in ovules of apomictic hybrids but in sexual ovules it normally grew through the centre of the cytoplasm. One wonders if it might not be possible to enhance fertilization by treatment with appropriate chemicals to eliminate repulsion. We certainly need basic research in this area to explore all possibilities for effective use of this potentially valuable breeding tool.

Mutation breeding

Theoretically, induced mutations should prove remarkably effective for improvement of apomictic species. The true-breeding nature of obligate apomicts assures that any true genetic change through mutagenesis will remain constant in succeeding generations. Elimination of the sexual female gametophyte in reproduction would allow seed propagation of plants with considerable chromosomal aberrations that might result from irradiation. One would expect that any characteristic present in the natural habitat, however remote, might possibly be accessible through artificial mutagenesis.

Unfortunately, there has not been enough experimentation to allow us to generalize about the opportunities possible through induced mutation. Results of the few studies conducted with apomictic grasses look encouraging but we approach this subject with caution. Radiation resulted in extensive morphological changes in apomictic *Paspalum dilatatum* (Bashaw & Hoff, 1962; Burton & Jackson, 1962). In neither set of experiments was any change in method of reproduction observed, but many mutants with significant modifications in growth habit and morphology bred true in succeeding generations. The particular objectives in these experiments, disease resistance and sexual reproduction, were not achieved but the results showed great promise for induction of morphological changes. Working with a different species, *Poa pratensis*, Hanson & Juska (1962) succeeded in producing potentially valuable mutants with disease resistance using ionizing radiation.

Julen (1960) reported changes in morphology and sexual reproduction among progeny of *Poa pratensis* following radiation. Out of 286 progeny, 38 had entirely sexual seed formation, 101 were apomictic and the remaining 147 were facultative. Progeny of sexual plants segregated into apomictic, sexual and partially sexual plants. 'The results indicated that it would be possible to induce sexuality by X-raying in an apomictic species, make crosses and in later generations select new apomictic types with favorable morphological characters.' Hanna & Powell (1973) succeeded in producing a facultative apomictic pearl millet by a dual treatment with thermal neutrons and diethyl sulphate.

Wide crosses – potential for transfer of apomixis

The genes controlling apomixis can most readily and effectively be utilized if found within the target species. Although genes controlling apomixis have been found and utilized in breeding such plants as *Citrus* (Parlevliet & Cameron, 1959), and certain forage (Bashaw, 1980) and turf (Pepin & Funk, 1971) grasses, genes controlling obligate or a high level of apomixis have not been discovered in the major cultivated grain crops. The transfer of genes controlling apomixis through interspecific and intergeneric crosses offers an opportunity for using apomixis in our cultivated crops. A number of factors such as ploidy of the cultivated and wild species, genetic nature of apomixis in the wild species, phylogenetic relationships and cross-compatibility of wild and cultivated species, genome relationships and fertility of wide crosses will determine the success of transferring genes controlling apomixis from a wild species by traditional methods.

Ploidy levels

Ploidy levels (e.g. diploid, tetraploid, etc.) of the cultivated sexual and wild apomictic species are an important consideration in an interspecific hybridization gene transfer programme. Higher fertility in the interspecific hybrid and backcross derivatives can be expected if tetraploid forms of the diploid cultivated sexual species are used in the crossings (Petrov *et al.*, 1976; Dujardin & Hanna, 1983, 1987*a*). Use of tetraploid forms of the cultivated species may even determine whether or not the original interspecific cross is successful (Dujardin & Hanna, 1989). Introduction of different genotypes of the tetraploid or polyploid cultivated species at various stages of the backcrossing programme is also important for enhancing fertility of the interspecific derivatives. A tetraploid female parent contributes at least a diploid set of chromosomes of the cultivated species to the interspecific hybrid and usually results in higher male and female fertility. This may require induction of tetraploids in normally diploid species if tetraploids are not available. Similar ploidy considerations are needed for the male apomictic wild species. One needs to select the wild species that will result in the highest fertility (if a number of apomictic species are available in a genus) and the highest level of apomictic reproduction in subsequent generations. This will probably require experimentation, because little is known about inheritance and expression of apomixis in most species. In *Pennisetum*, *P. squamulatum* (a hexaploid) was selected after experimenting with three different apomictic species and ploidy levels: triploid *P. setaceum*, tetraploid *P. orientale* and hexaploid *P. squamulatum* (Hanna & Dujardin, 1985).

Research in *Pennisetum* indicates that polyploidy is not necessary for expression of the genes controlling apomixis. Apomixis was expressed in pearl millet–*P. orientale* derivatives when the genes were either in the simplex or duplex condition. Obligate apomixis is expressed in pearl millet–*P. squamulatum* BC_3 derivatives when only one or two chromosomes from hexaploid ($2n = 54$) apomictic *P. squamulatum* are present.

Phylogenetic and genome relationships

Wild species that are taxonomically closely related to the cultivated species are better donors of genes controlling apomixis than distantly related species. The relationship may be due to residual homology and/or a common genome(s). Degree of relation may also exist within a gene pool. In the *Pennisetum* genus, triploid *P. setaceum*, tetraploid *P. orientale* and hexaploid *P. squamulatum* are all obligate apomicts and belong to the tertiary gene pool. However, *P. squamulatum* shows residual homology with the genome of cultivated pearl millet and has produced the most fertile interspecific derivatives.

One should always consider using bridging species or hybrids in an interspecific gene transfer programme even though they may be sexual. Bridging parents can be used to improve fertility in interspecific derivatives by creating a more desirable balance of the various genomes. Fertility was enhanced in apomictic interspecific hybrids between pearl millet and *P. squamulatum* crosses by crossing these hybrids with sexual hexaploid interspecific hybrids between pearl millet and *P. purpureum* (Dujardin & Hanna, 1985).

Genetics of apomixis

As discussed earlier, genetic systems controlling apomixis vary between and within plant genera and within species. Species with the simplest genetic control are the most desirable to use in interspecific gene transfer because one has to select simultaneously for a high level of apomixis and against undesirable characteristics of the wild species. Genes controlling apomixis should also condition obligate or a high level of apomixis under diverse environmental conditions. The degree of facultative reproduction tolerated for a species would depend on how it is utilized and/or harvested.

Progress in apomixis transfer

Progress is being made in transferring genes controlling apomixis from wild to cultivated species. Both tetraploid ($2n = 4x = 36$) and octoploid ($2n = 8x = 72$) apomictic *Tripsacum dactyloides* have been crossed with diploid ($2n = 2x = 20$) and tetraploid ($2n = 4x = 40$) *Zea mays* (de Wet *et al.*, 1973; Petrov *et al.*, 1976). Backcross derivatives were recovered with low seed fertility, varying levels of apomixis and expression of *Tripsacum* characteristics. Similar results were obtained when sexual diploid ($2n = 2x = 14$) and tetraploid ($2n = 4x = 28$) pearl millet (*Pennisetum glaucum*) were crossed with apomictic tetraploid ($2n = 4x = 28$) *P. orientale*. Although *P. orientale* is highly apomictic, the

Fig. 5.6. Pathway for developing a 29-chromosome apomictic BC_3 plant between sexual pearl millet ($2n = 28$) and *P. squamulatum* ($2n = 54$).

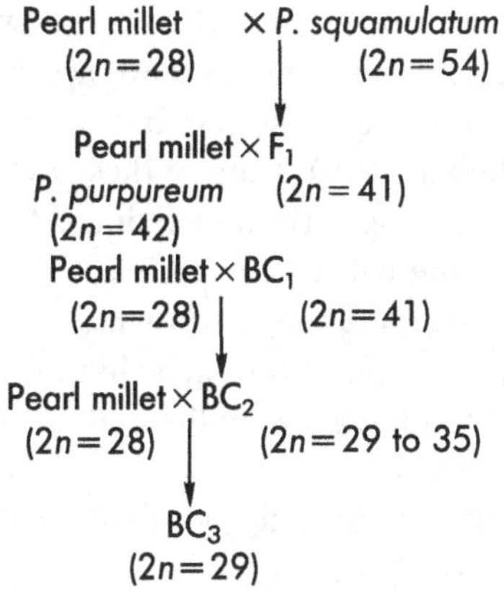

interspecific hybrids and backcross derivatives showed low levels of apomixis and were highly male and female sterile (Dujardin & Hanna, 1987*b*).

Crosses and backcrosses between tetraploid pearl millet and obligate apomictic hexaploid ($2n = 6x = 54$) *P. squamulatum* (Fig. 5.6) have produced some of the most encouraging results for interspecific transfer of genes controlling apomixis (Dujardin & Hanna, 1983; Hanna & Dujardin, 1985). Obligate apomixis has been maintained to the BC_3 generation due to a *P. squamulatum* chromosome with the gene(s) controlling apomixis. Plants are also partially male fertile. Male fertility was enhanced in the backcrosses by introducing *P. purpureum* as a bridging species (Dujardin & Hanna, 1985). Further progress towards developing an apomictic pearl millet for grain and forage production appears to depend on transferring a segment of the *P. squamulatum* chromosome with the gene controlling apomixis, recombination due to residual homology between the pearl millet and *P. squamulatum* chromosome, or producing a substitution line with tetraploid pearl millet.

The potential exists for interspecific transfer of genes controlling apomixis in wheat (*Triticum aestivum* L.). A hexaploid wild relative of wheat, *Elymus rectisetus*, has been reported to be apomictic (Hair, 1956; Crane & Carman, 1987). Progress in transferring genes controlling apomixis from *E. rectisetus* to wheat will depend on producing partially fertile interspecific hybrids and backcrosses and maintaining a high level of apomictic reproduction.

Some reported wide crosses between sexual and apomictic species have been highly sterile. An intergeneric cross between sexual diploid ($2n = 2x = 14$) pearl millet and apomictic tetraploid ($2n = 4x = 36$) *Cenchrus ciliaris* were completely sterile but showed apomictic embryo sac development (Read & Bashaw, 1974). Similarly, an interspecific hybrid ($2n = 3x = 25$) between diploid pearl millet and apomictic triploid ($2n = 3x = 27$) *P. setaceum* was highly sterile but showed apomictic embryo sac development (Hanna, 1979). Both of these hybrid combinations appeared to be dead ends using current techniques for transferring genes controlling apomixis to the cultivated species.

Apomixis is a strong force in perpetuating natural interspecific hybrids. Fountaingrass, *P. setaceum*, is a triploid apomictic hybrid that probably originated from a sexual diploid species pollinated by an apomictic tetraploid species. Although chromosome behaviour is very irregular (Simpson & Bashaw, 1969), the interspecific hybrid is able to perpetuate itself through apomictic reproduction.

Common dallisgrass, another natural apomictic hybrid, is an important pasture grass in the tropical and subtropical regions of the world. Efforts to improve its low seed fertility and ergot (*Claviceps paspali*) susceptibility through hybridization have been unsuccessful because of an apomictic barrier. Past research has shown that sexual tetraploid yellow anther dallisgrass has the same I and J genomes as apomictic common dallisgrass (Burson, 1983). Current research shows that common dallisgrass may be a hybrid between tetraploid yellow anther dallisgrass and a Uruguayan hexaploid dallisgrass with the I, J and X genomes (Burson, 1989). The X genome apparently carries the gene(s) for apomixis but the *Paspalum* species contributing the X genome has not been identified. Identification of all species contributing the common dallisgrass genomes should make improvement of the species more effective.

Screening for apomixis

Transferring genes controlling apomixis via wide crosses requires observing large populations of plants. Two major criteria influence plant selection at each crossing generation. First, plants are selected that reproduce by obligate or a high level of apomixis. Secondly, apomictic plants need to be partially male fertile because they can be used most efficiently as pollinators. Large populations need to be observed beginning at the initial cross because most apomicts are highly heterozygous and can produce a diversity of interspecific hybrids with varying levels of fertility.

Apomictic plants can be identified by both cytological methods and progeny tests of open-pollinated plants. Current ovule clearing techniques (Crane, 1978) allow the reproductive behaviour to be determined within 48 h of specimen collection and the selected plants can be used in further crosses before anthesis is completed. Cytological results should be confirmed with progeny tests.

Gene transfer methods

The most desirable gene transfer method would be genetic recombination due to residual homology between genomes of the cultivated and wild species. This type of gene transfer would result in the least number of undesirable genes being transferred. Transfer of small chromosome pieces through translocations may be the next best method of gene transfer if residual homology does not exist. Whether translocations are spontaneous or induced, this method requires screening very large populations. Another gene transfer consideration is the production of substitution lines. Substitution lines probably would be more effective at a polyploid level and could be used if the extra chromosome does not contribute too many undesirable genes.

In the future, new techniques such as the use of vectors, electroporation, protoplast fusion and micro-manipulation of chromosomes offer exciting potential for gene transfer not only between species but between genera and possibly between plant families.

Role of apomixis in evolution

There has been considerable disagreement among plant scientists in the past regarding the significance of apomixis in evolution. Early scientists seemed to overlook the existence or significance of facultative apomixis and concentrated mainly on describing the various manifestations of apomictic reproduction. The possibility of cross-compatible sexual and apomictic plants in the same species was generally ignored and there is little discussion in early literature regarding prospects for controlling and manipulating apomixis. Darlington (1939) viewed apomixis as 'an escape from sterility but an escape into a blind alley of evolution'. According to this idea, apomixis could lead only to extinction of the species. Babcock and Stebbins (1938) were somewhat more liberal in their observation that 'when separated from their sexual ancestors apomicts will in time become relic species and die out'. Stebbins (1941) conceded that apomicts were adapted to take great advantage of conditions prevailing when they come into existence, rather than to adjusting themselves to changing conditions over a long period of time. He concluded that apomixis is not a major factor in evolution, however important it is in increasing the polymorphism and the geographic distribution of the genus. It would seem that any force which enhances polymorphism must have a role in evolution.

We now recognize that sexual or partially sexual plants or strains probably exist in most, if not all, apomictic species and among relatives of the apomicts. It seems reasonable that facultative apomicts precede obligate apomicts in the development of the agamic complex. Thus, from the very beginning an apomictic species would not be restricted in the generation of sufficient variation for effective speciation. Clausen (1954) described facultative apomixis as an evolutionary equilibrium system in which 'the apomictic process is in balance with an almost dormant sexual process that can be invoked and can release a part of the stored sexual variability for a time'. In the location of origin of the species completely sexual, facultative and obligate plants may still be present. Sexual plants generally appear to be weaker than the apomicts and may themselves be inconspicuous. Over a large region one may observe considerable polymorphism indicating the presence and continuous role of the sexual or partially sexual members. However, under natural conditions obligate or nearly obligate apomicts tend to dominate individual ecosites. These conditions support Clausen's view of facultative apomixis as a seemingly dormant but effective source of variation. The best adapted apomictic strain colonizes an ecosite and serves as a permanent germplasm bank, preserving precious genes for future recombinations. Unless there are drastic changes in environmental conditions, the plants in local populations may all appear to be obligate apomicts but usually there is a latent sexual capacity available to generate sufficient diversity to meet changing conditions.

We now recognize, bearing in mind results of genetic studies and breeding experiments, that apomixis is a dynamic factor in the evolution of an agamic complex (Bashaw, Hovin & Holt, 1970). The heterozygous nature of apomictic plants ensures that a wide range of genotypes is produced each time sexual and apomictic plants hybridize, whether intra- or inter-specific. The presence of obligate apomixis in the hybrid population plays at least four important roles in the future of the species. The most obvious role is production of true-breeding hybrids with permanent heterosis. Natural selection then assures that any ideally adapted hybrids with exceptional vigour will soon dominate the scene. Secondly, obligate apomixis can assure fertility in derivatives of wide crosses that might not otherwise survive. Elimination of the sexual female gamete allows for seed reproduction except in those cases where the pollen is not sufficiently viable to stimulate pseudogamy or in some extreme crosses where the entire ovule aborts. Even in the cases of inviable self pollen, we have been able to stimulate good seed-set in apomictic *Pennisetum setaceum*, *P. flacidum* and other species with foreign pollen. Perhaps of equal significance is the fact that apomixis, either obligate or facultative, provides a method for preservation of genotypes over aeons of time. Finally, apomixis provides an effective method for polyploid build-up. As noted previously, fertilization of the unreduced egg is a relatively infrequent phenomenon in apomicts. The results of its occurrence are apparent in sexual as well as apomictic species. Harlan & de Wet (1975) state that 'the most common general and widespread form of spontaneous polyploidy is produced by $2n + n$ reproduction. It is a general phenomenon that probably takes place at a low but significant frequency in nearly all species of sexual plants.' The phenomenon takes on particular evolutionary significance in obligate apomicts where the new polyploid does not have to survive the gametic screen imposed on sexual organisms. The results can be a fertile, highly heterozygous super genotype with the best characteristics of both parents. Assuming an abundance of related species growing in close proximity, it is difficult to imagine a more effective mechanism for speciation.

Conclusion

Apomixis is a fascinating asexual method of reproduction widely distributed throughout the plant kingdom. It is especially prevalent in some of the most important forage grasses where it presents serious problems yet exciting opportunities for plant improvement. Apomixis has been identified in wild 'grassy' relatives of some grain crops and probably can be identified among the ancestors of most species of Gramineae. Apomixis is often encountered as an agamic complex of related sexual, partially sexual and obligate apomictic species and ecotypes. Intra- and inter-specific hybridization, polyploidy and a diversity of apomictic phenomena interact in the natural habitat resulting in plant populations which challenge the imagination and offer unique opportunities for a broad range of basic investigations. Development of successful methods for gene transfer requires a thorough understanding of all aspects of apomixis and demands novel approaches for synthesis of germplasm. The reward for these efforts can be a remarkably efficient system for production of true-breeding hybrid cultivars.

Development and control of apomictic mechanisms for gene transfer would facilitate commercial production of our major food, feed and fibre crops and could revolutionize world agriculture. The potential for control and manipulation of apomixis in plant breeding is just beginning to be realized and conventional breeding methods have resulted in development of a few apomictic hybrid cultivars of forage grasses. Recent advances in biotechnology and genetic engineering suggest new possibilities for introduction of apomixis into purely sexual species. Once transferred, the apomictic mechanism can be an effective natural method for cloning plants through seed propagation. Genes controlling apomixis are undoubtedly present in diverse gene pools, some more accessible than others, with many yet to be identified. Successful widespread commercial use of apomixis will most readily be realized from a team effort with all disciplines from agronomist to molecular biologist working together.

References

Babcock, E. B. & Stebbins, G. L. (1938). The American species of *Crepis*, their relationships and distribution as affected by polyploidy and apomixis. *Carnegie Inst. Wash. Publ.* 504.

Bashaw, E. C. (1962). Apomixis and sexuality in buffelgrass. *Crop Sci.* **2**, 412–15.

(1968). Registration of Higgins buffelgrass. *Crop Sci.* **8**, 397–8.

(1969). Registration of buffelgrass germplasm. *Crop Sci.* **9**, 396.

(1980*a*). Apomixis and its application in crop improvement. In *Hybridisation of Crop Plants*, ed. W. R. Fehr & H. H. Hadley, pp. 45–63. ASA Press: Madison, WI.

(1980*b*). Registration of Nueces and Llano buffelgrass. *Crop Sci.* **20**, 112.

Bashaw, E. C. & Hignight, K. W. (1990). Gene transfer in apomictic buffelgrass through fertilisation of an unreduced egg. *Crop Sci.* **30** (in press).

Bashaw, E. C. & Hoff, B. J. (1962). Effects of irradiation on apomictic common dallisgrass. *Crop Sci.* **2**, 501–4.

Bashaw, E. C. & Holt, E. C. (1958). Megasporogenesis, embryo-sac development and embryogenesis in dallisgrass, *Paspalum dilatatum* Poir. *Agron. J.* **50**, 753–6.

Bashaw, E. C., Hovin, A. W. & Holt, E. C. (1970). Apomixis, its evolutionary significance and utilization in plant breeding. In *Proc. 11th Int. Grassland Congr.*, pp. 245–8. Queensland, Australia.

Brown, W. V. & Emery, W. H. P. (1958). Apomixis in the Gramineae: Panicoideae. *Amer. J. Bot.* **45**, 253–63.

Burson, B. L. (1983). Phylogenetic investigations of *Paspalum dilatatum* and related species. In *Proc. 14th Int. Grassland Congr.*, ed. J. A. Smith & V. W. Hayes, pp. 170–3. Westview Press: Boulder, Colorado.

(1989). Phylogenetics of apomictic *Paspalum dilatatum.* In *Proc. 16th Int. Grassland Congr.* (in press). French Grassland Society: Nice.

Burton, G. W. & Forbes, I. (1960). The genetics and manipulation of obligate apomixis in common bahiagrass (*Paspalum notatum* Flugge). In *Proc. 8th Int. Grassland Congr.*, pp. 66–71. Alden Press: Oxford.

Burton, G. W. & Jackson, J. E. (1962). Radiation breeding of apomictic prostrate dallisgrass. *Paspalum dilatatum* var. *Pauciciliatum. Crop Sci.* **2**, 495–7.

Clausen, Jens. (1954). Partial apomixis as an equilibrium system in evolution. *Caryologia, Suppl. to Vol.* **6**, 469–79.

Crane, C. F. (1978). 'Apomixis and crossing incompatibilities in some Zephyranchaceae'. PhD thesis, University of Texas, Austin.

Crane, C. F. & Carman, J. G. (1987). Mechanisms of apomixis in *Elymus rectisetus* from Eastern Australia and New Zealand. *Amer. J. Bot.* **74**, 477–96.

Darlington, C. D. (1939). *The Evolution of Genetic Systems.* Cambridge University Press.

de Wet, J. M. J., Harlan, J. R., Engle, L. M. & Grant, C. A. (1973). Breeding behavior of maize–*Tripsacum* hybrids. *Crop Sci.* **13**, 254–6.

Dujardin, M. & Hanna, W. W. (1983). Apomictic and sexual pearl millet × *Pennisetum squamulatum* hybrids. *J. Hered.* **74**, 277–9.

(1985). Cytogenetics of double crosses between pearl millet × *Pennisetum purpureum* amphiploids and pearl millet × *P. squamulatum* interspecific hybrids. *Theor. Appl. Gen.* **69**, 97–100.

(1987*a*). Cytotaxonomy and evolutionary significance of two offtype millet plants derived from a pearl millet × (pearl millet × *Pennisetum squamulatum*) apomictic hybrid. *J. Hered.* **78**, 21–3.

(1987*b*). Inducing male fertility in crosses between pearl millet and *Pennisetum orientale* Rich. *Crop Sci.* **27**, 65–8.

(1989). Crossability of pearl millet with wild *Pennisetum* species. *Crop Sci.* **29**, 77–80.

Gustafsson, A. (1947). *Apomixis in Higher Plants.* Haken Ohlson's Boktryeker: Lund, Sweden.

Hair, J. B. (1956). Subsexual reproduction in *Agropyron. Heredity*, **10**, 129–60.

Hanna, W. W. (1979). Interspecific hybrids between pearl millet and fountaingrass. *J. Hered.* **70**, 425–7.

Hanna, W. W. & Bashaw, E. C. (1987). Apomixis: its identification and use in plant breeding. *Crop Sci.* **27**, 1136–9.

Hanna, W. W. & Dujardin, M. (1985). Interspecific transfer of apomixis in *Pennisetum.* In *Proc. 15th Int. Grassland Congr.*, pp. 249–50. Kyoto, Japan.

Hanna, W. W. & Powell, J. B. (1973). Stubby head, an induced facultative apomict in pearl millet. *Crop Sci.* **13**, 726–8.

Hanna, W. W., Schertz, K. F. & Bashaw, E. C. (1970). Apospory in *Sorghum bicolor* (L.) Moench. *Science*, **170**, 338–9.

Hanna, W. W., Powell, J. B., Millot, J. C. & Burton, G. .W. (1973). Cytology of obligate sexual plants in *Panicum maximum* Jacq. and their use in controlled hybrids. *Crop Sci.* **13**, 693–7.

Hanson, A. A. & Juska, F. V. (1962). Induced mutations in Kentucky bluegrass. *Crop Sci.* **2**, 369–71.

Harlan, J. R. & de Wet, J. M. J. (1963). Role of apomixis in the evolution of the *Bothriochloa–Dichanthium* complex. *Crop Sci.* **3**, 314–16.

(1975). On a wing and a prayer: the origins of polyploidy. *Bot. Rev.* **41**, 361–91.

Harlan, J. R., Brookes, M. R., Borgaonkar, D. S. & de Wet, J. M. J. (1964). The nature and inheritance of apomixis in *Bothriochloa* and *Dichanthium. Bot. Gaz.* **125**, 41–6.

Hovin, A. W., Berg, C. C., Bashaw, E. C., Buckner, R. C., Dewey, D. R., Dunn, G. M., Hoveland, C. S., Rineker, C. M. & Wood, G. M. (1976). Effects of geographic origin and seed production environments on apomixis in Kentucky bluegrass. *Crop Sci.* **16**, 635–8.

Julen, G. (1960). The effect of X-raying on the apomixis in *Poa pratensis. Genetica Agraria*, **13**, 60–65.

Knox, R. B. & Heslop-Harrison, J. (1963). Experimental control of aposporous apomixis in a grass of the Andropogoneae. *Bot. Notiser*, **116**, 127–41.
Muntzing, A. (1933). Apomictic and sexual seed formation in *Poa. Hereditas*, **26**, 115–90.
Nitzsche, W. (1983). Interspecific hybrids between apomictic forms of *Poa palustris* L. × *Poa pratensis* L. In *Proc. 14th Int. Grassland Congr.*, ed. J. A. Smith & V. W. Hayes, pp. 155–7. Westview Press: Boulder, Colorado.
Nogler, G. A. (1984). Gametophytic apomixis. In *Embryology of Angiosperms*, ed. B. M. Johri, pp. 475–518. Springer-Verlag: Berlin, Heidelberg, New York.
Nygren, A. (1954). Apomixis in angiosperms. *Bot. Rev.* **20**, 577–649.
Parlevliet, J. E. & Cameron, J. W. (1959). Evidence on the inheritance of embryony in citrus. *Proc. Amer. Soc. Hort. Sci.* **74**, 252–60.
Pepin, G. W. & Funk, C. R. (1971). Intraspecific hybridization as a method of breeding Kentucky bluegrass (*Poa pratensis* L.) for turf. *Crop Sci.* **11**, 445–8.
Petrov, D. F., Belousova, V. I., Laikova, L. I. & Yatsenko, R. M. (1973). First case of the transmission of an apomixis element from *Tripsacum* to corn. *Doklady Akademii Nauk SSR.* **208**, 222–4.
Petrov, D. F., Belousova, N. I., Fabina, E. S., Laikova, L. I., Yatsenko, R. M. & Sorokina, T. P. (1976). Transfer of some elements of apomixis from *Tripsacum* to maize. In *Apomixis and its Role in Evolution and Breeding*, ed. D. F. Petrov, pp. 9–73. Nauka Publishers, Siberian Division: Novosibirsk. (Translated to Amerind Pub. Co. Pot. Ltd, New Delhi, and published by USDA and NSF, Washington, DC, 1984.)
Read, J. C. & Bashaw, E. C. (1974). Intergenetic hybrid between pearl millet and buffelgrass. *Crop Sci.* **14**, 401–3.
Rutishauser, A. (1948). Pseudogamie und Polymorphie in der Gattung *Potentilla. Arch Julius Klaus-Stift Vererbungsforsch*, **23**, 267–424.
Savidan, Y. H. (1983). Genetics and utilization of apomixis for the improvement of guineagrass (*Panicum maximum* Jacq.). In *Proc. 14th Int. Grassland Congr.*, ed. J. A. Smith & V. W. Hayes, pp. 182–4. Westview Press: Boulder, Colorado.
Simpson, C. E. & Bashaw, E. C. (1969). Cytology and reproductive characteristics of *Pennisetum setaceum. Amer. J. Bot.* **56**(1), 31–6.
Stebbins, G. L. (1941). Apomixis in angiosperms. *Bot. Rev.* **7**, 507–42.
Taliaferro, C. M. & Bashaw, E. C. (1966). Inheritance and control of obligate apomixis in breeding buffelgrass, *Pennisetum ciliare. Crop Sci.* **6**, 473–6.
Tang, C. Y., Schertz, K. F. & Bashaw, E. C. (1980). Apomixis in sorghum lines and their F_1 progenies. *Bot. Gaz.* **141**, 294–9.
Tinney, F. W. (1940). Cytology of parthenogenesis in *Poa pratensis. J. Agr. Res.* **60**, 351–60.
Voigt, P. W. & Bashaw, E. C. (1972). Apomixis and sexuality in *Eragrostis curvula. Crop Sci.* **12**, 843–7.
(1976). Facultative apomixis in *Eragrostis curvula. Crop Sci.* **16**, 803–6.
Voigt, P. W. & Burson, B. L. (1983). Breeding of apomictic *Eragrostis curvula*. In *Proc. 14th Int. Grassland Congr.*, ed. J. A. Smith & V. W. Hayes, pp. 160–3. Westview Press: Boulder, Colorado.

6

The implications of reproductive versatility for the structure of grass populations

A. J. Richards

Introduction

Much of the adaptive radiation of the flowering plants reflects evolutionary interactions between animals and floral functions. Following Linnaeus, we recognize this in our classifications, giving weight to characters that influence patterns of haploid gametophyte dispersal and recombination (the flower), and diploid sporophyte dispersal (the fruit).

In the grasses, reproductive links with animals have largely disappeared. External influences on patterns of genetic recombination and gene migration are abiotic, or result from competitive interactions with other plants. Consequently, grasses are unable to employ many of the strategies by which zoophilous and zoochorous plants control reproductive resource allocation, influence mate choice and confer fitness to offspring. Pollinator specialization in time and space, monoecy, gynodioecy, dioecy, heteromorphy, herkogamy, ethological mechanisms conferring pre-pollination isolation, 'safe-site' and/or assortive dispersal of fruits and seeds: these are absent or very unusual in grasses.

In the grasses, two major strategies remain which influence genetic recombination and gene migration, often resulting in non-panmictic genotype distributions. First, many grasses are self-compatible (s-c). Although most s-c grasses are chasmogamous, and many are dichogamous, the sequential anthesis of florets within spikelets and inflorescences, and the vegetative spread of genetic individuals (genets), will often result in high levels of geitonogamous self-pollination and self-fertilization. Homogamy will render self-fertilization (autogamy) yet more likely, while some grasses are effectively cleistogamous, so that self-pollination becomes almost automatic.

Nevertheless, even normally cleistogamous selfers can sometimes undergo low levels of outcrossing, which will lead to bursts of heterozygosity and variability. *Festuca microstachys* is normally completely cleistogamous, Kannenberg & Allard (1967) showing extremely low levels of outcrossing. However, Adams & Allard (1982) report that in exceptional climatic circumstances, levels of outcrossing approach 7%. *Hordeum murinum* seems to be fully cleistogamous in British conditions, yet Booth & Richards (1978) show that evolution in this complex has resulted from polyphyletic amphidiploid events which must have resulted from outcrossing.

The second strategy is that of asexual reproduction which is probably more prevalent and is achieved by more mechanisms in the grasses than in any other plant family. Vegetative reproduction occurs through rhizomes, stolons, bulbs, bulbils at the stem base, culm or inflorescence internode, or through floral proliferation ('vivipary'). As we will see, the enormous capacity of fit and virtually immortal genets to spread asexually over large areas profoundly influences the structure of grass populations.

In certain grasses, agamospermy (chiefly by forms of apospory) is also common, although rarely obligate. Interactions at the population level between genes for apospory, polyploidy and the environmental control of asexuality lead to complex sexual/asexual cycles (Richards, 1986). Some arctic poas have uniquely adopted three asexual mechanisms within the same population, i.e. vegetative spread, vivipary and apospory.

The genetic architecture of grass populations depends on how competitive interfaces between panmixis and selfing on the one hand, and panmixis and asexuality on the other, equilibrate at thresholds determined by abiotic (density-independent) and biotic (density-dependent) selective constraints. The breeding strategy adopted by a given species can be figuratively placed within this 'eternal triangle' (Fig. 6.1).

Each strategy incorporates feedback. The genetic architecture resulting from the breeding system itself influences the variation and evolution of those breeding systems.

Breeding systems and genetic architecture

Self-fertilization

We can predict that populations experiencing high levels of self-fertilization will be genetically uniform relative to outcrossers, and genets will be homozygous at most loci. However, high levels of genetic differentiation *between* populations or subgroupings within populations may be encountered in selfers, so that rare migrational events should lead to bursts of genetic variability and the potential for evolutionary change.

These predictions are largely fulfilled by allozyme surveys of popula-

Fig. 6.1. The 'eternal triangle' of breeding system interfaces in grasses.
A.c. = *Agrostis capillaris*; B.max. = *Briza maxima*; B.me. = *Briza media*; D.c. = *Deschampsia caespitosa*; D.g. = *Dactylis glomerata*; E.f. = *Elymus farctus*; F.m. = *Festuca microstachys*; F.o. = *Festuca ovina*; F.r. = *Festuca rubra*; H.mo. = *Holcus mollis*; H.m. = *Hordeum murinum*; L.p. = *Lolium perenne*; P.m. = *Panicum maximum*; Ph.a. = *Phalaris arundinacea*; P.a. = *Poa annua*; P.j. = *Poa* × *jemtlandica*; P.n. = *Poa nervosa*; S.t. = *Spartina* × *townsendii*.

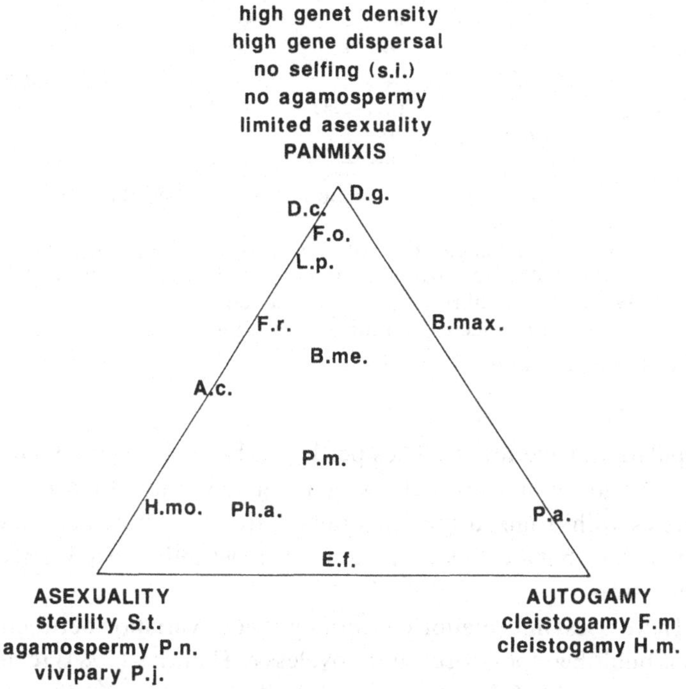

tions, including a number of grass species (reviewed in Hamrick, Linhart & Mitton, 1979; Brown & Marshall, 1981; Gottlieb, 1981; Barrett, 1982; Hamrick, 1983; Jain, 1983; Loveless & Hamrick, 1984; Barrett & Richardson, 1986).

Table 6.1 summarizes allozyme data within populations from 25 grass species, more thoroughly reviewed by Hamrick *et al.* (1979). Although there is some overlap in the distribution of genetic structure parameters between classes of breeding system (which are themselves not absolute), trends relating the genetic structure of populations to the breeding system are clear. Habitual selfers are on average polymorphic at fewer loci per

Table 6.1. *Characteristics of the genetic structure of grass populations*

Mean readings, range of readings and numbers of species/populations investigated are given in each cell.

Breeding strategy	PLP	A	PI	*H*
Outcrossers	84.8 (75–95) $n = 2$	2.9 (2.2–3.5) $n = 2$	0.264 (0.181–0.331) $n = 3$	0.267 $n = 1$
Mixed strategy	49.7 (15–100) $n = 7$	2.2 (1.6–2.7) $n = 4$	0.281 (0.216–0.377) $n = 4$	0.348 (0.276–0.420) $n = 2$
Selfers	36.4 (0–92) $n = 12$	1.7 (1.1–2.3) $n = 7$	0.138 (0.012–0.282) $n = 10$	0.028 (0.01–0.05) $n = 5$

PLP: proportion of all electrophoretic loci investigated that showed polymorphism within a population.
A: mean numbers of alleles present at a locus within a population.
PI: proportion of all electrophoretic loci that would be heterozygous under Hardy–Weinberg equilibrium genotype frequencies.
H: proportion of all electrophoretic loci found to be heterozygous.
After Hamrick *et al.* (1979).

population, have fewer alleles per locus, have lower potential levels of heterozygosity and are heterozygous at fewer loci than outbreeders. Species with a mixed mating strategy are at least as heterozygous as outcrossers, but are polymorphic at fewer loci and have fewer alleles than outcrossers.

There is less information concerning genetic variation between populations published for grasses, but Loveless & Hamrick (1984) review a few examples (Table 6.2). The attribute GST is a measure of the proportion of the total allozyme variability encountered in a species that is partitioned between populations. Unfortunately, for between-population comparisons, there are no data for outbreeding grasses to compare with those of selfing grasses. For data concerning population subdivisions, only two non-selfing grasses can be compared. However, for the data set as a whole (including other plants) as well as for the grass data, it is clear that allozyme variability between populations is as predicted generally greater for selfers than for outbreeders.

Genetic uniformity within populations resulting from habitual selfing will lead to breeding system uniformity within populations (for instance,

Table 6.2. *Amount of electrophoretic variation within a grass species partioned between populations (GST)*

Mean readings, range of readings and number of species investigated are given in each cell.

Breeding strategy	GST between populations	GST between population subgroups
Outcrossers		0.045 $n = 1$
Mixed strategy		0.010 $n = 1$
Selfers	0.371 (0.018–0.736) $n = 3$	0.240 (0.029–0.372) $n = 6$

After Loveless & Hamrick (1984).

for degrees of cleistogamy or homogamy); such populations should have a low potential for breeding system adaptation in response to changing environments.

Asexuality

For asexuals, our predictions concerning the genetic structure of populations will depend on the nature of the asexuality. Asexuality is characteristically successful in previously panmictic rather than non-panmictic populations, so post-asexual survivors should fix high levels of heterozygosity, and co-surviving genets should be highly differentiated genetically. In these features asexual populations differ markedly from autogamous populations.

Where sexuality has been lost through climatic change (northern populations of *Phragmites australis*, Richards, 1986), hybridity (*Spartina* × *townsendii*), uneven polyploidy (*Holcus mollis*, Harberd, 1967) or vivipary (*Poa* × *jemtlandica*), vegetative reproduction will predominate. It is likely that very few genets will eventually survive, often only one. Levels of genetic diversity *G* should be low, but levels of heterozygosity *H* should be high (e.g. Hughes & Richards, 1988).

In closed and stable habitats vegetative reproduction may largely replace sexual reproduction, even for sexually fertile populations, as seedlings fail under intense competition. As the community ages, the genetic diversity *G* will drop as fewer and fewer superfit genotypes predominate in the absence of genetic renewal from seed (Grime, 1973; Grubb, 1977).

Agamospermous grasses can be facultatively sexual

1 by virtue of environmentally influenced plasticity, e.g. *Bothriochloa* (Saran & de Wet, 1970), or *Dichanthium* (Knox, 1967);
2 as a result of polyembryony associated with apospory, e.g. *Poa pratensis* (Muntzing, 1940); or
3 through cytologically and genetically polymorphic populations with sexual/apomictic cycles, e.g. *Panicum maximum* (Asker, 1979).

Depending on the relative frequencies and fitnesses of sexually produced and asexually produced offspring, it is to be expected for facultative agamosperms that levels of *H* will approach or even exceed those in panmictic populations, but that levels of *G* will be rather lower than in panmictic populations. These predictions are largely fulfilled for *P. maximum* (Usberti & Jain, 1978). Although this example seems to be the only report of the population structure of agamospermous grasses, we can presume that facultatively viviparous grasses, such as *Festuca ovina* var. *vivipara*, will behave similarly.

For obligate agamospermy, a rare condition in grasses, but well known in several diplosporous poas, 'populations' should be highly heterozygous but genetically uniform. However, as discussed in Richards (1986), much depends on the semantics of the word population, which is difficult to apply to totally asexual populations. Where two or more 'biotypes' coexist, they will differ genetically, by definition, but they are not part of a common gene-pool, and they may have very distant phyletic connections. The only individuals in genetic contract are mother–daughter lines which should in theory be genetically uniform.

Facultatively asexual populations will retain some genetic variability between generations which reinforces the variable nature of the breeding system. Obligately asexual populations will have lost the variability between generations which allows breeding system versatility, and only mutational or hybridizational events have the potential of reintroducing sexuality. Here again, the breeding system is self-reinforcing.

Panmixis and gene flow

Departures from panmixis

Genotype frequencies for polymorphic loci, where alleles are selectively neutral and no gene migration occurs, should not depart from Hardy–Weinberg equilibria in panmictic populations. Panmixis is random genetic exchange between individuals in a population of infinite size. These conditions are probably never met in plants, but for self-incompatible grasses with limited vegetative spread, frequent genet replacement from seed, and seed and pollen travel with large variances, genotype frequencies may often not depart from panmictic expectancies. In fact, there seem to be little population genetic data from the grasses to substantiate this claim; most surveys have been of plants with 'interesting' (i.e. non-panmictic) breeding systems (Hamrick *et al.*, 1979, Table 6.1).

Even in suitable candidates for panmixis, such as *Cynosurus cristatus*, subgrouping effects may result in some non-panmictic gene flow. Ennos & Dodson (1987) demonstrate how variation in flowering time causes assortative mating between synchronous genets. Asymmetrical gene exchange of this type may serve to reinforce the genetically controlled subgrouping.

Reproductively controlled subgroupings of this kind may reflect Wallacean isolation encouraged by local adaptations to a heterogeneous environment. Well-known examples include flowering time differences between metal-tolerant genotypes of mine spoil and neighbouring non-tolerant populations (McNeilly & Antonovics, 1968). The array of non-random genotype and gene transport distributions discovered for populations of *Avena barbata* introduced into California provide another familiar example (Hamrick & Holden, 1979).

Experimental investigations into genetic exchange within populations are ideally based on studies of gene flow, but in most cases this is approximated by pollen travel and seed travel. The 'neighbourhood' concept, originated by Wright (1938) has been most recently revised by Crawford (1984*a*,*b*), who uses observations of variances of both seed and pollen travel to estimate likelihoods of parents of an individual at the centre of a circle originating within its radius.

The only work which reports neighbourhood calculations based on pollen and seed flow variances for a grass (the introduced annual *Avena barbata* in California: Rai & Jain, 1982), estimates neighbourhood numbers N of between 40 and 400 in artificially organized plots. Studies of the genetic structure of populations demonstrated extensive local subgrouping, which might at least in part be accounted for by this localized gene dispersal.

The estimate of the total number N of potential parents within this area A is probably most predictive of gene flow when considered in relation to the actual population size of n genets. When N equals or exceeds n, panmictic expectancies are likely to be met, particularly when n is greater than 100. In these conditions, gene dispersal is effectively random, and the population is of effectively infinite size. When *N* is much smaller than *n*, gene flow will be non-random within the population, and various subgroups may arise. Turner, Stephens & Anderson (1982) have modelled the effects of nearest-neighbour mating in infinitely large outbred populations, and show that non-random genotype distributions showing patchiness and high levels of homozygosis should result from this particular mating pattern.

Restriction of gene flow may also encourage, or be encouraged by, localized adaptation (e.g. Snaydon & Davies, 1972); within-'population' breeding barriers (e.g. McNeilly & Antonovics, 1968); and low *H* levels leading to non-adaptive allele fixation (Richards, 1986).

Pollen dispersal

There seems to be little data pertaining to pollen and seed dispersal in wild grass populations. Whitehead (1969) reviews theoretical considerations concerning the dispersal of air-borne pollen. As is typical for nearly all patterns of biological dispersal, grass pollen should travel on leptokurtic schedules. Gleaves (1973) successfully tested the model

$$f(x) = 1/x^{\mathrm{k}}$$

where *f* represents pollen flow, *x* distance and k a constant varying with the density of clustering of genets in the plant population, against Griffiths' (1950) data for gene dispersal in *Lolium perenne*, using a red tiller base marker. This suggests that the input of seed travel into the gene dispersal of this species is insignificant.

Environmental constraints on air-borne pollen travel are complex. Uninterrupted, grass pollen should fall at a constant rate, giving non-leptokurtic dispersal schedules. This rate has been measured for *Phleum pratense* in the range of between 10 and 14 cm s^{-1} (Raynor, Ogden & Hayes, 1971). In the presence of a steady wind the simple relationship

$$f(x) = 1/x$$

should result. However, wind gusts, and pollen will be released chiefly at gust peaks, the release being followed by a sharply declining gust speed. The distribution of gust speeds will itself be leptokurtic, and the effect of this leptokurtosis on pollen dispersal will be further reinforced by patterns of pollen deposition.

Pollen will most frequently lodge on obstructions in the immediate vicinity of the point of release, such as self stigmas. The further pollen travels, the less likely it is to lodge on a stigma per distance travelled. This secondary input into the leptokurtosis of pollen travel will be further reinforced by the density and clustering of genets which affect patterns of pollination (rather than pollen travel). Rai & Jain (1982) have shown that different experimental plot designs influence gene migration schedules in *Avena barbata* to the extent that platykurtic or even bimodal distributions can result from evenly spaced or highly clustered plant patterns respectively.

Pollen deposition schedules are such that genet size will influence patterns of geitonogamous or xenogamous pollen flow. Handel (1983) shows for the sedge *Carex platyphylla* that large genets with more than 10 culms are more likely to receive self pollen than smaller genets.

Estimates of pollen travel are reported for five grasses in Richards (1986, based on Altman & Dittmer, 1964). For *Phleum pratense*, *Dactylis glomerata* and *Lolium perenne*, more than 20% of pollen caught in traps occurred at 200 m or more from source. In contrast, for the domesticated subtropical grasses with large sticky pollen *Zea mays* (Paterniani & Short, 1974) and *Pennisetum glaucum*, less than 10% of pollen detected occurred at distances of more than 5 m from source. Silander & Antonovics (1979) show that for *Spartina patens*, pollen movement is halved over a distance of 1 m and at 4 m from source, only 5% of the maximum pollen flow is recorded.

Gene dispersal

Gene dispersal consists of both pollen dispersal to stigmas, and seed dispersal. Although dispersal of sporophytic seeds should have twice the influence on gene dispersal of gametophytic pollen, little seems to be known about seed dispersal schedules in grasses. However, a number of studies estimate rates of gene dispersal in grasses.

Genet density profoundly influences gene dispersal. Estimates of gene travel per generation for *Lolium perenne* and *Avena barbata* in dense artificial plots are low; 99% of events occurred within 20 m of source in the former, and within 6 m of the source in the latter.

In natural circumstances, especially where population densities are low, gene travel may occur over much longer distances. Densities are characteristically low on toxic mine spoil. McNeilly (1968), comparing seedling with adult tolerances, shows that the transport of copper-tolerant genes on to non-tolerant populations of *Agrostis capillaris* downwind from the Drws-y-Coed copper mine shows little diminution 160 m from the mine edge. Similarly, considerable gene flow of lead- and zinc-tolerant genes from the Trelogan mine on to non-tolerant populations of *Anthoxanthum odoratum* is also recorded despite some barriers to gene flow (McNeilly & Antonovics, 1968).

Conversely, little gene exchange is reported by Snaydon (1970) between adjacent plots of *Anthoxanthum* on the dense stands of Park Grass, Rothamsted, although some are separated by less than 20 m.

The effect of plant density and pollen flow on the genetic structure of grass populations is well shown by Vernet (pers. comm.) for mine and non-mine populations of *Arrhenatherum elatius* in France. Mine plants when selfed are about 70% as fertile as crossed plants, while the corresponding figure for non-mine plants is only about 12%. Heterozygote frequencies generally follow panmictic expectancies for the sparse but partially selfed mine populations, while heterozygote deficiencies are commonly observed in the dense non-mine populations, despite low selfing levels. Consequently, the genetic fixation index F shows a good relationship to panicle density, rather than to the breeding system.

From this very limited data we can conclude that for self-incompatible grasses, non-panmictic restrictions in gene flow will chiefly depend on the density and pattern of genets relative to dispersal schedules.

For a maize crop, with approximately 10 genets per square metre, nearly 1000 plants may be freely pollinated from a single source, but pollen is unlikely to travel freely throughout a large field.

For an ancient *Lolium* pasture, single genets might exceed an area of 20 m radius, and the number of genets available for most cross-pollinations may be in single figures; yet the whole population may be accessible to occasional pollination events.

In the former case, where $n > N$, but N is large, subgroupings may occur, but inside these genotype frequencies should maintain panmictic equilibria; in the latter, where $n < N$, but both may be small, subgroupings should not occur, but non-panmictic genotype frequencies and non-adaptive fixations could prevail.

The genetic structure of grass populations

Prior to the classical observations of Harberd (1961, 1962, 1967), almost no information existed as to the population structure of perennial grasses with vegetative, asexual, reproduction. For *Festuca rubra*, Harberd took a total of 1481 random isolates from an area of about 83.5 m^2 on a Scottish hill. Using a combination of morphological markers, and indications from self-incompatibility, he estimated that about 170 genets occurred in this area, but 90% of these were represented by a single sample (Fig. 6.2*a*). One individual, appropriately labelled 'W', was represented by no less than 51% of the ramets sampled. Sampling over a wider area revealed that 'W' occurred elsewhere on the moor over an

Fig. 6.2. (*a*) Frequency distribution of numbers of ramets identified as belonging to certain genets in a Scottish population of *Festuca rubra* (after Harberd, 1961). An example of a skewed genotype frequency distribution. (*b*) Survivorship curve for the annual *Vulpia fasciculata* in a Welsh sand-dune (after Watkinson & Harper, 1978). High genotype densities and low mortalities ensure almost even genotype frequency distributions in the absence of asexual reproduction. (*c*) Suggested genotype distribution for the two British populations of the introduced bigeneric hybrid × *Calammophila baltica* (after Rihan & Gray, 1982). Numbers are guestimates. In asexual populations genotype frequency distributions may be unimodal.

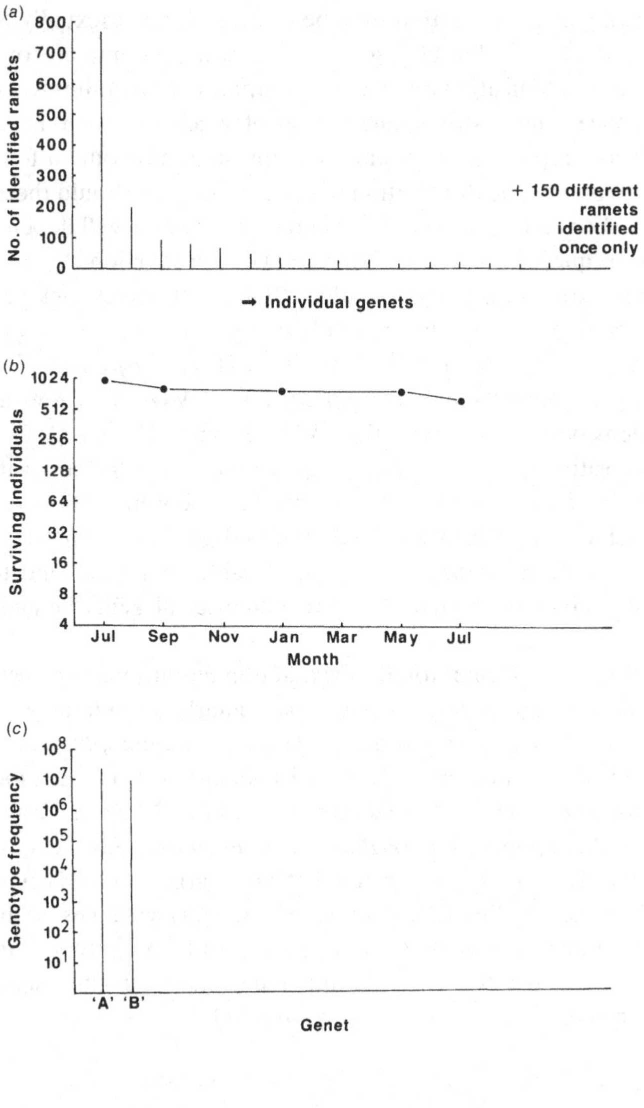

area of at least 200 m diameter. On known growth rates, Harberd suggests that 'W' may be at least 1000 years old.

Using the same quadrat, Harberd identified 42 genets of *F. ovina*, a few of which were widespread throughout the quadrat, although none had the predominance of *F. rubra* 'W'. *F. ovina* is not rhizomatous like *F. rubra*, and in the absence of interventions from outside agencies, it was again suggested that on known rates of spread, successful clones may be more than 1000 years old. However, the activity of grazing animals may result in enhanced genet motility.

More recent studies have suggested that for perennial grasses with effective asexual and sexual means of reproduction and dispersal, this pattern of genet distribution may be typical. A few asexually 'superfit' individuals of considerable age and size will predominate over large areas, while colonization by seed into temporarily open sites will result in the presence of many other genets, most of which are less fit, have a short half-life and rapid turnover, and are represented by only a few tillers.

In such plants, the distribution of genet frequency should therefore be strongly skewed (Fig. 6.2*a*). The degree of skewness will depend on the relative frequencies of asexual and sexual reproduction.

At one extreme, populations of small annuals, which lack vegetative reproduction and reproduce sexually every year, will have equal genet frequencies with a non-skewed distribution (Fig. 6.2*b*). Most of these are primarily autogamous, and thus genetic diversity *G* will be low, although genet density could be very high. Watkinson & Harper (1978) record genet densities of up to 5000 per square metre for the tiny sand-dune annual *Vulpia fasciculata*. Such plants have low mortalities and high survivorships. At densities in excess of 500 per square metre, density-dependent effects on fitness are considerable, but they tend to affect fecundity rather than viability, and influence all genets equally (Fig. 6.3*a*).

At the other extreme, totally asexual populations may consist of very few genets, even a single genet, particularly in relatively uniform habitats. The sterile intergeneric hybrid × *Calammophila baltica* has colonized the sand-dunes of eastern England twice. In each locality, a single (different) genet extends for over 1 km (Rihan & Gray, 1982). Many northern populations of *Phragmites australis* rarely if ever set seed, and Richards (1986) has suggested that some large reed beds may consist of single clones. Jerling (1985) quotes Sjors, who has suggested that some Swedish individuals may be 6000 years old. A similar pattern of predominance by a few very old individuals has been suggested for the sand-dune colonizer *Ammophila arenaria* in late seral stages by Huiskes (1979).

Fig. 6.3. (*a*) Regression line showing density-dependent effects on fecundity in *Vulpia fasciculata* (after Watkinson & Harper, 1978). Readings ($n = 44$) give an excellent fit to this line. (*b*) Vegetative growth of transplants of *Anthoxanthum odoratum* into a structured *Anthoxanthum* plot (after Antonovics & Ellstrand, 1984). Minority genotypes (hatched line) surrounded by foreign genotypes are asexually fitter than majority genotypes. Genotype heterogeneity as well as environmental heterogeneity will tend to favour genotype diversity by frequency-dependent selection for rare genotypes. This will tend to equilibrate genotype frequencies, especially at low densities. (*c*) Relationship between genet density and the proportion of sexual reproduction involved in tiller birth in a pioneer colony of the sand-dune grass *Elymus farctus* (after Harris & Davy, 1986). At very low densities, where sexual reproduction is not favoured, asexual reproduction predominates and genet diversity is low.

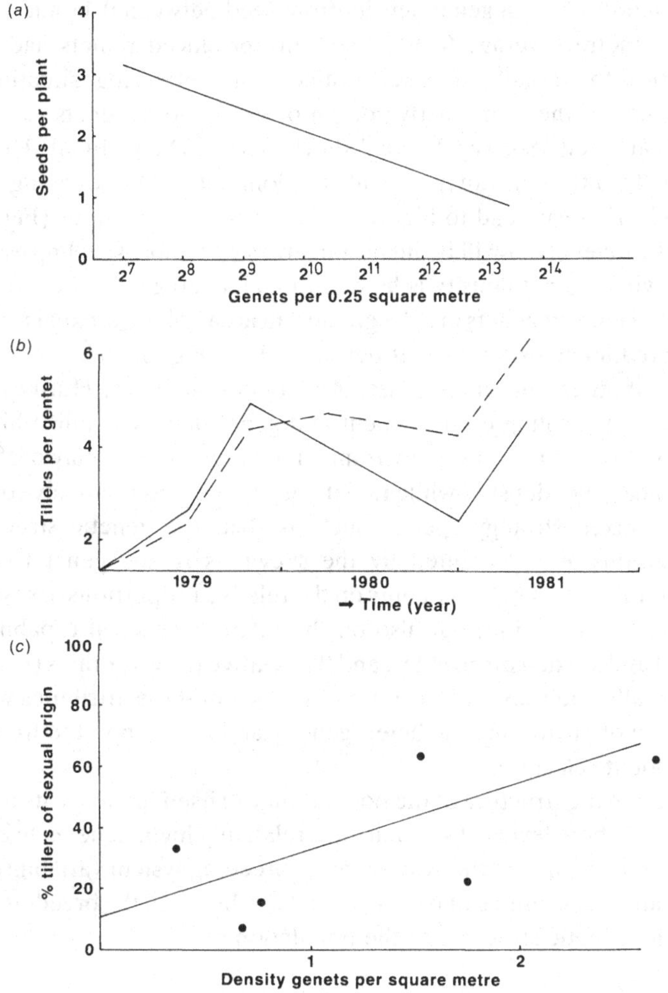

In these cases genet distributions may be not so much skewed as unimodal (Fig. 6.2*c*). Where more than one asexual genet has survived from a sexual past, selection will be density-independent. In relatively uniform habitats, such as a sand-dune (× *Calammophila*), or a reed bed, the genetic structure of a population will be a product solely of the relative asexual fitnesses of the genets.

Conversely, in heterogenous habitats, frequency-dependent selection (Antonovics & Ellstrand, 1984; Ellstrand & Antonovics, 1985) may tend to equilibrate genet frequency in asexuals, and to maximize genetic diversity *G* (Fig. 6.3*b*) (see next section).

Most grasses have flexible breeding systems which are a mixture of asexual, autogamous and xenogamous strategies (Fig. 6.1). The sand-dune strand-line colonizer *Elymus farctus* (= *Agropyron junceiforme*) was found to have a genet density from seed between 0.11 and 1.88 per square metre (average 0.69). Asexually produced ramets had similar densities to sexually produced ramets; in colonizing situations the frequency of the reproductive origin of newly born ramets was almost identical (seed 186, vegetative 181) (Harris & Davy, 1986). Here, the asexually fittest genotypes will predominate, but ongoing sexual reproduction will lead to highly skewed genet distributions (Fig. 6.2*a*). Genetic diversity *G* will be highest in areas with most seedling establishment where genet density is high, although average genetic differences between pairs of genets may be greatest in areas where asexual reproduction predominates, and genet density is low (Fig. 6.3*c*).

For another sand-dune grass, *Ammophila arenaria*, Huiskes (1977) shows that seedling establishment and genet density is much higher in wetter parts of the dunes; environmental heterogeneity may profoundly influence genet density, while ramet density remains relatively constant.

For mixed strategy species such as this, the genetic structure of populations as represented by the skewness of the genet frequency distribution will depend not only on the relative proportions of sexual and asexual reproduction, but also on the relative dispersal capabilities of sexual and asexual propagules, and the relative turnover rates (half-lives) of sexually and asexually born ramets. Each of these attributes will be a function of environmental heterogeneity and will respond to frequency-dependent selection.

The genetic structure of the population will itself influence its breeding system. Where levels of sexuality are relatively high, genetic diversity *G* will also be high, and this will reinforce breeding system variability. Here is yet another example of two-way feedback between the breeding system and the genetic structure of the population.

Density-dependent and frequency-dependent selection

Law, Bradshaw & Putwain (1977) show that plants of *Poa annua* from pasture situations, said to be under density-dependent regulation, are longer lived, show more vegetative growth and take much longer to reach a sexual phase than those selected under open (density-independent) conditions. Pasture plants invest more resource in asexual reproduction, and are said to be 'K' strategists (MacArthur & Wilson, 1967), while those from open habitats invest more in sexual reproduction and are thought of as 'r' strategists. We may expect that the 'K' strategy populations, under density-dependent selection, will have more strongly skewed genet frequency distributions than those from 'r' strategy populations which have a higher level of genet turnover.

As pioneer communities develop, colonizers typically progress from an open, selectively density-independent condition in which 'r' strategists (predominantly sexual) might be expected to be successful, to a closed, selectively density-dependent condition where 'K' strategists (predominantly asexual) should predominate. According to Grime's 'hump-backed' model (1973), once the population enters the predominantly asexual, density-dependent phase, relatively few successful genets will predominate. Consequently, the genet frequency distribution should become progressively more skewed as genet replacement by seed becomes rarer to a very low equilibrium point. Even at this low equilibrium, however, Soane & Watkinson (1979) show that reproduction by seed plays a vital role in maintaining some genetic diversity in populations.

The progressive genetic impoverishment of colonizing populations under density-dependent selection was first demonstrated for grasses by Charles (1961, 1964). More recent studies have been able to utilize zymographic techniques. Gray (1987) followed populations of the heathland grass *Agrostis curtisii* after fire. Population sizes peaked at 4 years at 1.4 genets per square metre and after 8 years had declined gradually to 1.0 genets per square metre. Survivorships of the original cohort were very high, but reduced dramatically and progressively for succeeding cohorts, so that approximately 95% of colonizing genets 4 years after the fire failed to survive 4 years more. Eight years after the fire, 80% of surviving genets were at least 7 years old.

Gray has also studied the salt-marsh grass *Puccinellia maritima*, discovering that pioneer ungrazed populations, with much bare mud, averaged 4.5 genets per square metre, but in a grazed marsh with a closed turf largely composed of *P. maritima*, genets were much larger, and the density only 2.3 per square metre.

McNeilly & Roose (1984) compare genet densities in 10-year-old swards with those in a pasture which had remained unploughed for at least 40 years. Estimates of genet density for *Lolium perenne* varied from 36 to 43 per 0.25 square metre for the younger populations, and only 5 genets in the same area in the older pasture. Studies on *Agrostis stolonifera* by Wu, Bradshaw & Thurman (1975) where genet densities are compared between 10-year-old copper-polluted sown lawns, and old pastures, show genet densities similar to those in *A. curtisii*, but levels of genet impoverishment comparable to those in *L. perenne*. Many features which vary between species, perhaps especially rates of vegetative dispersal and relative investments into sexual and asexual reproduction, will strongly influence genet density and genet impoverishment.

Aarssen & Turkington (1985) demonstrate an interesting comparison between *Lolium perenne* and *Holcus lanatus* in sown Canadian pastures of varying ages. While *L. perenne* shows a marked decrease in genetic variability with age, almost no differences in variability are detectable in *Holcus*. *Holcus* occurs as an invading weed in these areas, and the authors suggest that it arises from few relatively fit founders, its genetic variability being further boosted by continuing invasion. In comparison with *L. perenne*, the greater investment in sexual reproduction, and the larger seeds in more freely dispersed fruits shown for *Holcus* may account for this behaviour.

The extent to which frequency-dependent selection maintains genetic variability in a grass through seral succession contrary to increasing density-dependent forces, will be a function of the heterogeneity of the habitat, the mobility of the grass and the intensity of intraspecific competition. By introducing genotypically novel tillers of *Anthoxanthum odoratum* into wild populations, Antonovics & Ellstrand (1984) demonstrate the fitness advantage often experienced by minority genotypes. In a later experiment they introduced asexually derived, invariable tillers, and sexually derived, variable tillers into parts of the wild population from which the propagants originated, at varying densities (Ellstrand & Antonovics, 1985). Sexually derived tillers proved fitter on average than asexually derived tillers. As there was a tendency for this advantage to be maximized at low densities, it was concluded that frequency-dependent effects outweigh density-dependent effects in maintaining genetic variation within populations. It was noteworthy, however, that some frequency-dependent effects on genetic variability could be observed at high (e.g. late seral stage) densities.

As suggested by Law *et al.* (1977) for *Poa annua*, and Watkinson & Harper (1978) for *Vulpia fasciculata*, density-dependent pressures tend to influence reproductive strategy and resource allocation, rather than genetic diversity. Changes in density-dependent effects through seral succession may encourage changes to reproductive syndromes which will fundamentally affect genet distribution patterns within a population, but they are less likely to select for genetic variability *per se*. In contrast, frequency-dependent pressures responding to levels of environmental heterogeneity will favour genetic variability, and so may indirectly select for mothers possessing breeding systems which will maximize this variability.

It follows that attributes such as self-incompatibility, effective pollen and seed travel, and low investment into asexual reproduction which favour panmixis should be selected for at low population densities, for instance in pioneer habitats or colonizing situations. These breeding systems are then more likely to remain at a selective advantage in innately heterogeneous environments.

Conversely, in stable late seral habitats with closed communities, particularly those such as a sand-dune, salt-marsh or reed bed which are relatively environmentally homogeneous, density-dependent selection should select for a low sexual fecundity and a large investment into asexual reproduction and vegetative growth. (Williams, 1975, proposes the converse with his 'strawberry-coral model', namely that colonizers should first establish through an asexual phase, and that sexuality will be favoured in later density-dependent phases in community development. Although this model may hold for certain growth forms and habitats, it seems not to be true for the grasses, and later data argues against the universality of this dogma.)

This change of reproductive strategy as communities develop may occur simply by the replacement of a primarily sexual colonizing species by a mostly asexual species of stable communities. However, many grass species exhibit a high level of plasticity with respect to reproductive strategy. Much of the variability in reproductive allocation expressed between populations of different densities and successful stages in the American grass *Andropogon scoparius* (Roos & Quinn, 1977) proved to be phenotypic and was lost after culture in standard conditions.

Nevertheless, there are clear examples of species exhibiting evolutionary adaptations of breeding systems to seral progressions from density-independent to density-dependent selection. *Poa annua*, already quoted (Law *et al.*, 1977), forms an excellent example. The coastal north American *Spartina patens* displays a remarkably catholic habitat range. Silander & Antonovics (1979) examined sand-dune, salt-marsh and meadow populations over a transect of only 200 m. After cultivation in constant conditions, the plants from unstable dunes showed two- and four-fold increases in seed production compared with the other more permanent populations, and also invested slightly more resource in vegetative reproduction. Here, selection had apparently occurred in relatively pioneer habitats for greater investment in both sexual and asexual reproduction.

Gray (1987) interpreted differences between pioneer low-marsh and stable upper-marsh populations of *Puccinellia maritima* in terms of the intensity of selection by competition. He found that the low-density pioneer populations are genetically more variable in all attributes examined, including reproductive characters. Not only did the stable populations alone show evidence of strong ecotypic adaptation, but seedlings from the upper marsh were notably more variable than their parents, evincing strong stabilizing selection on the adults in this area. This may explain the finding that upper-marsh plants not only invested proportionally more resource into asexual reproduction, as predicted, but also invested more resource into sexual reproduction. It may be that when density effects disappear completely, as on the lower marsh, relaxed selection of a pioneer specialist lowers constraints on to sexual as well as asexual reproduction.

This intriguing finding draws our attention to the necessity to differentiate on the one hand selection for resource allocation between sexual and asexual reproduction, and on the other hand selection between panmictic (variable offspring) and non-panmictic, e.g. subgrouping effect (less variable offspring), breeding systems. Both elements have strong feedback characteristics (Fig. 6.4), and these interact. Grasses from early seral stages will undergo frequency-dependent selection for genetic variability, and this should select for sexual outbreeders with near-panmictic systems. They should also be efficient colonizers, and this should select for high resource allocations to reproduction. However, their variability will militate against constancy both in the quantity of reproductive resource allocation, and its nature (sexual against asexual). This is clearly shown for *Puccinellia maritima*.

Conversely, grasses from late seral stages will undergo density-dependent selection which will result in low genetic variability and consequent non-panmictic breeding systems, which should reinforce the low variability. They should also be efficient survivors, and this should select for high resource allocations to vegetative growth (arguably including vegetative reproduction). This asexuality will also encourage low populational levels of genetic variability, thus reinforcing not only the

Fig. 6.4. Feedback relationships within and between selection for resource allocation strategies, and selection for breeding system strategies, under frequency-dependent (early seral stage) and density-dependent (late seral stage) selection in perennial grasses with mixed breeding strategies.

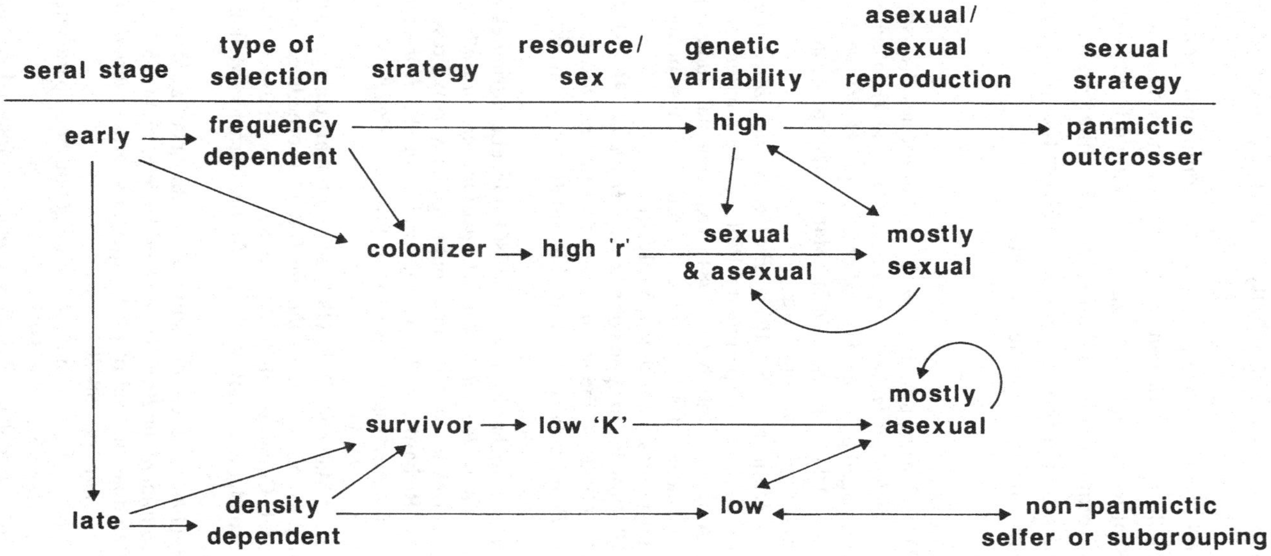

asexuality, but also the low variability resultant upon density-dependent selection.

In each seral context, systems are not only mutually reinforcing, but may also be evolutionarily dissonant. Colonizers may select for reproductively inefficient individuals through hitch-hiking effects associated with high levels of frequency-dependent selection, while grasses from closed swards may be sufficiently invariable as to restrict the colonizing potential of their offspring into new pioneer habitats.

References

Aarssen, L. W. & Turkington, R. (1985). Within-species diversity in natural populations of *Holcus lanatus*, *Lolium perenne* and *Trifolium repens* from four different-aged pastures. *Journal of Ecology*, **73**, 869–86.

Adams, W.T. & Allard, R.W. (1982). Mating system variation in *Festuca microstachys*. *Evolution*, **36**, 591–5.

Altman, P. L. & Dittmer, D. S. (1964). *Biological Data Book*. Federation of American Societies for Experimental Biology: Washington DC.

Antonovics, J. A. & Ellstrand, N. C. (1984). Experimental studies of the evolutionary significance of sexual reproduction. 1. A test of the frequency-dependent hypothesis. *Evolution*, **38**, 103–15.

Asker, S. (1979). Progress in apomixis research. *Hereditas*, **91**, 231–40.

Barrett, S. C. H. (1982). Genetic variation in weeds. In *Biological Control of Weeds with Plant Pathogens*, ed. R. Charudattan & H. Walker, pp. 73–98. John Wiley: New York.

Barrett, S. C. H. & Richardson, B. J. (1986). Genetic attributes of invading species. In *The Ecology of Biological Invasions: an Australian perspective*, ed. R. H. Groves & J. J. Burdon. Cambridge University Press.

Booth, T. A. & Richards, A. J. (1978). Studies in the *Hordeum murinum* aggregate: disc electrophoresis of seed proteins. *Botanical Journal of the Linnean Society*, **76**, 115–25.

Brown, A. H. D. & Marshall, D. R. (1981). Evolutionary changes accompanying colonization in plants. In *Evolution Today*, Proceedings of the Second International Congress of Systematic and Evolutionary Biology, ed. G. G. E. Scodder & J. L. Reveal, pp. 351–63. Carnegie–Mellon University: Pittsburg.

Charles, A. H. (1961). Differential survival of cultivars of *Lolium*, *Dactylis* and *Phleum*. *Journal of the British Grassland Society*, **16**, 69–75.

(1964). Differential survival of plant types in swards. *Journal of the British Grassland Society*, **19**, 198–204.

Crawford, T. J. (1984*a*). The estimation of neighbourhood parameters for plant populations. *Heredity*, **52**, 273–83.

(1984*b*). What is a population? In *Evolutionary Ecology*, ed. B. Shorrocks, pp. 135–73. Blackwell Scientific Publications: Oxford.

Ellstrand, N. C. & Antonovics, J. A. (1985). Experimental studies of the evolutionary significance of sexual reproduction II. A test of the density dependent hypothesis. *Evolution*, **39**, 657–66.

Ennos, R. A. & Dodson, R. K. (1987). Pollen success, functional gender and assortative mating in an experimental plant population. *Heredity*, **58**, 119–26.

Gleaves, J. T. (1973). Gene flow mediated by wind borne pollen. *Heredity*, **31**, 355–66.

Gottlieb, L. D. (1981). Electrophoretic evidence and plant populations. *Progress in Phytochemistry*, **7**, 1–46.

Gray, A. J. (1987). Genetic change during succession. In *Colonization, Succession and Stability*, ed. A. J. Gray, M. J. Crawley & P. J. Edwards, pp. 274–93. Blackwell Scientific Publications: Oxford.

Griffiths, D. J. (1950). The liability of seed-crops of perennial rye grass (*Lolium perenne*) to contamination by wind-borne pollen. *Journal of Agricultural Science*, **4**, 19–38.

Grime, J. P. (1973). Competition and diversity in herbaceous vegetation. *Nature*, **244**, 311.

Grubb, P. J. (1977). The maintenance of species-richness in plant communities: the importance of the regeneration niche. *Biological Reviews*, **52**, 107–45.

Hamrick, J. L. (1983). The distribution of genetic variation within and among natural plant populations. In *Genetics and Conservation*, ed. C. M. Schoenewald-Cox, S. M. Chambers, B. MacBride & L. Thomas, pp. 335–48. Benjamin/Cummings: Menlo Park.

Hamrick, J. L. & Holden, L. R. (1979). Influence of microhabitat heterogeneity on gene frequency distribution and gametic phase disequilibrium in *Avena barbata*. *Evolution*, **33**, 521–33.

Hamrick, J. L., Linhart, Y. B. & Mitton, J. B. (1979). Relationships between life history characteristics and electrophoretically-detectable genetic variation in plants. *Annual Review of Ecology and Systematics*, **10**, 173–200.

Handel, S. N. (1983). Pollination ecology, plant population structure, and gene flow. In *Pollination Biology*, ed. L. Real, pp. 163–211. Academic Press: New York.

Harberd, D. J. (1961). Observations on population structure and longevity in *Festuca rubra* L. *New Phytologist*, **60**, 184–206.

(1962). Some observations on natural clones in *Festuca ovina*. *New Phytologist*, **61**, 85–100.

(1967). Observations on natural clones of *Holcus mollis*. *New Phytologist*, **66**, 401–8.

Harris, D. & Davy, A. J. (1986). Strandline colonization by *Elymus farctus* in relation to sand mobility and rabbit grazing. *Journal of Ecology*, **74**, 1045–56.

Hughes, J. & Richards, A. J. (1988). The genetic structure of populations of sexual and asexual *Taraxacum*. *Heredity*, **60**, 161–71.

Huiskes, A. H. L. (1977). The natural establishment of *Ammophila arenaria* from seed. *Oikos*, **29**, 133–6.

(1979). *Ammophila arenaria* (L.) Link (*Psamma arenaria* (L.) Roth.). (Biological Flora of the British Isles.) *Journal of Ecology*, **67**, 363–82.

Jain, S. B. (1983). Genetic characteristics of populations. In *Ecological Studies: analysis and synthesis*, ed. H. A. Mooney, pp. 240–58. Springer-Verlag: Berlin.

Jerling, L. (1985). Are plants and animals alike? A note on evolutionary plant population ecology. *Oikos*, **45**, 150–2.
Kannenberg, L. W. & Allard, R. W. (1967). Population studies in predominantly self-pollinated species. VIII. Genetic variability in the *Festuca microstachys* complex. *Evolution*, **21**, 227–40.
Knox, R. B. (1967). Apomixis: seasonal and population differences in a grass. *Science*, **157**, 325–6.
Law, R., Bradshaw, A. D. & Putwain, P. D. (1977). Life-history variation in *Poa annua*. *Evolution*, **31**, 233–46.
Loveless, M. D. & Hamrick, J. L. (1984). Ecological determinants of genetic structure in plant populations. *Annual Review of Ecology and Systematics*, **15**, 65–95.
MacArthur, R. H. & Wilson, E. O. (1967). *The Theory of Island Biogeography*. Princeton University Press.
McNeilly, T. (1968). Evolution in closely adjacent plant populations. III. *Agrostis tenuis* on a small copper mine. *Heredity*, **23**, 99–108.
McNeilly, T. & Antonovics, J. A. (1968). Evolution in closely adjacent plant populations. IV. Barriers to gene flow. *Heredity*, **23**, 205–18.
McNeilly, T. & Roose, M. L. (1984). The distribution of perennial rye grass genotypes in swards. *New Phytologist*, **98**, 503–13.
Muntzing, A. (1940). Further studies on apomixis and sexuality in *Poa*. *Hereditas*, **26**, 115–90.
Paterniani, E. & Short, A. C. (1974). Effective maize pollen dispersal in the field. *Euphytica*, **23**, 129–34.
Rai, K. N. & Jain, S. K. (1982). Population biology of *Avena*. IX. Gene flow and neighbourhood size in relation to microgeographic variation in *Avena barbata*. *Oecologia*, **53**, 399–405.
Raynor, G. S., Ogden, E. C. & Hayes, J. V. (1971). Dispersion and deposition of timothy pollen from experimental sources. *Agricultural Meteorology*, **9**, 347–66.
Richards, A. J. (1986). *Plant Breeding Systems*. Allen & Unwin; London.
Rihan, J. R. & Gray, A. J. (1982). The hybrid marram × *Calammophila baltica* in Britain. *Annual Report of the Institute of Terrestrial Ecology 1982*, 78–80.
Roos, F. H. & Quinn, J. A. (1977). Phenology and reproductive allocation in *Andropogon scoparius* (Gramineae) populations in communities of different successional stages. *American Journal of Botany*, **64**, 535–40.
Saran, S. & de Wet, J. M. J. (1970). The mode of reproduction of *Dicanthium* (sic) *intermedium* (Gramineae). *Bulletin of the Torrey Botanical Club*, **97**, 6–13.
Silander, J. A. & Antonovics, J. A. (1979). The genetic basis of the ecological amplitude of *Spartina patens*. I. Morphometric and physiological traits. *Evolution*, **33**, 1114–27.
Snaydon, R. W. (1970). Rapid population differentiation in a mosaic environment. I. The response of *Anthoxanthum odoratum* populations to soils. *Evolution*, **24**, 257–69.
Snaydon, R. W. & Davies, M. S. (1972). Rapid population differentiation in a mosaic environment. II. Morphological variation in *Anthoxanthum odoratum*. *Evolution*, **26**, 390–405.

Soane, I. D. & Watkinson, A. R. (1979). Clonal variation in a population of *Ranunculus repens*. *New Phytologist*, **82**, 557–73.

Turner, M. E., Stephens, J. C. & Anderson, W. W. (1982). Homozygosity and patch structure in plant populations as a result of nearest-neighbor pollination. *Proceedings of the National Academy of Sciences, USA*, **79**, 203–7.

Usberti, J. A. & Jain, S. K. (1978). Variation in *Panicum maximum*; a comparison of sexual and asexual populations. *Botanical Gazette*, **139**, 112–16.

Watkinson, A. R. & Harper, J. L. (1978). The demography of a sand dune annual: *Vulpia fasciculata*. I. The natural regulation of populations. *Journal of Ecology*, **66**, 15–33.

Whitehead, D. R. (1969). Wind pollination in the angiosperms: evolutionary and environmental considerations. *Evolution*, **23**, 28–35.

Williams, G. C. (1975). *Sex and Evolution*. Princeton University Press.

Wright, S. (1938). Size of population and breeding structure in relation to evolution. *Science*, **87**, 430–1.

Wu, L., Bradshaw, A. D. & Thurman, D. A. (1975). The potential for evolution of heavy metal tolerance in plants. III. The rapid evolution of metal tolerance in *Agrostis stolonifera*. *Heredity*, **34**, 165–87.

7

An assessment of grass succession, utilization and development in the arid zone

M. D. Kernick

Introduction

The worldwide distribution of the grass family, Poaceae, and its central role in providing a large part of the world's most important food crops – the cereals – is well known. Rather less importance is sometimes accorded to the pivotal role of grasses as a major component of the majority of grazing lands that cover more than a third of the world's land surface. Not only do these grassland ecosystems provide the major feed resource for pastoral livestock, particularly in Africa and Asia, but where they are resistant to fire and overgrazing they provide an effective and essential soil cover that aids infiltration, reduces run-off and prevents soil erosion.

The purpose of this chapter is to review, against a background of global food imbalances, increasing desertification and depletion of grazing lands and grass cover, especially in arid and semi-arid regions, progress in developing grasses for difficult environments, and in particular to draw attention specifically to those grasses that can exploit disturbed ground and difficult soils. With this aspect at the forefront, progress in reseeding depleted grazing lands with suitable grasses is reviewed and future grass development priorities are then outlined.

Global food imbalances and desertification

Despite overproduction in agriculture, particularly in Europe and North America, the world as a whole continues to face serious food imbalances. Nowhere is the situation more apparent than in Africa, where many countries are hard-pressed to keep food production ahead of population growth. Indeed, FAO paints a rather dismal picture of Third World hunger in the next two decades with some 64 countries, 29 of them in Africa, being projected as unable to meet their population needs by the end of the century using present technologies, even assuming all cultivable land is under production (Dover & Talbot, 1987).

The situation is not, however, all gloomy. The 'Green Revolution' with its use of high-yielding crop varieties allied to high inputs of fertilizer and pesticides has brought huge increases in agricultural production in many parts of Asia and also Latin America. In Asia countries that have become largely self-sufficient in staple food crops as a result of applying the technology of the 'Green Revolution' include India, Indonesia, Malaysia, the Philippines and Thailand (Booth *et al.*, 1986). But in most countries the total land area benefiting from the 'Green Revolution' has generally been small because the high technological inputs involved work only under the best economic, social and ecological conditions. In South-east Asia the core areas devoted to intensive Green Revolution food production account for less than 5% of the total land area.

Inevitably overemphasis on the 'Green Revolution' has led to neglect of the less-favoured agricultural areas which comprise the bulk of the cultivable land. Where traditional farming methods are still practised, food production, in many instances, has actually fallen, by as much as 15% in some African countries during the past decade (Lawrence, 1986). This has led to cereal imports rising by 117% and food aid by 172% (World Bank, 1985). But now the very food surpluses which in the past have made such large-scale supplies of cereals and other food aid to the Third World possible are threatened. A series of poor harvests in 1988 starting with the great US–Canada drought and including crop failures in China, USSR, Australia and South America has left cereal stocks at their lowest for 40 years (Lichfield, 1989).

While the serious run-down in world food stocks could undoubtedly cause North America and Western Europe to review their present food production policies, more worrying is the continued loss of cultivable land, particularly in the Third World, due to massive environmental disturbance. Continuing drought, floods and bad land management practices have all contributed to this situation, which is now frequently referred to in the drier areas of the world as desertification. However, the term 'desertification' is becoming more widely used to describe not only desert-like conditions but also other forms of degradation such as salination and deforestation.

Desertification can be described as a process of ecological degradation by which economically bioproductive land becomes less productive (Kassas, 1988). In extreme instances the final scene is a barren landscape incapable of sustaining communities that once depended on it. The United Nations Environmental Programme (UNEP) estimates that globally 3475 million hectares of dry land (arid and semi-arid) are at least moderately desertified (a loss of some 25% of potential productivity);

1500 million hectares are severely desertified (a loss in excess of 50% of potential productivity); and some 850 million persons are affected (Tolba, 1986). The impact of desertification was felt to be particularly severe in sub-Saharan Africa probably because UNEP's assessment estimates (Mabbutt, 1984) were made after a 15-year rainfall decline (Nelson, 1988). Studies have shown that rainfall in sub-Saharan Africa has steadily decreased since the late 1960s (Todorov, 1985; Nicholson, 1986) and that each drought has been more severe than the last (Rasmusson, 1987).

While the effects of continuous and intermittent drought are considered to be far-reaching, recent studies (Luk, 1983; Songqiao, 1988) in the arid regions of northern China reveal that drought merely accentuates the deleterious impact of human overexploitation, i.e. clearance of vegetation for agriculture, the cutting and uprooting of woody species for fuel and the burning of vegetation for pasture and charcoal.

It is now becoming clearer that desertification is often a localized phenomenon, which in the Sahel is generally more prevalent around year-long water supplies and where pastoral and agricultural land use overlap (Breman & Uithol, 1984). A recent study (Olsson, 1984) now throws doubt on the earlier finding that the desert in northern Sudan had moved 90–100 km south during the period 1958–79 (Goudie, 1981). It is also apparent in the African Sahel that the devastating drought of 1968–73 caused a higher order of ecological stress than the broadly comparable droughts of 1910–15 and 1944–8, largely because of increasing anthropogenic pressures (Goudie, 1981). For the immediate future it would seem that a lower rainfall than that experienced between 1930 and 1960 is also likely to be experienced in the Sahel for several more decades (Lamb, 1988).

Faced with the threat of continuing droughts, further desertification in the world's arid zones can probably be arrested only by adopting ecologically sustainable forms of agriculture (Dover & Talbot, 1987) that also make maximum use of local plant species, among which the pioneering grasses must play a prominent part.

Grazing land and grass cover depletion

More than 200 million people use the world's grazing lands, often referred to as rangelands, for some form of pastoral production and 30 to 40 million of these people are wholly dependent on livestock (World Resources, 1988).

Over the last 30 years pastoralists in many of the world's arid and semi-arid regions have had to contend with a gradual loss of grazing lands as they have been converted to cropland. The scale of the conversion has often been high, particularly in Africa and Asia where rapidly increasing human and livestock populations, especially since the 1950s, have greatly increased the pressure on the land. For example, 20 million hectares have been converted to cropland in northern China (Blaikrie & Brookfield, 1987), some 10 million hectares in Iran (Pearce, 1968), as much as 3 million hectares in northern Iraq, 11 million hectares in Turkey (Kernick, 1979) and over 2.7 million hectares in Tunisia (UNESCO/UNEP/UNDP, 1980).

It is in the semi-arid zones of sub-Saharan Africa, however, that the highest loss of rangelands has occurred. The region particularly affected is the sand-dune belt south of the Sahara – a band about 400 km wide stretching some 6000 km from the Atlantic to the Red Sea, receiving 100–600 mm mean annual rainfall and known as the Sahel – that developed during the Pleistocene (2000–12 000 BP) as a thorn scrub savannah (Mensching, 1988). Loss of rangeland in the Sahelian countries has led to serious wind-induced soil erosion and repeated failure of millet crops in recurrent droughts. This has led in some cases to the abandonment of agricultural settlements, with Niger and the Sudan being particularly affected (Zaroug, 1985). In the Sudan in the more humid phase following the droughts of the 1970s rain-fed cultivation shifted as far north as the 200 mm isohyet (El Moghraby & Ali, 1987).

Serious loss of grazing and forest land through conversion to cropland has also occurred in the Ethiopian and Kenyan highlands, the Atlas mountains of Algeria and Morocco, the mountains of eastern Turkey and northern Iraq, the Elburz and Zagros mountains of Iran and the Himalayas. This has led to increased soil erosion and a higher incidence of flash flooding, which is of particular concern because many of these highland zones are vitally important watersheds.

In the world's arid and semi-arid regions depletion of the grass cover is increasing mainly as a result of the overexploitation of the land resource. For the African continent, *The Grass Cover of Africa* (Rattray, 1960) provides useful preliminary indications of the existing dominant grass cover that has evolved under the combined influence of climate, soils and land use. Two findings of the study have special importance for the arid and semi-arid areas of Africa: first, the secondary nature of the present grass composition with *Aristida* species becoming very common as a result of the overexploitation of the previous grass cover, and secondly, the difficulty of assessing what the more permanent grass species should be. More often than not, the evidence available seems to point to some form of disclimax grassland leaving the nature of the true climax in doubt

(Clayton, 1986). However, in the light of the serious and increasing degradation of both the soil and the microclimate that is now apparent in many arid zones, it seems very unlikely that the original grassland climax, where known, could ever be restored.

The most serious and often permanent loss of the original grass cover occurs as a result of the large-scale expansion of dryland farming. A good example is *Stipa tenacissima* grassland in the North African steppe, whose lower limit is the 200 mm isohyet, and which has been greatly reduced in this way (Le Houerou, 1969). Rapid expansion of dryland farming in the Near East has also caused the loss of valuable perennial grasses. In southern Iraq the important tufted perennial grasses *Aristida plumosa*, *Chrysopogon gryllus*, *Cymbopogon olivieri* and *Hyparrhenia hirta* are now present only as relics in the moist steppic zone (Guest, 1966). Similarly, in Iran massive disturbance to the steppic vegetation since time immemorial has left only relics of important grasses such as *Aristida caerulescens*, *A. plumosa*, *Cenchrus ciliaris*, *Cymbopogon laniger*, *Hyparrhenia hirta*, *Stipa barbata* and *Tricholaena teneriffae* (Pabot, 1961, 1967). Such relics in Iran's steppic zone have, however, been sufficient to lead to an excellent recovery of *Aristida plumosa* and *Stipa barbata* after a minimum of 5 years' protection (Nemati, 1986); similar regeneration after 5 years' protection has also been recorded for *Stipa barbata*, *S. lagascae* and *S. parviflora* in the North African steppe (Le Houerou, 1970). In contrast, on the southern margins of the Sahara, perennial *Aristida* and *Cenchrus* grass species once cleared through cultivation have little or no chance to re-establish themselves and the ground becomes dominated by annual herbs and grasses (Wickens & White, 1979). In fact, particularly after cultivation, re-establishment of the original grass cover in many arid zones is now likely to prove much more difficult due to the widespread soil degradation involving progressive loss of organic matter, structure, water-holding capacity and even the A soil horizon.

Overgrazing, periodic drought and both the low and high frequency of fires have also led to widespread loss of perennial grasses in sub-Saharan Africa resulting in their replacement by annual grasses and largely unpalatable shrubs (Skovlin, 1986; Toutain, 1986; World Resources, 1986). Perennial grasses particulary affected have been *Andropogon gayanus*, *Aristida* spp., *Cenchrus ciliaris* and *Hyparrhenia dissoluta* in the western Sahel (Rossetti, 1965) and *Chrysopogon aucheri* in the eastern Sahel, and especially in northern Somalia (Hemming, 1966, 1973).

In the arid zone of northern Kenya with around 250 mm of annual rainfall a cycle of dry years resulted in a change from perennial to annual grassland irrespective of grazing pressure, while perennial grassland further south with around 650 mm of annual rainfall shows remarkable regeneration (Pratt, 1984). However, in northern Nigeria close to the 600 mm isohyet the 1970–80 drought period caused the replacement of the perennial *Andropogon gayanus* grassland by the annual grass *Cenchrus biflorus* (Mortimore, 1989).

In the Sahel recurrent droughts over the past 20 years have undoubtedly been the major cause of the widespread changeover from a perennial to an annual grass cover. This situation is dramatically confirmed by the results of ecological studies carried out in Mali during the period 1976–80, covering a transect between the 50 mm and 600 mm rainfall isohyets, which showed that the entire grass cover biomass was produced by annual grasses, chiefly *Cenchrus biflorus*, *Diheteropogon hagerupii* and *Schoenefeldia gracilis* (Penning de Vries & Djitaye, 1982).

Unfortunately, in the western Sahel *Cenchrus biflorus*, a rather unpalatable species, is replacing the more valuable annual grass *Aristida mutabilis* whose seed stocks in the soil have become seriously depleted after successive droughts (Peyre de Fabreque, 1989). So far as is known, this phenomenon has not been encountered among the important annual grasses in North Africa, the Near East or Asia.

In the western Baluchistan Province of Pakistan many millennia of overgrazing have largely destroyed the natural vegetation with *Chrysopogon aucheri* now predominating in the arid southern portion and *C. fulvus* in the more humid northern part. There is reason to believe that *Stipa*, *Pennisetum* and *Enneapogon* were important grass genera at one time and that *Chrysopogon* represents a stage in deterioration. Partial protection for 10 years at a site in the arid southern region resulted in regeneration of the palatable grasses *Chrysopogon aucheri*, *Stipa szowitsiana*, *S. linearis*, *Enneapogon persicus* and *Oryzopsis aequiglumis* (Whyte, 1968).

Persistent long-term overgrazing has also caused widespread loss of grass cover in the *Dichanthium/Cenchrus/Lasiurus* type of disclimax grassland that dominates the sub-tropical and semi-arid regions of India and Pakistan, and which are characterized under less grazing pressure by the valuable perennial grasses *Cenchrus ciliaris*, *C. setigerus*, *Cynodon dactylon*, *Dichanthium annulatum*, *Eleusine compressa*, *Panicum antidotale* and *Sporobolus marginatus* (Dabadghao, 1960; Shankarnaryan, 1977).

In the Thar desert of western Rajasthan in India progressive degeneration of the *Dichanthium/Cenchrus/Lasiurus* grassland has led to the progressive loss of the palatable perennial grasses *Cenchrus ciliaris*, *C. setigerus* and *Eleusine compressa* and their replacement by poor annual forage species, especially *Aristida* spp. and *Cenchrus biflorus*, and ultimately bare ground (Shankarnaryan, 1985). A further concern is that intensive seed collection from *Cenchrus biflorus*, *Echinochloa colonum*, *Panicum antidotale* and *Panicum turgidum* for human consumption in times of drought may also be seriously hindering natural regeneration in the Thar desert (Saxena, 1977).

In Asia, particularly in northern China, where the grasslands are very similar to those in the Near East with respect to their flora, vegetation and physiogonomy, annual grasses also appear to be increasing their ground cover under the progressive extension of man-induced desiccation (Numata, 1979). In north-eastern China good *Leymus* (*Aneurolepidium*) *chinense* grassland tends to deteriorate towards *Stipa* dominance under overgrazing (Zhu Ting-Cheng, Li Jiandong & Yang Dianchen, 1981), with further grazing pressure leading to the replacement of the more valuable *Stipa grandis* by the coarser perennial grass *Calamagrostis squarrosa* (Zhao Ji, Zhou Giming & Fan Weihong, 1985).

The continuing degradation of the world's arid grazing lands and particularly the depletion of their grass cover should prompt both ecologists and plant breeders to pay more attention to those indigenous plant species, especially the grasses, that can both stand the harsh environment and recolonize the degraded lands. It should be recognized that it is probably with these grasses that the continuing ecological stability of the arid zone will increasingly lie.

Role of grasses in maintaining ecological stability

It has been suggested that the success of the grasses lies primarily in the evolution of a versatile life-style adapted to unstable or fluctuating environments, particularly those associated with strongly seasonal rainfall regimes or the early stages of succession following disturbance (Clayton, 1986).

This constant ability to adapt to changing environmental conditions and biotic factors also underlines the essential role that grasses are playing in maintaining ecological stability in disturbed and fragile ecosystems, particularly in the arid regions of the world. Here, greater attention to those grass genera and species presently recolonizing disturbed sites could hold real promise for obtaining further ecological stability in the future.

In the Sahel and in the drier areas of southern Africa both *Aristida* and *Sporobolus* species are important colonizers of disturbed sites, and in southern African grasslands under continuous grazing pressure they also displace a secondary grass cover of *Eragrostis* species, except for those areas where *Eragrostis curvula* is the dominant pioneer species (Rattray, 1960; Bosch, 1989).

In the eastern Sahel *Aristida sieberiana*, *A. pallida* and *A. papposa* (accepted name now *Stipagrostis uniplumis*), along with the desert grasses *Cymbopogon schoeanthus*, *Lasiurus hirsutus* and *Panicum turgidum*, are also common colonizers on sandy soils throughout the Saharo-Sahelian Zone (Rattray, 1960; Rossetti, 1965; Le Houerou, 1980). In the dry Sahel *Cymbopogon schoeanthus* ssp. *proximus*, a tufted perennial which occurs from Mauritania to Ethiopia, is also a common colonizer on sandy soils and appears to be increasing as a result of overgrazing (Rattray, 1960; Ibrahim & Kabuye, 1987).

Both *Aristida pungens* (accepted name now *Stipagrostis pungens*) and *Panicum turgidum* are important colonizers of sand-dunes in the Sahel, where the southern limit of the former species in Mali and Mauritania appears to be close to the 150 mm rainfall isohyet. At a lower rainfall limit further north *Aristida pungens* tends to replace *Panicum turgidum*, although this depends to some extent on relief and soil conditions (Rossetti, 1965).

Annual species of *Aristida*, particularly *A. adscensionis* and *A. mutabilis*, both widespread in the northern Sahara, are common colonizers of disturbed land in the Sahel. *A. adscensionis* commonly occurs in association with the widespread and valuable annual forage species *Schoenefeldia gracilis* on clay soils; while *A. mutabilis* frequently occurs in association with the widespread annual grasses *Cenchrus biflorus* and *Eragrostis tremula* on sandy soils, especially in heavily grazed areas near watering points.

Throughout the Sahel perennial species of *Sporobolus*, particularly *S. ioclados* and *S. helvolus*, which provide excellent salt grazing for cattle in Mauritania, Senegal and the Sudan, are important stability components of the grass cover of alluvial, clay and gypsum soils. In north-eastern Somalia *Sporobolus ruspolinus* is the characteristic perennial grass cover of gypsum soils. In this area *S. ruspolinus* and *S. ioclados*, together with *Andropogon kelleri*, *Chrysopogon aucheri* and *Dactyloctenium robecchii*, are said to arrest run-off and assist infiltration, which allows *Acacia bussei* trees with their superficial root system to survive the arid conditions (Hemming, 1973). *Dactyloctenium robecchii*, which is characteristic of overgrazed areas in north-eastern Somalia, provides valuable stability to the grass cover, since this shrubby grass, though grazed in the wet season when it is green, is unpalatable in the dry season due to its sharp stiff leaves.

Tetrapogon villosus is another perennial tussock grass which plays a valuable stability role on poor stony and loose sandy soils in the drier areas of the eastern Sahel, extending also into the Somali–Masai region, where the annual rainfall is at least 350 mm, and in the Sudan it is said to provide good fodder (IBPGR/Kew, 1984). A similar stability role is afforded by the perennial tussock grass *Pennisetum schimperi* which is increasing in overgrazed areas of the Ethiopian highlands (Gebrehiwot & Tadesse, 1985; Woldu, 1985).

In the Mediterranean and Irano-Turanian regions of North Africa, the Near and Middle East and also in Asia, particularly northern China, annual and perennial species of *Stipa* play a similar pioneering role in disturbed arid environments to that of *Aristida* species in the drier zones of sub-Saharan Africa.

The perennial steppic grasses *Stipa barbata*, *S. lagascae* and *S. parviflora* are important stability components on overgrazed areas and shallow soils in North Africa and the Near East where the annual rainfall varies from 150 to 250 mm. *Stipa barbata* is of special interest because of its wide ecological amplitude, with an altitude range of 450–2300 m in Iraq, and of 400–3500 m in Iran. In contrast, *S. lagascae* has a more restricted elevation range of 600–800 m in Iraq and 1500–2600 m in Iran (Pabot, 1961; Bor & Guest, 1968).

In the arid regions of China perennial species of *Stipa* also play a prominent ecological role as major stability components of the present steppic/grassland system. The most widely distributed species in this grassland ecosystem is *Stipa gobica* which belongs to a group of short-growing, xerophytic *Stipa* species including *S. breviflora*, *S. glareosa*, *S. orientalis* and *S. caucasia* that have a well-developed fibrous root system and vegetative tillering buds that are perfectly adapted to the extreme climate where they have existed for centuries. As such, they can be called inert plants and are considered to form one of the most stable subsystems in grassland ecosystems today (Ren Jizhou, Hu Zizhui & Fu Yikun, 1985).

The more typical steppe in northern China is dominated by taller-growing species of *Stipa*, particularly *S. grandis* and *S. krylovii*, with *S. baicalensis* predominating on meadow-steppe at elevations greater than 1250 m (Shu Jiang, 1985; Zhu Ting-Cheng, Li Jiandong & Zu Yuangang, 1985). While its ecological role is important, it appears that the taller *Stipa* steppe is not as stable as the shorter *Stipa* steppe with overgrazing leading to the gradual disappearance of the former dominant grass species (Zhao Ji *et al.*, 1985).

Perennial species of *Stipa*, including *S. speciosa*, *S. chrysophylla* and *S. humilis*, are also widespread and important stability components of the overgrazed desert grassland steppes of Patagonia and the Andean highlands of South America (Soriano, 1979).

In the Mediterranean and Near East region the perennial grasses *Poa bulbosa* and *P. sinaica* play an especially important role in providing ecological stability to overgrazed steppes and deserts. Both of these grasses, which are considered by some scientists to be a single agamic complex of *P. bulbosa* (Heyn, 1962), can reproduce themselves by above- and below-ground bulbils as well as forming a dense soil-binding surface mat of fibrous roots. In many desert areas in the Near East they now form the only dominant grass cover, which in many cases constitutes the last line of defence against soil erosion.

P. bulbosa has a very wide ecological amplitude in the Mediterranean region with ecotypes adapted to 100 mm of annual rainfall at the southern extent of its distribution and to more than 1000 mm at its northern extremity and with an altitude range of 50–2000 m (Sukopp & Scholtz, 1965; Bor & Guest, 1968). Both *P. bulbosa* and *P. sinaica* are commonly found growing on sandy/gravelly soils, often in association with *Carex stenophylla*, with *P. sinaica* predominating in the drier desert areas of Sinai, southern Iraq and Arabia. But *P. bulbosa* is not restricted to the Near East in its distribution since it also occurs on overgrazed sandy desert areas in north-western China. There, it grows in association with *Carex physoides*.

Aristida plumosa, whose curved plumose awns apparently aid seed burial in the soil (Pabot, 1961) is another valuable colonizer of sandy or gravelly deserts and steppes with a geographical distribution almost without interruption from the western Sahara to Tibet. Other caespitose perennial grasses spanning the same region are *Cymbopogon laniger*, a valuable colonizer of stony, sandstone and gypsum soils, and *Hyparrhenia hirta* and *Tricholeana teneriffae*, both colonizers of rocky soils and having a strong rooting habit with tussocks that remain green throughout the summer.

Again, of special value in the Near East region are *Aeluropus littoralis* and *A. repens*, creeping stoloniferous perennial grasses that colonize desert saline soils, and the shrubby perennial grasses *Lasiurus hirsutus*, *Panicum turgidum* and *Pennisetum dichotomum* that colonize loose sandy wadis, plains and dunes (Vesey-Fitzgerald, 1957*a*,*b*; Kernick, 1966; FAO, 1975; Fatahallah, 1976).

In the steppes and foothills of the Near East, where dryland cultivation has disturbed the natural vegetation, the annual grass *Hordeum spontaneum* and the short-lived perennial grasses *Hordeum bulbosum* and *Secale montanum* have shown themselves to be excellent colonizers of abandoned land. Again, in the disturbed environments of the Taurus mountains of southern Turkey and the Zagros and Elburz mountains of Iran the native perennial grasses *Agropyron caespitosum*, *Bromus persicus*, *B. tomentellus* and *Hordeum fragile* play a major ecological role in providing stability to the mountain grassland ecosystem because of their strong rooting and resistance to overgrazing (Pabot, 1967).

Similarly, in the Himalayan foothills in the Terai region the unpalatable perennial sub-tropical grass *Vetivera zizaniodes* can be considered the last bastion of ecological stability in the final stage of regression of the tall savannah grassland as further grazing and burning lead to the complete removal of the vegetative cover (Whyte, 1964). Elsewhere in the lower Himalayan region the perennial grasses *Chrysopogon gryllus*, *Ch. montanus* and *Cymbopogon montanus* are important stability components of the dry, hilly tracts in the sub-tropical zone.

In north-eastern China *Leymus* (*Aneurolepidium*) *chinense* is an important colonizer of abandoned fields in the steppic region with *Achnatherum splendens* also an important conservation species on the margins of the meadow steppe and cultivated lands (Numata, 1979).

In the search for ecological stability in the arid zone, however, the pivotal role of the pioneering grasses must be balanced against the relative contribution of other well-adapted native plants, particularly shrubs and trees (Valenza & Diallo, 1980). On depleted grazing and agricultural lands this means promoting a balanced ecological approach whereby conservation/fodder grasses, along with appropriate shrubs and trees, are introduced alone or in mixture with food crops in order to strengthen and sustain the traditional agricultural system.

Use of grasses for reseeding depleted grazing lands

Over a quarter of a century ago comment was made that although many of the 10000 grass species in the world contribute to natural grazing, only some 40 of them had any significance as sown pasture plants (Hartley, 1964). Today, this comment still largely holds true. A further factor of interest is that the principal source of useful forage grasses is Africa, where some 45 species have been listed as being important (Bogdan, 1977). However, it is also significant that only nine native perennial grass species from this listing – namely *Andropogon gayanus*, *Cenchrus ciliaris*, *C. setigerus*, *Chloris gayana*, *Cynodon dactylon*, *Dichanthium annulatum*, *Eragrostis curvula*, *Lasiurus hirsutus* and *Panicum antidotale* – are well adapted for use in the arid zone.

Although relatively little progress has yet been made in developing grasses for the rehabilitation of the degraded grazing lands in the dry Sahelian zone of Africa, a number of indigenous grasses show promise. Foremost among these are the perennial species *Andropogon gayanus* and *Cenchrus ciliaris*. *Andropogon gayanus* is, in fact, a complex of four separate varieties (Clayton, 1972). *A. gayanus* var. *gayanus* and var. *bisquamulatus* are the most interesting for seeding in the drier parts of the Sahel with a mean annual rainfall of 400 mm, with the latter variety being the most widely used in dryland seeding.

In the Sudanian–Sahelian belt of Burkina Fasso, Mali and Niger both *Andropogon gayanus* and *Cenchrus ciliaris* have been seeded successfully on denuded village rangelands with 350–450 mm of annual rainfall (Sikora, 1980). In regions of Senegal (Bambey) with more than 600 mm of annual rainfall dryland seedings of *Andropogon gayanus* and *Cenchrus ciliaris* have produced more than 5 tonnes of dry matter per hectare per year (5 tonnes dm/ha/year) (Naegele, 1977).

In the dry Sahelian zone of Mauritania, Niger and the Sudan with 100–350 mm of annual rainfall some success has also been achieved with the direct seeding and/or vegetative propagation of the perennial desert grasses *Aristida pungens* and *Panicum turgidum* for fixing sand-dunes (Baumer, 1961; Naegele, 1977) and with *Cymbopogon schoeanthus* for direct seeding in sandy depressions (Shenker & Goumandakoye, 1987).

Because of the widespread changeover from perennial to annual grassland in the Sahel, the possibilities that exist for seed multiplication of the valuable indigenous annual grasses *Aristida mutabilis*, *Dactyloctenium aegyptium*, *Eragrostis tremula* and *Schoenefeldia gracilis*, with a view to their future use for reseeding denuded rangeland areas, have also been recognized (Naegele, 1977).

Eragrostis curvula, which is an important native grass in the semi-arid grazing lands of southern Africa, has proved valuable for stabilizing sandy soils and sand-dunes in the Patagonian foothills in Argentina where it has been mechanically seeded with *Secale cereale*; as a soil builder and living fallow it has not been surpassed by any other cultivated plant in the region (Coras, 1960). The success of *Eragrostis curvula* in the cold Patagonian environment is surprising, coming as it does from the relatively warmer areas of southern Africa. Its cold tolerance may have evolved during the Pleistocene, when southern Africa was much colder than it is now, and this attribute has not subsequently been lost (Harlan, 1983).

In the cold winter areas (250–350 mm annual rainfall) of North Africa and the Near East introduced varieties of *Agropyron cristatum*, *A. desertorum* and *Elymus hispidus* subsp. *hispidus* (= *A. intermedium*) from the USA were successfully reseeded on depleted rangelands in northern Iran and Iraq, while *A. cristatum* and *E. hispidus* subsp. *hispidus* did well in the cold Midelt region of Morocco with *E. elongatus* (= *A. elongatum*) has also given encouraging results in the mild semi-arid region of Tunisia (Hussain, 1975; Le Houerou, 1985; Nemati, 1986).

In the early 1960s in northern Iran more emphasis was given to including native perennial grasses in range reseeding trials, particularly *Bromus tomentellus*, *B. cappadocicus*, *Elymus tauri* (= *A. tauri*), *Hordeum bulbosum* and *Secale montanum*. At the lower rainfall limit of 200–250 mm these grasses proved more drought-resistant than some of the introduced *Agropyron* and *Elymus* species. In particular, the large-seeded *Hordeum bulbosum* and *Secale montanum* germinated more quickly at lower temperatures (10–15°C) and with as little as 20–30 mm of rainfall, and also emerged better in vesicular soils (FAO, 1970).

All the grasses mentioned as being successfully seeded in small-scale range reseeding trials in Algeria, Iran, Iraq, Morocco and Tunisia have to be planted in rows spaced at least 70 cm apart for the best results, since they persist well only at low plant densities. Forage yields, the bulk of which are produced in the spring period (March–June), have averaged 1000–2000 g dm/ha/year (Kernick, 1978). In the Near East grass reseeding has proved most practical on depleted village grazing lands (Kernick, 1985).

In the sub-steppic zone of northern Iran (average annual rainfall 350 mm) most of the introduced and some of the native grasses were summer dormant, except for some species such as *Agropyron cristatum*, *A. tauri*, *Festuca arundinacea*, *Hordeum fragile* and *Secale montanum* which produced substantial re-growth in July, averaging 500–900 kg/dm/ha; the percentage moisture in the first 30 cm of the soil usually drops from a peak of 20–30% in March to 2–6% by the end of July, with the root system of all the perennial grasses being concentrated in the top 30–50 cm of the soil and only a few roots reaching or exceeding 1 m in depth (Kernick, 1967). Some of the introduced and native perennial grasses included in the range reseeding trials in the sub-steppic zone of northern Iran such as *Agropyron desertorum*, *Elymus elongatum*, *Oryzopsis holciformis*, *Stipa barbata*, *S. lagascae* and *S. orientalis* also exhibited seed dormancy which, if drought prevailed, permitted germination to occur 2 to 3 years after seeding, a particularly valuable characteristic.

In the cold steppic zone of the central plateau region of Iran, where there is 200 mm of annual rainfall, limited adaptability plantings of salt-tolerant introduced and native perennial grasses on saline–alkaline soils in the Ghazvin plain with a high water table (1–2 m deep) indicated that the most promising species were *Agropyron elongatum*, *Elymus cinereus*, *Festuca arundinacea*, *Hordeum bulbosum*, *Puccinellia capillaris* and *P. distans* (Kernick, 1986).

On sandy/gravelly soils in the cold central plateau region of Iran with 150–200 mm of annual rainfall, the native perennial grass *Stipa barbata* has proved to be the most successful species for reseeding depleted steppic rangeland. A similar result was also obtained in the steppic zone of Syria, where *Stipa barbata* gave the best establishment in good rainfall years with more than 170 mm when compared with other seeded grasses including *Agropyron libanoticum*, *Poa bulbosa* var. *vivipara* and *P. sinaica* (Sankary, 1979). However, *Poa bulbosa* var. *vivipara* has been successfully reseeded on depleted rangeland in the steppic zone of Afghanistan and in the Samarkand desert of southern USSR (Kernick, 1978) and also *P. sinaica* on depleted range near Quetta in Pakistan (Said, 1960).

In the arid steppe of Algeria (average annual rainfall of 200 mm) introduced varieties of both *Elymus elongatum* and *Phalaris aquatica* (= *P. tuberosa*) have been reseeded successfully on depleted rangelands, with the latter species also doing well in range reseedings in the high plateau zone where it has proved very resistant to overgrazing (Gallagher, 1972). *P. aquatica* has also performed well in dryland range plantings in the steppic zone (300–400 mm annual rainfall) of Jordan and Syria (Bailey, 1967; Draz, 1974).

In the southern arid zone (100–200 mm annual rainfall) of Morocco *Elymus elongatum* has so far proved to be the most successful perennial grass for reseeding depleted rangelands (Omar, 1985), although earlier studies (Foury, 1956) had underlined the intrinsic value of the native perennial grasses *Cenchrus ciliaris*, *Cymbopogon schoeanthus* ssp. *laniger*, *Hyparrhenia hirta*, as well as *Aristida* and *Stipagrostis* species. In follow-up range reseeding trials only *Cenchrus ciliaris*, *Hyparrhenia hirta*, *Stipa lagascae* and *Digitaria commutata* ssp. *nodosa*, an important native perennial grass now rare on shallow sandy soils in southern Algeria, Morocco and Tunisia, have received attention.

In the southern arid zone of Morocco, recent dryland range plantings of *Cenchrus ciliaris* and *Digitaria commutata* ssp. *nodosa* have apparently not persisted well due to drought and overgrazing (Omar, 1985). Difficulty has also been experienced in establishing *Cymbopogon schoeanthus* ssp. *laniger* in southern Morocco where the temperatures are not mild enough and the moisture is insufficient during the seedling stage (Omar & Kabak, 1986).

The poor germination sometimes occurring in the seeds of *Hyparrhenia hirta* and *Aristida pungens* and the presence of long or branched awns in *Aristida*, *Stipagrostis* and *Stipa* species causing an impediment to the mechanical handling of the seeds presently restricts the greater use of these grass species for reseeding depleted rangelands (Schoenenberger, 1982; Le Houerou, 1985).

The problem of low germination in some perennial native grasses such as *Panicum antidotale*, *P. turgidum* and *Stipa lagascae* has encouraged the vegetative propagation of these species which have been successfully established as transplants on depleted rangelands in the north-west coastal region of Egypt (Migahid & El Shourbagui, 1961; Ibrahim, 1968; Abouguendia, 1985).

In the Rajasthan desert of India and the Thal and Sind deserts of Pakistan, which receive from 150 to 450 mm annual rainfall, principally in the monsoonal summer period, considerable success has been achieved through dryland seeding of improved and unimproved native and also introduced varieties of *Cenchrus ciliaris*, *C. setigerus*, *Lasiurus hirsutus* and *L. sindicus* on depleted rangelands (French, 1968; Chakravarty & Kackar, 1970; Khan, 1970; Puri & Paliwal, 1976).

Forage yields of *Cenchrus ciliaris*, *C. setigerus* and *Lasiurus sindicus* have averaged 1000–1600 kg/dm/ha/year on sandy soils in the Rajasthan desert (Jodhpur) in a year with 300 mm of rainfall; in a year with 248 mm the yield of *Cenchrus ciliaris* was reduced to 590 kg/dm/ha and *Lasiurus sindicus* to 990 kg/dm/ha, indicating the greater drought resistance of the latter species (Chakravarty, 1970). In the Thal desert of Pakistan individual dryland seedings of *Cenchrus ciliaris* and *Lasiurus hirsutus* have increased the forage yield of depleted native range from 29 kg/dm/ha before seeding to 1750 kg/dm/ha after seeding in a year with 225 mm of rainfall (Chakravarty, 1970; Khan, 1971).

In the cold winter arid zone of north-east China early experience with the seeding of introduced *Agropyron*, *Bromus* and *Elymus* species from the USA and Canada on depleted rangelands showed that these species had some difficulty in adapting to a monsoonal summer environment where most of the annual rainfall (250–350 mm) comes in the period June to August. As a result, attention was switched to native perennial grasses that were considered to be better adapted and possibly more productive in the long term (Ma Zhi Guang, 1985).

So far, the most successful native perennial grasses for seeding in the north-east arid zone of China have proved to be *Leymus* (= *Aneurolepidium*) *chinense* and *Clinelymus dahuricus* (Li Chonghao *et al.*, 1981). However, other native perennial grasses, such as *Leymus*

(= *Aneurolepidium*) *dasystachys*, *Elymus nutans*, *E. sibiricus* and *Hordeum brevisubulatum* have also shown considerable promise, and could become much more widely used in future if some of the large tracts of degraded and marginal cultivated land were removed from cropping and returned to pasture (Lu Lian, 1985; Zhu Ting-Cheng *et al.*, 1985).

What is clear from the foregoing review is the fact that in almost all cases it is the native perennial grasses that have generally given the best results in range reseeding trials in the arid zone. This should stimulate further efforts towards the collection and development of valuable ecotypes and varieties of arid zone native grasses that could help to extend their effective use for reseeding depleted rangelands in the future.

Future grass development priorities

Particularly in the arid zone of the Sahel, North Africa and the Near East, serious concern has been expressed about the possibilities of increasing genetic erosion of the forage resource occurring due to widespread rangeland degradation and desertification (Harlan, 1983; Knight, 1983). In so far as the grasses are concerned, however, little is yet known of the extent of the genetic erosion that may be taking place since the species really meriting attention have not been studied adequately. For instance, the genetic variability in those arid zone grasses which could prove useful, such as *Cenchrus ciliaris*, *Panicum antidotale*, *P. turgidum* as well as other valuable species of *Eragrostis*, *Pennisetum* and *Sporobolus*, has yet to be fully determined.

Among the wide range of indigenous perennial grasses that are important pioneers and stability components in the Saharo-Sahelian zone, priority attention now needs to be given to the collection and screening of a wide range of ecotypes of *Cenchrus ciliaris*, *Cymbopogon schoeanthus*, *Dichanthium annulatum*, *Hyparrhenia hirta*, *Lasiurus scindicus* (= *L. hirsutus*), *Panicum turgidum* and *Stipagrostis plumosa* (Fig. 7.1) which occur in both the northern and southern Sahara (IBPGR/Kew, 1984; Kernick, 1989). Because of their wide ecological amplitude, all these grass species are of special interest for future development with regard to their possible large-scale use for reseeding depleted Saharan rangeland.

It is important to understand that some of these desert grasses present special problems and challenges for the plant breeder. For instance, *Cenchrus ciliaris* in its various ecotypes is predominantly apomictic which may in Nature limit diversity unless apomixis is interrupted by occasional instances of sexual reproduction; alternatively, it is possible to obtain sexual strains of *Cenchrus ciliaris* which if pollinated with apomictic

Fig. 7.1. ***Stipagrostis plumosa*** **(L.) Munro ex T. Anders. 1. Habit, ×2/3; 2. spikelet, ×2; 3. lower glume, ×8; 4. upper glume, ×8; 5. floret with base of awn, ×8; 6. lemma, ×8; 7. palea, ×8; 8. flower, ×8; 9. grain, ×8; 10. ligule, ×4. (After** ***Flora of Iraq, vol. 9, Gramineae*** **(1968), p. 390, plate 148.)**

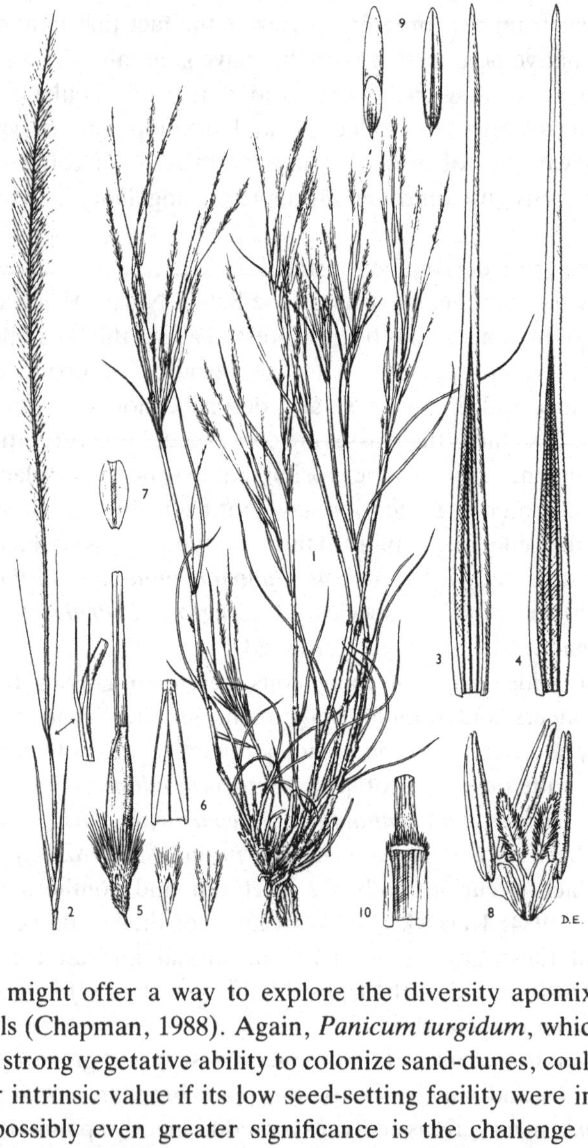

strains might offer a way to explore the diversity apomixis presently conceals (Chapman, 1988). Again, *Panicum turgidum*, which has developed a strong vegetative ability to colonize sand-dunes, could become of greater intrinsic value if its low seed-setting facility were increased.

Of possibly even greater significance is the challenge of trying to produce awnless cultivars of species of *Aristida*, *Stipagrostis* and *Stipa*, which are so widespread in the arid zone of Africa, the Near East and Asia, thus allowing normal mechanical handling of the seeds. With the right inbreeding and backcrossing programme, it might be possible to produce awnless mutants which could then be stabilized for further agronomic use. Earlier success in locating and utilizing natural mutants for overcoming the seed-retention problems in *Phalaris aquatica* suggests that this approach could be employed successfully for overcoming major problems inhibiting domestication in other important grasses (McWilliam & Gibbon, 1981).

A further challenge facing plant breeders and agronomists concerned with range improvement is to develop native grasses for reseeding saline and impoverished soils. In the Sahel attention needs to be given to the collection and screening of perennial species of *Sporobolus*, particularly *S. helvolus*, *S. iocladus* and *S. ruspolinus* which are well adapted to saline and alkaline soils. These grasses seem to spread mainly by stolons but they also produce very small seeds. In some species of *Sporobolus* the pericarp, which is not firmly attached to the seed coat as it is in other grasses, swells and becomes sticky, aiding seed dispersal and possibly also seed germination on hard soil surfaces (IBPGR/Kew, 1984).

Leptochloa (*Diplachne*) *fusca* is another salt-tolerant perennial grass that merits further development for use on saline or sodic seasonally flooded rangeland in the Sahel and southern Africa as well as in the arid sub-tropical zone of India and Pakistan. In the Punjab Province of Pakistan it colonizes salt-affected soils where its fodder value is recognized by local farmers who depend on it to feed their buffaloes (Kernick, 1986). The seeds are small and poor germination has been reported on salt-affected soils, so there is clearly a need for further screening and selection to find more salt- and soda-tolerant strains.

Since soils over much of Africa are especially low in phosphorus, greater attention in the future may also need to be paid to the selection of ecotypes of arid zone grasses that perform well under this limitation. In the arid zone of Australia it has been shown that *Cenchrus ciliaris*, an introduced species, requires about 25 ppm of available phosphorus – a level rarely occurring naturally in soils – for satisfactory seedling growth rate, root development and subsequent drought survival (Christie, 1975).

In the Mediterranean region of North Africa and the Near East the question has been raised as to whether the many native annual grasses represent an important genetic resource (Knight, 1983). While there is very little information on which to base an answer, it can be postulated that any native perennial grass could be enriched from its annual cereal relative (Chapman, 1988). For instance, in the Near East *Hordeum spontaneum*, which is the progenitor of cultivated barley, has been shown to possess populations that differentiate over short distances suggesting the operation of edaphic natural selection (Nevo, 1981). Because of the excellent spring forage growth of *Hordeum spontaneum* even in dry seasons it would be worth while to collect genetic material from a wide range of habitats for crossing with other perennial *Hordeum* species, including *H. bulbosum*, *H. fragile* and *H. violaceum* (Kernick, 1978). In the Sahel similar benefits might be obtained from crossing the annual *Pennisetum americanum* with other related perennial species of *Pennisetum* such as *P. squamulatum* (Dujardin & Hanna, 1983).

In view of the extensive distribution and important ecological role that *Poa bulbosa* plays throughout North Africa and the Near East, extending even into the arid zone of China, priority attention also needs to be given to wide-scale collection of ecotypes of both *P. bulbosa* and *P. sincaica* in order to exploit the very considerable reserve of germplasm that exists (Fig. 7.2). Selection among the material collected is likely to prove very rewarding and lead to the development of more productive forms for both species.

Chrysopogon aucheri is another arid zone perennial grass of particular importance which, because of its very wide distribution from the Sahel through Arabia, southern Iraq and Iran, to Afghanistan and Pakistan, merits increased study in the future. This grass has been recommended for the regeneration of degraded desert rangelands and for fixing sands (Naegele, 1977), but so far no agronomic seedings have yet been made. Ecotypic collections are an urgent necessity and should provide valuable introduction and breeding material for extending the possibilities for successfully reseeding this species in the arid zone.

It is also opportune to give some attention to those less palatable, sometimes inert, perennial grasses that have already been identified as playing an essential stability role in degraded grassland ecosystems. In particular, more study should be devoted to *Andropogon kelleri*, *Cymbopogon schoeanthus* var. *proximus*, *Dactyloctenium robecchii*, *Pennisetum schimperi* and *Tetrapogon villosus* in the Sahel; to *Cymbopogon olivieri*, *C. parkeri* and *Tricholeana teneriffae* in the Near East; to *Chrysopogon gryllus*, *Ch. montanus* and *Cymbopogon montanus* in the dry hill tracts in the lower Himalayan region; and to *Achnatherum splendens* and *Calamagrostis squarrosa* in the north-east arid zone of China. Some of these grasses could prove valuable for erosion control on abandoned or very degraded cultivated land in the arid zone if they were planted in widely spaced strips that permitted cropping in between.

Although extensive forage plant collections have already been made in the arid zone, particularly in North Africa and the Near East, one region where natural evolution and hybridization may still be occurring is in the

Fig. 7.2. (*a*) Variation of the hairs at the base of the lower pales of flowers of different plants of *Poa bulbusa*. (*b*) Underground bulbils of *Poa* from different environments. (i) *P. sinaica* (Negev, about 100 mm annual rainfall); (ii) *P. bulbosa* (Jerusalem, about 500 mm annual rainfall). (Redrawn from *Bull. Res. Council of Israel*, 11D, 117–26 (1962).)

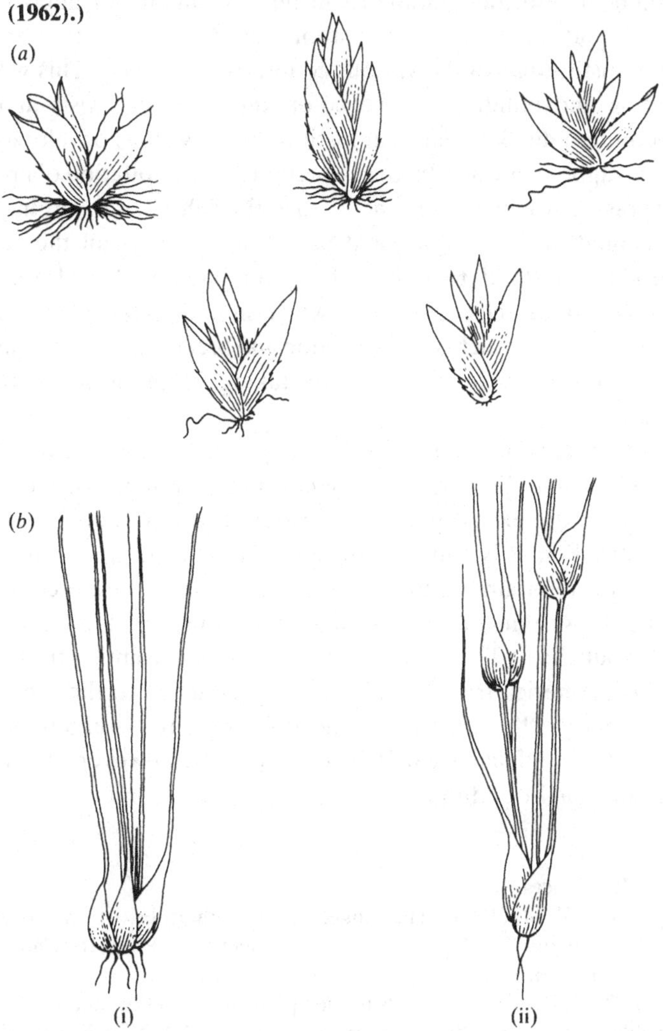

mountains of Baluchistan, and further north in the Karakoram mountains, which form the border between the dry Irano-Turanian climate and the characteristic monsoonal climate of Pakistan (Meher-Homji, 1963). Where annual and perennial temperate species of *Bromus*, *Poa* and *Stipa* meet the monsoonal grass species of *Cenchrus*, *Chrysopogon* and *Cymbopogon* (Kitamura, 1964), it could prove of great interest to make extensive ecotypic collections which, when properly screened, could provide very valuable material for reseeding depleted rangelands in the Near East and elsewhere.

Finally, because of the vastness of the arid zone, most of the grass development priorities outlined will be difficult, if not impossible, to attain without maximum co-operation between the existing arid zone research institutions in the various countries concerned. This will mean expanding and building on the co-operation which exists at present between the individual national institutions as well as developing a co-operative approach whereby the collection and exploitation of important desert grasses in a large arid zone, such as the Sahara, is co-ordinated and implemented on a trans-national basis. In this way, all the countries concerned will stand to benefit from an integrated arid zone grass development research programme, which should, as far as is possible, be designed so that as much of the selection and breeding work is done *in situ* to avoid any possible genetic shift taking place in the germplasm collected.

For the genera mentioned in this chapter except *Cenchrus*, *Dichanthium*, *Hordeum*, *Poa* and *Pennisetum*, only the most meagre, formal genetic literature exists apart from scattered chromosome counts. Such genetic variation as is known is from botanists' collections, and systematic hybridization among these is often beyond the resources of many institutes to sustain or even seriously to contemplate. One possible answer would be to divert the attention of trained scientists from Europe and North America from cereals in overproduction to those colonizing grasses where, self-evidently, the need for research is urgent across an immense geographical area. It is not apparent, however, in what way such a shift might be funded.

References

Abouguendia, Z. M. (1985). The rangeland of northern Egypt, the potential for restoration and development. Tokten Project UNDP/ASRF: Cairo, Egypt. 35 pp., mimeo.

Bailey, E. T. (1967). Pasture and fodder plant introduction and establishment problems. Report to the Government of Jordan. FAO TA Report No. 2405, Rome. 9 pp., mimeo.

Baumer, M. (1961). Re-seeding trials in Kordofan. *The Sudan Journal of Veterinary Science and Husbandry*, **2**(1), 68–77.

Blaikrie, P. & Brookfield, H. H. (1987). *Land Degradation and Society*. Methuen: London. 296 pp.

Bogdan, A. V. (1977). *Tropical Pasture and Fodder Plants* (grasses and legumes). Longmans: London & New York.

Booth, A., David, C. C., Syarifuddin Baharsyah, Meyanathan, S., Sivalingam, G., Chan, F., Dow Mongkolsmai, Bennet, A. G. & Mauldon, R. G. (1986). *Food Trade and Food Security in Asean and Australia.* Asean–Australia Joint Research Project: Kuala Lumpur & Canberra. 269 pp.

Bor, N. L. & Guest, E. (1968). *Flora of Iraq. Vol. 9: Gramineae.* Baghdad, Ministry of Agriculture: Republic of Iraq. 587 pp.

Bosch, O. J. H. (1989). Degradation of the semi-arid grasslands of southern Africa. *Journal of Arid Environments*, **16**(2), 165–75.

Bremen, I. H. & Uithol, P. N. J. (1984). *The Primary Productivity of the Sahel (PPS) Project: a Bird's Eye View.* Centre for Agrobiological Research: Wageningen, The Netherlands.

Chakravarty, A. K. (1970). Forage production from arid deserts. *Indian Farming*, December, 15–17.

Chakravarty, A. K. & Kackar, M. L. (1970). Selection of grasses and legumes for pastures in the arid zone. 1. Variation of morphological and physiological characters in *Lasiurus scindicus* Henr. *Indian Forester*, **96**(6), 433–6.

Chapman, G. P. (1988). An approach to desert containment and retrieval. *Biologist*, **35**(4), 217–20.

Christie, E. K. (1975). A study of phosphorus nutrition and water supply on the early growth and survival of buffel grass on sandy red earth from south-west Queensland. *Australian Journal of Experimental Agriculture and Animal Husbandry*, **15**, 239–49.

Clayton, W. D. (1972). *Graminae in French West Tropical Africa* (2nd edn), Vol. 2, Part 2. London Crown Agents for Overseas Governments and Administrations.

Clayton, W. D. & Renvoize, S. A. (1986). *Genera Graminum: grasses of the world.* Kew Bull. Addit. Ser. 13. Royal Botanic Gardens, Kew: London.

Coras, G. (1960). Performance of Weeping Lovegrass on the semi-arid Argentine Pampa. *Proceedings VIII International Grassland Congress*, pp. 231–3.

Dabadghao, P. M. (1960). Types of grass cover in India and their management. *Proceedings XIII International Grassland Congress*, pp. 226–30.

Dover, M. & Talbot, L. M. (1987). *To Feed the Earth: agro-ecology for sustainable development.* World Resources Institute: Washington DC.

Draz, O. (1974). *Range Management and Fodder Development.* Draft report to the Government of Syria, FAO/SYR/68/011. 74 pp.

Dujardin, M. & Hanna, W. W. (1983). Apomictic and sexual pearl millet × *Pennisetum squamulatum* hybrids. *The Journal of Heredity*, **74**, 277–9.

El Moghraby, A. I. & Ali, O. M. M. (1987). Desertification in western Sudan and strategies for rehabilitation. *Environmental Conservation*, **14**(3) autumn, 227–31.

FAO (1970). *Pasture and Fodder Crop Investigations, Iran.* Technical Report 1. AGP: SF/IRA. UNDP/FAO: Rome. 94 pp.

(1975). *Range Management in Sind, Pakistan: project findings and recommendations.* AG DP/PAK/71/001. Terminal report. 58 pp.

Fatahallah, M. M. (1976). *Range Management and Fodder Crops in the Northern Areas.* Report to the People's Democratic Republic of Yemen. UNOTC PDY/75/R40. FAO: Rome. 98 pp.

Foury, A. (1956). Les plantes fourragères, les plus recommendables au Maroc

dans le bassin méditerranean. *Cahiers Recherche Agronomique, Rabat, Maroc*, **7**, 47 pp.
French, N. H. (1968). Grass seeding in the Thal, West Pakistan. *Annals Arid Zone, Jodhpur, India*, **7**(2), 221–9.
Gallagher, R. (1972). *Fodder Crops and Pasture Lands in Algeria.* AGP: SF/ALG16. FAO: Rome.
Gebrehiwot, L. & Tadesse, A. (1985). Pasture research and development in Ethiopia. In *Pasture Improvement Research in Eastern and Southern Africa*, ed. J. A. Kategile, pp. 77–91. Proceedings Workshop: Harare, Zimbabwe.
Goudie, A. (1981, 1986). *The Human Impact on the Natural Environment* (1st & 2nd edn). Basil Blackwell: Oxford. 338 pp.
Guest, E. (1966). *Flora of Iraq. Vol. 1: Introduction.* Baghdad, Ministry of Agriculture: Republic of Iraq. 213 pp.
Harlan, J. R. (1983). The scope for collection and improvement of forage plants. In *Genetic Resources of Forage Plants*, ed. J. G. McIvor & R. A. Bray, pp. 3–14. CSIRO: Melbourne, Australia.
Hartley, W. (1964). The distribution of the grasses. In *Grasses and Grasslands*, ed. C. Barnard, p. 29. Macmillan: London.
Hemming, C. F. (1966). The vegetation of the northern region of the Somali Republic. *Proceedings of the Linnean Society, London*, **177**, 173–270.
(1973). *An Ecological Classification of the Vegetation of the Bosaso Region.* AGP: SF/SOM/70/512, working paper. FAO: Rome. 67 pp.
Heyn, C. C. (1962). Studies on bulbous *Poa* in Palestine. 1. The agamic complex of *Poa bulbosa. Bulletin Research Council, Israel D*, **11**(2), 117–26.
Hussain, I. (1975). *Range Management in Northern Iraq.* FO:D8/IRQ 168/518 Technical Report 3. FAO: Rome. 98 pp.
Ibrahim, K. N. (1968). Some important native forage grasses in northern Egypt: their value in range management. *Pakistan Journal of Forestry*, **1**, 57–74.
Ibrahim, K. N. & Kabuye, C. H. S. (1987). *An Illustrated Manual of Kenya Grasses.* FAO: Rome. 765 pp.
International Board for Plant Genetic Resources & Royal Botanic Gardens, Kew (1984). *Forage and Browse Plants for Arid and Semi-arid Africa.* IBPGR Crop Genetic Resources Centre, AGP, FAO: Rome. 293 pp.
Kassas, M. A. F. (1988). Ecology and management of desertification. In *Changing Geographic Perspectives*, ed. H. J. de Blij, pp. 195–211. *Proceedings Centennial Symposium National Geographic Society, Washington DC.*
Kernick, M. D. (1966). *Plant Resources, Range Ecology and Fodder Plant Introduction.* FAO TA Report No. 2181. FAO: Rome. 95 pp.
(1967). Results of tests on dryland forage crops at Homand Station 1960–66. *Pasture and Fodder Crops Investigation Project IRA/10.* Ministry of Natural Resources, Tehran, Technical Progress Report No. 3. 57 pp.
(1978). *Indigenous Arid and Semi-arid Forage Plants of North Africa, the Near and Middle East.* EMASAR Phase II, Volume IV, FAO/UNEP Technical Data. FAO: Rome. 689 pp.
(1979). *Rangelands and Their Use in the Near East Region.* ICARDA Forage Training Course, Aleppo, Syria. 32 pp.
(1985). Development of village grazing perimeters by seeding pasture species and implementing proper grazing management. *FAO Expert Consultation*

on Rangeland Rehabilitation and Development in the Near East. FAO: Rome, October, AGP 810. 14 pp., mimeo.

(1986). Forage plants for salt-affected areas in developing countries. *Reclamation and Revegetation Research*, **5**, 451–9.

(1989). Pasture conditions and development on the northern fringes of the Sahara. In *Changing Sahara*, ed. P. Darling. Earthscan: London (in press).

Khan, C. M. A. (1970). Range management – a challenge in West Pakistan. *Journal of Forestry*, **2**(1), 329–50.

(1971). Rainfall pattern and monthly forage yields in the Thal ranges of Pakistan. *Journal of Range Management*, **24**, 66–70.

Kitamura, S. (1964). *Plants of West Pakistan and Afghanistan: results of the Kyoto University Scientific Expedition to the Karokoram and Hindukush 1955. Vol. II.* Kyoto University: Japan. 283 pp.

Knight, R. (1983). Mediterranean and temperate grasses. In *Genetic Resources of Forage Plants*, ed. J. G. McIvor & R. A. Bray, pp. 47–61. CSIRO: Melbourne, Australia.

Lamb, H. H. (1988). *Weather, Climate and Human Affairs: a book of essays and other papers.* Routledge: London. 354 pp.

Lawrence P. (1986). *World Recession and the Food Crisis in Africa.* James Currey Ltd. 314 pp.

Le Houerou, H. N. (1969). La vegetation de la Tunisie Steppique (avec reference à l'Algerie à la Libye et au Maroc). *Recherches Physiogonomiques, Ecologiques, Sociologique et Dynamique.* Institut National de la Recherche Agronomique: Tunis. 622 pp.

(1970). North Africa: past, present and future. In *Arid Lands in Transition*, ed. H. E. Dregne, pp. 227–78. Symposium XIII, American Association for the Advancement of Science: Washington, DC. 524 pp.

(1980). The rangelands of the Sahel. *Journal of Range Management*, **33**(1) January, 41–5.

(1985). Forage and fuel plants of North Africa and the Near and Middle East. In *Plants for Arid Lands*, ed. W. E. Wickens, J. R. Goodin & D. N. Field, pp. 117–41. Proceedings of Kew International Conference on Economic Plants for Arid Lands, Royal Botanic Gardens, Kew, 23–7 July 1984.

Lichfield, J. (1989). Rethinking the politics of plenty. *Independent* (newspaper), London, 29 January 1989.

Li Chonghao, Zheng Xuanfeng, Zhao Kuiyi & Ye Juxin (1981). Basic types of pasture vegetation in the Songnen Plain. *Proceedings XIV International Grassland Congress, Kentucky, USA*, pp. 432–4.

Luk, S. H. (1983). Recent trends of desertification in the Maowusu Desert, China. *Environmental Conservation*, **10**(3) autumn, 213–24.

Lu Lian (1985). Steppe vegetation resources on the plateau of Hebei Province and the direction of their utilization. *Proceedings XV International Grassland Congress, Kyoto, Japan*, pp. 554–5.

Mabbutt, J. A. (1984). A new global assessment of the status and trends of desertification. *Environmental Conservation*, **2**(2) summer, 103–13.

McWilliam, J. R. & Gibbon, C. V. (1981). Selecting for seed retention in *Phalaris aquatica. Proceedings XIV International Grassland Congress, Kentucky.*

Ma Zhi Guang (1985). The results in loosening the soil to improve fringed

sagebush grassland in Inner Mongolia. *Proceedings XV International Grassland Congress, Kyoto, Japan*, pp. 615–16.
Meher-Homji, V. M. (1963). Les bioclimats du sub-continent Indiens et leurs types analogues dans la monde. *Travaux de la Section Scientifique et Technique Institut Français de Pondichery*, **7**, 254 pp.
Mensching, H. G. (1988). Land degradation and desertification in the Sahelian zone. In *Arid Lands – today and tomorrow*, ed. E. E. Whitehead, C. F. Hutchinson, B. T. Timmermann & R. G. Varady, pp. 605–13. Proceedings of an International Research and Development Conference, Tucson, Arizona.
Migahid, A. & El Shourbagui, M. N. (1961). The ecological amplitude of the desert fodder grass *Panicum turgidum*. III Transplantation of *Panicum turgidum* in Ras El Hekma and Fuka. *Bulletin Institut du Desert d'Egypte*, **8**(1), 68–98.
Mortimore, M. J. (1989). *Adapting to Drought: farmers, famine and desertification in West Africa*. Cambridge University Press. 299 pp.
Naegele, A. (1977). *Plantes Fourragères Spontanées de l'Afrique tropicale Seche*. EMASAR Phase II, Vol. 3, Données Techniques. FAO: Rome, 509 pp.
Nelson, R. (1988). *Dryland Management: the 'desertification' problem*. Environment Department Working Paper No. 8. World Bank: Washington DC.
Nemati, N. (1986). Pasture improvement in Iran. *Journal of Arid Environment*, **1**, 27–35.
Nevo, E. (1981). Microgeographic differentiation in allozyme polymorphisms of wild barley (*Hordeum spontaneum*), Poaceae. *Plant Systematic Evolution*, **138**, 287–92.
Nicholson, S. E. (1986). Climate, drought and famine in Africa. *Journal of Climate and Applied Meteorology*, **24**, 1388–99.
Numata, M. (1979). Distribution of grasses and grasslands in Asia. In *Ecology of Grasslands and Bamboolands in the World*, ed. M. Numata. Junk: The Hague.
Olsson, K. (1984). *Long-term Changes in the Woody Vegetation in North Kordofan, the Sudan: study with the emphasis on* Acacia senegal. Linds Universitet Naturgeographiska Instituten in co-operation with the Institute of Environmental Studies, University of Khartoum.
Omar, B. (1985). *Les Parcours hors Forêts*. Ministère de l'Agriculture et de la Referme Agraire, Projet FAO/MOR/TCP/4802. FAO: Rome. 124 pp.
Omar, B. & Kabak, A. (1986). *Cymbopogon schoeanthus* (L) Spreng: germination and seedling development under different temperature and water potential conditions. *Proceedings 2nd International Rangeland Congress, Adelaide*.
Pabot, H. (1961). Interim report on the natural vegetation of the Khuzistan region and headwaters. IRA/TE/PC. RAO: Rome. 86 pp., mimeo.
(1967). Pasture development and range improvement through botanical and ecological studies. Report to the Government of Iran UNDP/FAO TA Rep. No. 2311, 129 pp., mimeo.
Pearce, C. K. (1968). A range, pasture and fodder crop programme for Iran: a problem analysis and working plan. IRA 10. Ministry of Natural Resources and FAO: Tehran. 54 pp., mimeo.

Penning de Vries, F. W. T. & Djitaye, M. A. (1982). *La Productivité des Paturages Saheliens: une étude des sols, des vegetation et de l'éxploitation de cette resource naturelle.* Centre de Recherche Agrobiologique: Wageningen, The Netherlands.

Peyre de Fabreque, M. B. (1989). Ecological studies in the Sahel. In *Changing Sahara*, ed. P. Darling. Earthscan: London (in press).

Pratt, D. (1984). Ecology and livestock. In *Livestock Development in Sub-Saharan Africa – constraints, prospects and policy*, ed. J. R. Simpson & J. V. Evangelou. Westview Press: Boulder, Colorado.

Puri, D. C. & Paliwal, M. K. (1976). Selection of promising grass strains for the Kota region. I. *Cenchrus ciliaris* L. *Annals Arid Zone, Jodhpur, India*, **16**(2), 85–8.

Rasmusson, E. M. (1987). Global climate change and variability: effects on drought and desertification in Africa. In *Droughts and Hunger in Africa: denying famine a future*, ed. M. H. Glantz. Cambridge University Press.

Rattray, J. M. (1960). *The Grass Cover of Africa.* FAO Agric. Study No. 49. FAO: Rome. 168 pp. with map.

Ren Jizhou, Hu Zizhui & Fu Yikun (1985). The ecological role of plant resources in the arid regions of China. In *Plants for Arid Lands*, ed. W. E. Wickens, J. R. Goodin & D. N. Field, pp. 277–87. Proceedings of Kew International Conference on Economic Plants for Arid Lands, Royal Botanic Gardens, Kew, 237 July 1984.

Rossetti, C. (1965). Ecological survey mission to West Africa: studies on the vegetation (1959 and 1961): discussions and conclusions. UNSF/DL/ES/5. FAO: Rome. 77 pp., mimeo.

Said, M. (1960). Development of rangelands in Quette-Kalat region of West Pakistan. *Proceedings VIII International Grassland Congress, Montevideo, Brazil*, pp. 220–3.

Sankary, M. N. (1979). Autoecology of *Stipa barbata* Desf. from the Syrian arid zone in comparison with several Mediterranean type arid zone grass species. *Journal of Arid Environments*, **2**(3) September, 251–62.

Saxena, S. K. (1977). Vegetation and its succession in the Indian desert. In *Desertification and its Control*, ed. R. S. Jaiswal, pp. 176–92. Indian Centre for Agricultural Research (ICAR): New Delhi.

Schoenenberger, A. (1982). Les groupements vegetaux de la zone Saharienne: écologie des espèces intéressants pour les dunes et des sols érosives. FOR: DP/MOR/781/017, Document de Travail No. 4. FAO: Rome, 81 pp.

Shankarnaryan, K. A. (1977). Impact of overgrazing on the grasslands. *Annals Arid Zone, Jodhpur, India*, **16**, 349–59.

(1985). Ecological degradation of the Thar desert and eco-regeneration. *Scientific Reviews on Arid Research, New Delhi, India*, **5**, 1–13.

Shenker, A. & Goumandakoye, M. (1987). Water-point rehabilitation and management in the pastoral zone of Mali, West Africa. *2nd International Desert Development Conference, Cairo, Egypt, January.* 15 pp., mimeo.

Shu Jiang, S. (1985). Research on grassland ecosystems in the Inner Mongolia Region of China and the Strategy on Utilization. *Proceedings XV International Grassland Congress, Kyoto, Japan*, pp. 766–8.

Sikora, I. (1980). *Perspectives et developpement de la production semences espèces fourragères a mileu rurale dans la zone Soudano-Sahelienne.* Projet Regionale au Mali, Niger et Burkina Fasso pour l'Amélioration et Production des Cultures Fourragères, Rapport Technique. FAO: Rome.

Skovlin, F. M. (1986). Long-term trends from selected range sites throughout Kenya. *Proceedings 2nd International Congress, Adelaide, South Australia*, pp. 537–8.

Songqiao, Z. (1988). Human impacts on China's arid lands: desertification or de-desertification. In *Arid Lands – today and tomorrow*, ed. E. E. Whitehead, C. F. Hutchinson, B. T. Timmermann & R. G. Varady, pp. 1127–35. Proceedings of an International Research and Development Conference, Tucson, Arizona.

Soriano, A. (1979). Desert grasslands and steppes of Patagonia and the Andean highlands. In *Ecology of Grasslands and Bamboolands in the World.* Junk: The Hague.

Sukopp, H. & Scholtz, H. (1965). *Poa bulbosa*, an archaeophyte in the flora of central Europe. *Flora, Jena (B)*, **157**(4), 494–526.

Todorov, A. V. (1985). Sahel: the changing rainfall regimes and the 'normals' used for assessment. *Journal of Climate and Applied Meteorology*, **24**(2), 97–107.

Tolba, M. K. (1986). Desertification in Africa. *Land Use Policy*, **3**, 260–8.

Toutain, B. (1986). Recent vegetation changes and degradation in some Sahelian pastoral ecosystems of western Africa. *Proceedings 2nd International Rangeland Congress, Adelaide, South Australia*, pp. 73–4.

UNESCO/UNEP/UNDP (1980). Desertification in the Oglat Merteba Region, Tunisia: case study presented by the Government of Tunisia. In *Natural Resources Research XVIII Case Studies on Desertification.* UNESCO: Paris. 279 pp.

Valenza, J. & Diallo, A. K. (1980). Towards an animal/tree/grass combination. In *Browse in Africa*, ed. H. H. Le Houerou, pp. 387–8. ILCA Study: Addis Ababa.

Vesey-Fitzgerald, D. F. (1957*a*). The vegetation of the sea coast north of Jeddahyh, Saudi Arabia. *Journal of Ecology*, **45**, 547–62.

(1957*b*). The vegetation of central and eastern Arabia. *Journal of Ecology*, **45**, 779–98.

Whyte, R. O. (1964). *Grassland and Fodder Resources of India.* ICAR: New Delhi.

(1968). *Grasslands of the Monsoon.* Faber & Faber: London. 325 pp.

Wickens, G. W. & White, L. P. (1979). Land use in the southern margins of the Sahara. In *Management of Semi-arid Ecosystems*, ed. B. H. Walker. Elsevier: Amsterdam.

Woldu, Z. (1985). Variation in grassland vegetation on the central plateau of Shewa, Ethiopia, in relation to edaphic factors and grazing conditions. *Dissertationes Botanicae Band 84.* J. Cramer: Vaduz. 114 pp.

World Bank (1985). *World Development Report, 1985.* Washington DC.

World Resources 1986, 1988 and 1988–9. *An Assessment of the Resource Base that Supports the Global Economy.* World Resources Institute and International Institute for Environment and Development in collaboration with UNEP. Basic Books Inc.: New York.

Zaroug, M. G. (1985). Rangelands of Sub-Saharan Africa: their present use and prospects for their improvement. FAO Workshop on livestock policy, range and feed utilization guidelines for drought prone African countries, Khartoum, Sudan, November. 17 pp., mimeo.

Zhao Ji, Zhou Giming & Fan Weihong (1985). Remote sensing analysis of the grassland degradation in the Hailar Area. *Proceedings XV International Grassland Congress, Kyoto, Japan*, pp. 717–18.

Zhu Ting-Cheng, Li Jiandong & Yang Dianchen (1981). A study of the ecology of Yang-cao (*Leymus chinensis*) grassland in northern China. *Proceedings XIV International Grassland Congress, Kentucky, USA*, pp. 429–31.

Zhu Ting-Cheng, Li Jiandong & Zu Yuangang (1985). Grassland resources and the future development of grassland farming in temperate China. *Proceedings XV International Grassland Congress, Kyoto, Japan*, pp. 33–8.

8

In vitro *technology*

P. A. Lazzeri, J. Kollmorgen and H. Lörz

In vitro multiplication and regeneration

Introduction

The major principles underlying the use of *in vitro* techniques, whether for fundamental studies on plant cell biology or for crop improvement, are that plant tissues may be maintained in culture under defined conditions, and that individual cells are totipotent and have the capacity to regenerate new plants. These properties make it possible to make at the cell level genetic manipulations which are not possible at the whole plant level and then to return to whole plants via *in vitro* regeneration. In addition to using *in vitro* culture to produce modified plants, there is also the possibility of using these techniques for multiplication purposes, to propagate desired individual plants.

Such applications of *in vitro* techniques are now well established for many herbaceous dicotyledonous species, but the Poaceae are generally more difficult to culture *in vitro* and the development of usable systems has been slower. In recent years, however, considerable progress has been made (Vasil, 1987; Lazzeri & Lörz, 1988) and there are good prospects for the use of grass cell cultures as experimental and practical tools.

In the following sections we will discuss the basic principles and practices, and the current status of grass tissue and cell culture.

General characteristics of grass in vitro *cultures*

In grasses *in vitro* cultures must be initiated from immature, meristematic tissues, as mature differentiated tissues can generally not easily be induced to re-enter cell division and proliferate (Wernicke & Milkovits, 1987*a*,*b*). Further, in order to induce proliferation in grass tissues potent synthetic auxins such as 2,4-D, Dicamba or Picloram are required (Wernicke, Gorst & Milkovits, 1986; Vasil, 1987). Exogenous auxin is also needed to maintain the growth of established cultures and is usually only removed to allow differentiation and regeneration to take place.

A number of different explants may be used as sources of meristematic tissue for culture initiation, including shoot meristems, immature inflorescences, leaf bases and immature embryos. Of these different explant types, embryos and inflorescences are usually the most responsive and are most frequently used.

In their early stages grass tissue cultures are typically heterogeneous, containing several different callus types with differing morphogenetic capacities. The fastest-growing callus types are often non-regenerative and tend to overgrow morphogenic callus so that at subculture visual selection is made for the desired tissue types.

Two pathways of regeneration are recognized in tissue cultures: somatic embryogenesis and organogenesis. In the former bipolar structures with morphology similar to zygotic embryos are produced; these somatic embryos germinate to form plantlets like seed embryos do, via outgrowth of shoot and root apices. In the organogenic pathway an apical meristem is formed superficially on callus tissues, and this meristem then grows out to form shoots. These shoots may later form adventitious roots at their bases, allowing whole plants to be recovered. Of the two regeneration pathways, somatic embryogenesis affords more opportunities for manipulation, as it is possible from embryogenic callus cultures to produce suspension cultures which contain large numbers of easily accessible, potentially regenerable cells (see below).

Organogenic cultures are frequently produced when the explants used for initiation already contain shoot meristems (e.g. seedling shoot apices, entire immature embryos). In such cases shoots produced most likely result from *in vitro* 'microtillering' (Dunstan *et al.*, 1979), although in other cases meristems may arise *de novo* from somatic tissues. The ontogeny of regeneration has important implications for the occurrence of genetic alterations (somaclonal variation) among regenerant plants (Benzion, Phillips & Rines, 1986; Lörz & Brown, 1986; Rajasekaran *et al.*, 1986) and therefore for applications of the culture system. Plants regenerated from *in vitro* multiplication of pre-existing meristems are expected to be genetically indentical, whereas a degree of variation can be expected among plants derived from somatic cells, whether they are regenerated via embryogenesis or organogenesis. It is suggested that plants produced from embryogenic cultures show lower frequencies of genetic variation than those regenerated from organogenic cultures (Vasil, 1987), but there are very few studies in which plants derived from the two regeneration pathways have been compared. In one such study Armstrong & Phillips (1988) compared plants regenerated from organogenic or friable, embryogenic maize cultures. They found marginally higher frequencies of abnormalities in plants from embryogenic cultures, although these plants were less frequently chimaeric.

Independent of explant source or regeneration pathway, a major factor influencing both regeneration capacity and the genetic fidelity of regenerants is culture age. It is almost universally observed that the morphogenic capacity of tissue cultures decreases with time and, in addition, that the frequency of variant plants among regenerants increases (Fukui, 1983; Müller *et al.*, 1990). This process may, to some extent, be offset by efficient visual selection for cell types with long-term regenerative capacity (Vasil, 1987), by manipulations of culture conditions (Kavi Kishor & Reddy, 1986) or even by cryopreservation of regenerative cultures (Shillito *et al.*, 1989), but at present it must be accepted that plant cell cultures are evolving systems and that there will be a gradual loss of regenerative competence over time.

Regeneration from complex tissue cultures

The induction of morphogenetic cultures and subsequent plant regeneration has now been reported for all of the major cereal and grass species (for detailed bibliographies see Bright & Jones, 1985; Vasil, 1986). While this gives cause for optimism for the general application of *in vitro* manipulation techniques to the Poaceae, there is considerable inter- and intra-specific variation in response in culture, and truly efficient and reproducible systems exist for relatively few members of the family.

Of the major cereal species, rice is considered to be most amenable *in vitro*, although indica varieties are usually less easy to work with than japonicas. After rice, millet, triticale, maize, rye, barley, oats and wheat have generally proved progressively more difficult to manipulate in culture. While considerable interspecific variation in tissue culture response of grasses could be expected due to their diversity of morphology and environmental adaption, intraspecific differences might be expected to be much more limited. However, in most grasses intraspecific variation in culture response is marked, even within crop species which have undergone intensive selection for similar agronomic characteristics. The basis for this variation in response *in vitro* is little understood, but a number of studies have shown there to be heritable genetic components, with either nuclear genes or cytoplasmic factors being involved (Tomes & Smith, 1985; Hodges *et al.*, 1986; Petolino & Thompson, 1987). The transfer of regenerative capacity via crossing has been demonstrated in a number of cases (Duncan *et al.*, 1985), but to date no specific genes controlling regeneration capacity have been identified.

Cell suspension culture

Cell suspensions have a number of features making them useful for physiological and biochemical studies and for *in vitro* manipulations. They are relatively homogeneous by comparison with 'solid' tissue cultures, they have faster growth rates, can more easily be handled in bulk and can provide large numbers of relatively uniform experimental units (cell aggregates). In addition, in grasses suspension cultures are a convenient source of protoplasts and, in most cases, are the only source of dividing protoplasts (see below).

For plant genetic manipulation morphogenic suspensions are necessary. However, in grasses friable callus which adapts easily to liquid medium is generally non-morphogenic and most early work on grass suspensions resulted in cultures composed of meristematic centres which continually sloughed-off vacuolated cells (King, 1980). Such cultures are almost invariably non-morphogenic, or are capable only of root production.

With the recognition that maintainable embryogenic callus cultures could be selected for in grass species (Vasil & Vasil, 1981*a*) came the possibility of establishing embryogenic suspension cultures. Such cultures, capable of regenerating plants, were first produced with *Pennisetum* and *Panicum* species in the early 1980s (Lu & Vasil, 1981; Vasil & Vasil, 1981*b*) and since that time embryogenic suspensions have been reported for some 15 grass species (see Table 8.1). While these reports demonstrate that most grasses have the potential to form embryogenic suspension cultures, it is only in a few species that the process is truly repeatable (Kyozuka, Mayasmi & Shimamoto, 1987; Dalton, 1988*a*,*b*; Shillito *et al.*, 1989), and even in these species suspension establishment is strongly genotype-dependent.

The major difficulty in the establishment of embryogenic grass suspensions is that in most species only a very specific callus type is suitable. This callus is typically fast-growing and friable and disperses easily on transfer to liquid, but also maintains morphogenetic capacity. This particular callus phenotype was first described in maize and denoted 'Type II' callus (Armstrong & Green, 1985), but has subsequently been recognized in a number of other species (Harris *et al.*, 1988; Kyozuka, Otoo & Shimamoto, 1988; Lazzeri & Lörz, 1989). Established embryogenic suspensions of different grass species share a number of characteristics: they are typically composed of relatively small aggregates (10–100 cells)

Table 8.1. *Regeneration of plants from cell suspensions of grasses*

Species	Level of regeneration	Reference
Bothriochloa ischaemum	Plants	Johnson & Worthington (1987)
Bromus inermis	Albino plantlets	Gamborg *et al.* (1970)
Dactylis glomerata	Plants	Gray *et al.* (1984)
Festuca arudinacea	Plants	Dalton (1988*a*)
Hordeum vulgare	Albino plantlets	Kott & Kasha (1984)
	Albino plantlets	Lührs & Lörz (1988)
	Plants	Lazzeri & Lörz (1989)
Lolium multiflorum	Plants	Jones & Dale (1982)
	Plants	Dalton (1988*a*)
Lolium perenne	Plants	Dalton (1988*b*)
Oryza sativa	Plants	Fujimura *et al.* (1985)
	Plants	Abe & Futsuhara (1986)
	Plants	Zimny & Lörz (1986)
Panicum maximum	Plants	Lu & Vasil (1981)
Pennisetum americanum	Plants	Vasil & Vasil (1981*a*)
Pennisetum purpureum	Plants	Vasil *et al.* (1983)
Polypogon fugax	Plants	Chen & Xia (1987)
Saccharum officinarum	Plants	Ho & Vasil (1983)
Triticosecale	Plants	Stolarz & Lörz (1986)
Triticum aestivum	Plantlets	Maddock (1987)
Zea mays	Plants	Green & Rhodes (1982)
	Plants	Vasil & Vasil (1986)
	Plants	Kamo & Hodges (1986)

of small (approximately 15–30 μm diameter) isodiametric, cytoplasm-rich cells, which contain visible starch grains. They have relatively fast growth rates, with doubling times around 2–7 days and are subcultured at intervals between 3 and 10 days. In contrast with embryogenic suspensions of dicotyledonous species, embryogenic grass suspensions do not usually produce embryos in the liquid medium, but must be plated on to solid medium for morphogenesis.

Applications of in vitro *regeneration/multiplication in grasses*

Although the majority of grasses of economic interest are normally propagated via seed, there are situations where vegetative multiplication techniques are advantageous, and in these cases *in vitro* culture systems may offer alternatives to conventional propagation methods (Dale & Webb, 1985).

For clonal propagation *in vitro* tiller multiplication (Dalton & Dale, 1985) is the most commonly used technique as this method ensures genetic uniformity among the plants produced. Other multiplication systems, such as embryogenic callus cultures (Chandler & Vasil, 1984; Rajasekaran *et al.*, 1986) or inflorescence culture (Lo, Chen & Ross, 1980; Dale & Dalton, 1983) may have the capacity to deliver large numbers of propagules, but, as with all cultures in which there is a callus phase before the organization of the regenerant's apical meristem, there is the possibility of genetic variation occurring, so these methods may not always be appropriate.

Grass germplasm is routinely maintained in seedbanks, although it is also possible to use *in vitro* cultures for germplasm storage. Genetic stocks may be kept as *in vitro* tiller cultures, possibly at lowered temperatures to reduce growth rates (Dale & Webb, 1985), or cultures may be cryopreserved, theoretically allowing indefinite storage (Withers, 1985).

Advantages of maintaining germplasm as slow-growing *in vitro* cultures are that material is immediately available for multiplication as required and that sterile, axenic cultures enjoy reduced quarantine restrictions and may easily be transported internationally. Additionally, techniques such as meristem culture may be used to eliminate viruses (Leu, 1972; Kartha, 1985) and the cultures subsequently maintained *in vitro* as virus-free stocks.

Production of haploids *in vitro*

Of the various techniques of *in vitro* plant manipulation haploid production has the most obvious and direct application in plant breeding as it allows the achievement of homozygosity in a much shorter time (Dunwell, 1986). As a result, much effort has been devoted to developing protocols for haploid production in the major crop species, including those of the Poaceae. A number of recent reviews have covered the subject (Wenzel & Foroughi-Wehr, 1984; Dunwell, 1985; Snape *et al.*, 1986).

The two main techniques of *in vitro* haploid production are those of anther and ovary culture. The principle behind both techniques is that when isolated from the parent plant and cultured under defined conditions *in vitro*, haploid gametophytic cells can be induced to proliferate and to regenerate into plants. In some species a high frequency of spontaneous chromosome doubling occurs during the proliferation process to yield homozygous 'dihaploid' individuals, whereas in others haploid plants are usually produced. Such plants are sterile and chromosome doubling must be induced at the plantlet stage with colchicine or other treatments to produce fertile dihaploid individuals.

Anther culture

Successful anther culture requires that microspores be directed from their normal pathway of development to a pathway producing callus tissue or embryos. For this process to occur the microspores must be at a specific developmental stage and specific cultural requirements must be fulfilled.

In general, microspores of poaceous species are less responsive *in vitro* than those of the solanaceous species with which much of the original anther culture experiments were made. However, much work has been done in grasses to define the parameters influencing microspore division and development and in several species yields of haploid plants are sufficient for the practical application of the technique (Dunwell, 1986; Snape *et al.*, 1986).

As in most *in vitro* regeneration processes, genotype has an overriding influence on cereal anther culture response (Foroughi-Wehr, Friedt & Wenzel, 1982; Deaton *et al.*, 1987; Petolino & Thompson, 1987). However, while the transfer of anther culture response via crossing is quite frequently performed (Petolino and Thompson, 1987) there is little understanding of the nature of the genes influencing performance in culture.

The physiological status of donor plants has great influence on anther culture response, and in cereals this is often the single most important factor (Kuhlmann & Foroughi-Wehr, 1989). Generally, donor plants must be protected from physiological stress, and treatments such as pesticide applications must be kept to a minimum. Conditions under which sufficient mature pollen is produced to achieve full seed-set may not be appropriate for the production of microspores able to respond *in vitro*.

The stage of pollen development is another critical component of the anther culture process. Developmental stage is usually estimated by cytological analysis of anthers selected from different positions in the maturity gradient within the ear. In grasses microspores are usually cultured at the uninucleate stage.

Following the selection of anthers at the correct developmental stage it is common for a pretreatment to be applied before culture. In grasses this usually entails a period of storage (4–28 days) at low temperature under humid conditions. The actual mechanism of the pretreatment in increasing the number of responding microspores is not known, but it may involve changes in the tapetal layer or some synchronization of the microspores in a more responsive state.

Media for anther culture have received much attention as this is a variable which can be manipulated and controlled relatively easily. The media commonly used are basically similar to those used for culture of somatic tissues, although the phytohormone levels, and particularly those of 2,4-D, tend to be lower. Carbohydrate supply is one aspect of current interest as dramatic improvement in barley anther culture yield has been achieved by the use of maltose instead of sucrose (Hunter, 1987).

Microspore culture

Regeneration from isolated microspore cultures is an attractive system from a number of standpoints: it would facilitate the study of development from a single cell to a whole plant without the influence of surrounding tissue, it could provide cells for transformation via methods such as microinjection or particle bombardment (see Genetic transformation below) and it could allow *in vitro* selection at the gamete level (Ye *et al.*, 1987; and below). Regeneration from isolated microspores has now been reported for several cereal species (Köhler & Wenzel, 1985; Datta & Wenzel, 1987; Cho & Zapata, 1988; Coumans *et al.*, 1989) but at present efficiency levels are generally lower than for anther cultures.

Ovary culture

For mainly technical reasons ovary culture as a means of haploid production has had less attention than anther culture. In cereals haploid plants have been produced from ovary cultures of barley, wheat, rice and maize (Yang & Zhou, 1982; Dunwell, 1985), but to date there is too little information to judge the real potential or limitations of the technique.

Variation among regenerants

The genetic stability of plants from anther or ovary cultures is obviously of importance, particularly when dihaploid plants are to be incorporated in breeding programmes. The phenomenon of variation in gamete-derived plants has been termed 'gametoclonal variation', and was recently reviewed by Morrison & Evans (1987).

In grasses the most frequently observed variation among microspore-derived plants is albinism. Since relatively few plants have been produced from ovary cultures, it is hard to judge their relative stability, but it appears that the ratio of green: albino plants may be higher than for anther cultures (Yang & Zhou, 1982). In wheat it has been shown that albino regenerants have deletions in their plastid genomes (Day & Ellis, 1984), but it is not known if this condition is pre-existing or culture-induced. Although albinism has previously been a serious problem in barley anther culture, improved culture media and methods have markedly increased the frequency of green plantlets produced (Hunter, 1987; Kuhlmann & Foroughi-Wehr, 1989) and similar improvements may be possible in other species.

Application of in vitro *haploid production*

The major barrier to the wider use of *in vitro*-produced haploids in plant breeding is that of low yield. In grass species there is the additional problem of albinism. Refinement of techniques is, however, steadily increasing culture productivity and in species such as rice and barley many breeding programmes now routinely use anther culture to bring lines to homozygosity. The question of genetic stability of culture-derived material is not yet fully resolved, but it appears that when high numbers of doubled haploid can be produced, lines with good or even superior agronomic performance may be selected (Morrison & Evans, 1988).

In vitro selection

World food supply is heavily reliant on the successful cultivation of many members of the Poaceae. With substantial population increases in many countries there has been an increasing requirement for cultivars that are high in yield potential, yield stability, disease resistance and nutritional value. For many decades conventional plant breeding methods have been very successful in meeting these requirements. This success has been due to a large extent to the ability of astute and experienced plant breeders to select improved genotypes *in vivo*.

Selection is the breeder's main tool. A successful breeder must know what to select, have material from which to select and have the facilities to grow and to test the selected material (Mayo, 1988). In traditional breeding programmes selection is for the most part based on field-grown plants, although some traits such as disease resistance may initially be evaluated under controlled environmental conditions.

In the last decade there have been major advances in *in vitro* culture of many economically important members of the Poaceae, many of which were originally recalcitrant to cell culture. The present situation with respect to *in vitro* culture of the more important species has been summarized earlier and the rapid progress in the past few years has been highlighted. This progress has led at times to heated debate on the likely effects of strategies such as *in vitro* selection on conventional methods. Some have speculated that the new methods will replace the existing ones, others that they will complement them, while others have predicted little, if any, effect.

Several recent reviews address the strengths, weaknesses and limitations of *in vitro* selection and its likely contribution to the development of improved crop cultivars (for example, Chaleff, 1983; Maliga, 1984; Bright, 1985; Wenzel, 1985; Daub, 1986; Duncan & Widholm, 1986). Some of the perceived advantages of *in vitro* selection are that a large population of cells can be screened for variants in a controlled, pest-free environment (Duncan & Widholm, 1986), that it provides a strategy to solve problems intractable to conventional methods (Daub, 1986), that it creates a high degree of genetic variability and that it provides an opportunity to exploit the simpler genomes of microspores and haploids (Wenzel, 1985).

Cell cultures for in vitro *selection*

Cell cultures that can be initiated from various members of the Poaceae are described in the first section of this chapter. Table 8.2 gives an indication of the relative merits of the various cultures from the standpoint of *in vitro* selection. From a pragmatic view, only those cultures that can yield fertile regenerants are of use and in this respect calli on solid media are attractive. However, such calli represent rather complex systems and it is difficult to achieve uniform exposure of them to a selective agent. With aggregates in suspension, uniformity of exposure is not such a problem but regeneration is more difficult. Protoplasts and

Table 8.2. *The relative merits of different generalized scheme of cells and cell cultures for* in vitro *selection*

Cells/cell culture	Regenerative capacity	Similarity to plants *in vivo*	Number of cells or cell units per test	Uniformity of exposure to selective agent
Calli on solid media	++++	++++	+	−
Aggregates in suspension	++	++	+++	++
Protoplasts	+	+	++++	++++
Microspores	+	−	++++	++++

microspores provide systems that start with single cells and, as stated by Wenzel (1985) for microspores, they help to circumvent problems of cross-feeding and uncontrolled interaction of cells.

There are at least three possible ways in which microspores can be used for *in vitro* selection. First, to challenge them with a selective agent (e.g. herbicide) during induction and from the resulting culture to produce doubled haploid plants. Swanson *et al.* (1988) used this method to produce a rapeseed plant (*Brassica napus*) tolerant to the herbicide chlorsulfuron. They did, however, use a mutagen treatment which is an additional option if there is inadequate variation in the natural population. Although regeneration of cereals and grasses from microspore-derived cultures is still generally difficult, recent successes indicate that this may soon cease to be a stricture. From a practical viewpoint, microspore-derived plants are attractive as they are homozygous for both recessive and dominant traits and can be exploited immediately in conventional plant breeding programmes. A second method in which microspores can be utilized for *in vitro* selection is to challenge them during *in vitro* maturation and to use the resulting mature pollen grains for *in situ* pollination. Benito Moreno *et al.* (1988) described *in situ* seed production in tobacco after pollination with microspores matured *in vitro*. A third strategy is to apply selection to pollen during a pregermination (incubation) phase and to use surviving cells for pollination. Hodgkin (1988) used this method to select for resistance in *B. napus* to a toxic extract from *Alternaria brassicola* (Schw.) Wilts.

Microspores have, as yet, rarely been used for deliberate *in vitro* selection in the cereals and grasses (Ye *et al.*, 1987) but it is clear that they have many characteristics that make them well suited to such investigations.

Resistance/tolerance to pathogens

Phytopathologists have invested considerable time and effort in the development of refined techniques for *in vitro* culture of many of the pathogens that attack the Poaceae. In many instances they have also obtained information on the factors – for example, toxins and enzymes – involved in pathogenesis. This knowledge, combined with the recent developments in cell culture of the Poaceae, provides a sound basis for *in vitro* studies involving plant–pathogen interactions.

Wenzel (1985) and Daub (1986) provided comprehensive reviews on the use of cell culture to select for resistance to plant pathogens. They summarized many earlier studies, evaluated methodologies and discussed future prospects. However, at the time of their reviews there were few studies involving the Poaceae and the generally unresponsive nature of many of the crop members of that group limited the application of *in vitro* selection.

Gengenbach & Green (1975) grew maize callus on media amended with the HmT toxin of *Drechslera maydis* (Nisikado) Subram, and Jain (= *Helminthosporium maydis* Nisikado and Miyake) and demonstrated that resistant clones retained that trait for at least 127 days. In an essential extension of that work Gengenbach, Green & Donovan (1977) showed that maize regenerants from toxin-resistant callus were resistant to the toxin and the disease (southern corn leaf blight) it causes. However, Brettell & Thomas (1980) showed that toxin-resistant regenerants could also be regenerated from callus which had not been challenged with the HmT toxin. A similar event was noted by Larkin & Scowcroft (1983) in studies with *Helminthosporium sacchari* (Van Breda de Haan) Butler, the cause of eyespot disease of sugarcane. Nevertheless, in both the experiments of Brettell & Thomas (1980) and Larkin & Scowcroft (1983) selection in the presence of toxin increased the frequency of resistant plants.

Victorin, a highly specific pathotoxin from *Helminthosporium victoriae*, only affects oat lines with the dominant nuclear allele *Vb*. Rines & Luke (1985) challenged oat calli with victorin and among 175 calli of heterozyous *Vbvb* cultures grown in the presence of the toxin, 16 calli (13 culture lines) produced surviving callus sectors or shoots. Nine culture lines produced toxin-insensitive plants, two had toxin insensitivity similar to the parent and two lines did not regenerate. Plants regenerated from calli not exposed to the toxin were all sensitive to it. This report therefore contrasts with the results obtained with *D. maydis* and *H. sacchari* where some unchallenged callus was resistant. Rines & Luke (1985) also showed that the insensitive reaction to victorin was heritable, although progeny had become susceptible to crown rust (*Puccinia coronata* Cda. var. 'avenae', Fraser and E. Led.). They therefore demonstrated the possibility for *in vitro* selection to contribute to new cultivars but also highlighted a potential problem of the system. Ling *et al.* (1985) attempted *in vitro* screening of rice germplasm for resistance to brown spot (caused by *Helminthosporium oryzae*). Their methodology differed from other workers in that they crushed rice callus into small pieces before shaking it in toxin-containing medium. This therefore ensured that a large number of cells came into direct contact with the selective agent. There is a concern in studies with calli grown on agar amended with toxin because some cells would tend to be insulated from the chemical. Although Ling *et al.* (1985) found that two plants from the toxin treatment were resistant to the chemical, one plant from a control treatment was also unaffected by inoculation with *H. oryzae*. They reported that there was segregation for resistance to brown spot in the second generation, indicative of a dominant mutation.

Pauly, Shane & Gengenbach (1987) attempted to select for resistance to the toxin syringomycin (produced by the bacterium *Pseudomonas syringae* pv *syringae*) in wheat. Callus growth was inhibited by partially purified extracts from a syringomycin^{+}-mutant but not by a syringomycin^{-}-mutant. This therefore provided strong evidence of a positive correlation between *in vivo* and *in vitro* responses to the toxin. Although 5 out of 88 progeny plants showed a somewhat resistant reaction when inoculated with the bacterium, it is not known if the resistance was transmitted to subsequent generations or if it would have occurred without selection.

When cell cultures are challenged with a pathogen *per se* or its metabolites, there is concern that the cultures could be so 'overwhelmed' by the selection pressure that any resistant cells or sectors might not be detected. For this reason, Chawla & Wenzel (1987) used a discontinuous method in parallel with a continuous method in their search for barley and wheat cultures resistant to purified culture filtrates of *Helminthosporium sativum* P. K. and B. In the continuous method there were four successive cycles on toxic medium, while in the discontinuous method there was a 'pause' on non-toxic medium after the second or third cycle. Overall, there were 6–17% surviving calli and regenerants from these which were generally insensitive to the pathogen *in vivo*, whereas unselected plants were susceptible. It is not known, however, if the resistance was inherited.

To date, the only published study where a pathogen *per se* has been used to select for disease resistance in a cereal cell culture is that of Lihua, Jianming & Xuefeng (1986). In a well-planned series of experiments they investigated *in vitro* selection for bacterial blight (caused by *Xanthomonas oryzae* Dowson) resistance in rice. By using known susceptible and resistant cultivars, they first showed that *in vivo* resistance and suceptibility were expressed by callus cultures. They then inoculated a susceptible cultivar with *X. oryzae* and after 40–50 days' incubation 'newly formed' calli emerged from the old bacteria-coated calli. Out of 365 inoculated calli, 63 showed 'sectional proliferation', and of 45 regenerants, 44 had resistance to the disease. This resistance was stable and inherited for three generations. However, when nine plants were regenerated from uninoculated (control calli), four had resistance to the disease. Thus, although cell culture *per se* generated resistant plants, the effect was enhanced by a selection process.

Many of the findings described above indicate that *in vitro* selection will be a valuable tool in the development of disease-resistant cultivars in the Poaceae. There is, however, a need to extend much of the work to determine whether 'selected traits' are inherited and if they are, whether they are in a useful form. That most of the work has involved toxins is perhaps understandable as they provide a simple system in which cultured cells can be challenged. However, toxins are produced by only a few major pathogens in the Poaceae. Furthermore, host-specific toxins are not good candidates for *in vitro* selections because resistance to them already exists in nature and this resistance can be incorporated by conventional means (Daub, 1986). The work of Lihua *et al.* (1986) gives confidence that resistance to other pathogens may also be selected in tissue cultures. Suspension and protoplast cultures and microspores are less complex systems than calli on solid media and may therefore be better candidates for *in vitro* selection. Gendloff, Scheffler & Somerville (1987) showed that protoplasts from oat cultivars resistant to *H. victoriae* were insensitive to victorin, whereas protoplasts from susceptible cultivars were sensitive. This therefore indicates that protoplasts could be used for *in vitro* selection to *H. victoriae*. However, before suspension and protoplast cultures of the grasses and cereals can be used effectively for this purpose there must be further refinements in plant regeneration techniques. The finding that the resistant plants are sometimes regenerated in the absence of selection underlines the importance of adequate control treatments and may detract in part from *in vitro* selection, but it provides yet another method for obtaining disease resistance.

In studies with the wheat take-all fungus *Gaeumannomyces graminis* Arx and Olivier var. *tritici* Walker., Kollmorgen (unpublished) allowed the fungus to colonize callus and then attempted to rescue the callus with a fungicide treatment. However, resistant sectors were not detected, possibly because the fungus was highly virulent. As an alternative, seed progeny of regenerants are now being screened as seedlings *in vitro* using a system which permits rapid assessment of large numbers of plants. This system does not rely on a correlation between reaction of the fungus to calli and to plants but it is reliant on a simple bioassay which permits rapid screening of seedlings.

Amino acid overproduction

Nelson (1969) stressed that serious attention should be given to increasing the content of limiting amino acids in cereals. Methods have now been developed to utilize selection *in vitro* to increase production of nutritionally important amino acids in species such as pearl millet, barley, rice and maize. This is difficult to achieve by conventional methods as, in contrast to many diseases, it is difficult to apply selection pressure to select *in vivo*.

The method used to increase amino acid production involves subjecting callus to toxic amino acid analogues in either the presence or absence of a mutagen. As a survival mechanism, the callus responds by producing the analogue's corresponding amino acid. Miao, Duncan & Widholm (1988) used this method to develop threonine- and tryptophan- overproducing maize variants. Their data show that resistance to lethal levels of equimolar *L*-lysine plus *L*-threonine results in threonine levels between two and nine times greater in leaves and kernels of resistant plants than in wild-type plants. They showed that inheritance was via a single dominant gene which should therefore be able to be exploited in conventional breeding programmes. Resistance to 5-methyl-DL-tryptophan resulted from resistant calli having levels of free tryptophan 133–161 times greater than wild-type tissues. In regenerants tryptophan levels were at least 2000 times greater in resistant than wild-type plants but the regenerants were sterile.

Boyes & Vasil (1987) used a similar procedure to select for lysine overproduction in pearl millet. When embryogenic calli were cultured on a selective medium containing the amino acid analogue of lysine, tolerant calli resulted, which upon regeneration gave plants with seven times as much lysine in their leaves as control plants. Unfortunately, inheritance of the trait was not studied because lack of seed-set in the self-pollinated regenerants prevented progeny analysis.

Wakasa & Widholm (1987) studied *in vitro* selection of rice calli against a tryptophan analogue. In a selected callus line resistance to the analogue was due to the accumulation of high levels of free tryptophan (87-fold) that was associated with an increase of free phenylalanine content (9-fold). Regenerants also contained elevated levels of these amino acids. The authors pointed out, however, that for the trait to be useful commercially, homozygous plants must be obtained.

Kueh & Bright (1981) challenged mature barley embryos from mutagenized seed with proline analogues in agar media. They obtained a mutant with a six-fold increase in soluble proline in the leaves of its selfed progeny. Resistance to the analogue was due to a single, partially dominant nuclear gene.

Salt tolerance

In arid and semi-arid regions of the world insufficient rainfall results in a reliance on irrigation and a concentration of salts (mainly sodium) in soils and water supplies that is high enough to impair plant growth (Epstein *et al.*, 1980).

Selection for salt tolerance *in vitro* has been attempted in many species but because of metabolic adaptation of cultured cells, non-regeneration of selected cell types, drastic phenotypic changes and non-expression of salt-tolerant phenotypes in regenerants the results have been generally disappointing (Bright, 1985). Selection of callus cultures resistant to sodium chloride has been reported for sorghum (Bhaskaren, Smith & Schertz, 1983), oats (Nabors, Kroskey & McHugh, 1982), rice (Reddy & Vaidyanath, 1986), pearl millet (Rangan & Vasil, 1983) and maize (Lupotto, Mongodi & Lusardi, 1988). However, none of these workers demonstrated transmission of salt tolerance to field-grown plants or progeny of the regenerants. Lupotto *et al.* did notice, however, that regenerants from NaCl-selected plants had much larger root systems than control plants and suggested that this may be associated with increased salt tolerance.

Herbicides

In present-day agriculture herbicides are essential for economic weed control. They have become indispensable to the successful cultivation of cereals where high fuel costs and land degradation have become important issues. Although many of the recently developed herbicides have desirable attributes such as low toxicity to humans and other animals, rapid degradation in soil and effectiveness at low doses, they often lack selectivity (de Greef *et al.*, 1989).

Cell cultures have been used to select for tolerance to a range of herbicides, for example glufosinate in alfalfa (Donn *et al.*, 1984), 2,4-dichlorphenoxyacetic acid in perennial white clover (Oswald, Smith & Phillips, 1977) and chlorsulphuron and sulphometuron in tobacco (Chaleff & Ray, 1984). There is, however, a paucity of published information on selection for herbicide tolerance in the Poaceae. Shaner & Anderson (1985) reported selection of maize cell lines expressing greater than 100-fold enhanced tolerance to the imidazolinones. This resistance was also expressed in regenerants and was inherited in a manner suggesting that it is controlled by a single dominant gene.

In this particular instance resistance expressed *in vitro* was also expressed *in vivo*. However, in preliminary studies Racchi *et al.* (1989) showed that maize cell cultures consisting of undifferentiated cells were much less sensitive to glyphosate than differentiated tissues. Thus, for that particular herbicide, cell selection *in vitro* may be inappropriate. Selection *in vitro* will be effective only where selective agents affect cell cultures and *in vivo* plants in a similar manner.

Although *in vitro* selection has been the subject of many research projects in the Poaceae and several successes have been reported, it has yet to be shown that it will make a significant contribution to the development of new cultivars. Many studies have shown that novel cell cultures can be selected *in vitro* but few have demonstrated transmission of useful traits to the progeny of regenerants. Research is now required to extend the knowledge already gained, to a point where traits selected in culture can be incorporated in new cultivars. There is, perhaps, little merit in showing for yet another system simply that resistance to a particular factor can be selected for *in vitro*.

Protoplasts and somatic cell genetics

Protoplast culture and regeneration

Culture and regeneration of protoplasts are seen as the basic prerequisites for the use of higher plant cells in somatic cell genetics and for direct DNA-mediated gene transfer. Mesophyll protoplasts are most commonly isolated and cultured in case of dicotyledonous species. However, culture and regeneration experiments with mesophyll protoplasts isolated from grasses have so far failed. It has been concluded that most (if not all) cereal mesophyll cells are irreversibly differentiated and have lost their totipotency and the capacity to undergo sustained cell divisions in culture. Protoplasts isolated from meristematic (and dividing) tissue are seen as an alternative, but experimentally this material is not easily accessible.

Attempts to isolate protoplasts from meristematic grass tissues are also problematical in that these tissues are present in only small amounts and are compact, yielding only a few protoplasts. Where adequate yields are obtained, these protoplasts seldom divide. Similar problems apply to protoplast isolation from callus cultures, although there are a few reports of the isolation of dividing protoplasts from callus (Imbrie-Milligan & Hodges, 1986; Kyozuka *et al.*, 1987). In these cases the cultures used were fast-growing and friable.

Instead of isolating protoplasts directly from the plant, *in vitro* cultures can also be used as a source for the preparation of grass protoplasts. As it is assumed that protoplasts express the same competence in respect to regeneration as the cells from which they have been isolated, embryogenic suspensions are therefore the most promising material.

Since the mid-1970s there have been numerous reports of isolation and regeneration to callus of protoplasts from non-morphogenic suspensions (Koblitz, 1976; Chourey & Zuwarski, 1981), but in all cases the protoplast-derived calli remained non-morphogenic. During the process of establishing a cell culture suitable for the isolation of protoplasts, the original embryogenic cultures frequently lose their embryogenic capacity.

Meanwhile, several more stable embryogenic suspensions have become available and they have provided sources of fast-dividing, competent cells for protoplast isolation. Somatic embryos and plantlets were obtained from such isolations in *Pennisetum* and *Panicum* species (Vasil & Vasil, 1980; Lu, Vasil & Vasil, 1981), but these plantlets did not survive transfer to soil. A major breakthrough was the regeneration of rice protoplasts to green plants, reported by several laboratories over a short period (see Table 8.3). More recently, green plants have been obtained from sugarcane (Srinivasan & Vasil, 1986) and from maize. The first protoplast-derived maize plants turned out to be sterile as a consequence of somaclonal variation (Rhodes, Lowe & Ruby, 1988*a*), but subsequent publications have reported the regeneration of fertile plants from different maize genotypes (Prioli & Söndahl, 1989; Shillito *et al.*, 1989).

The principle common to all these reports of regeneration is the use of fast-growing embryogenic suspensions for protoplast isolation, suggesting that, as this type of culture is produced in other grass species, regeneration will be achieved.

Somatic hybridization

The technology of protoplast fusion and somatic hybridization holds considerable promise for new crop variety development. Any two protoplasts, no matter how distantly related, can be fused. Consequently, in principle at least, the barriers of incompatibility can be overcome without difficulty. In practice, however, regeneration capacity is totally absent or drastically reduced in very wide hybridizations. In such wide combinations selective elimination of one or the other parental chromosome type occurs rapidly. The problems of regeneration associated with wide somatic hybrids have stimulated researchers to concentrate recently on the combination of more closely related species.

Table 8.3. *Regeneration from protoplasts of grasses*

Species	Level of regeneration	Reference
Bromus inermis	Albino plantlets	Kao *et al.* (1973)
Dactylis glomerata	Plants	Horn *et al.* (1988)
Festuca arundinacea	Plants	Dalton (1988*a*,*b*)
Hordeum vulgare	Albino plantlets	Lührs & Lörz (1988)
	Green plantlets	Lazzeri & Lörz (1989)
Lolium perenne	Plants	Dalton (1988*a*,*b*)
Oryza sativa	Plants	Fujimura *et al.* (1985)
	Plants	Abdullah *et al.* (1986)
	Plants	Toriyama *et al.* (1986)
	Plants	Yamada *et al.* (1986)
Panicum maximum	Plantlets	Lu *et al.* (1981)
Panicum miliaceum	Albino plantlets	Heyser (1984)
Pennisetum americanum	Plantlets	Vasil & Vasil (1980)
Pennisetum purpureum	Plantlets	Vasil *et al.* (1983)
Poa pratensis	Albino plantlets	van der Valk *et al.* (1988)
Saccharum officinarum	Plants	Srinivasan & Vasil (1986)
Zea mays	Sterile plants	Rhodes *et al.* (1988*a*)
	Plants	Cai *et al.* (1988)
	Plants	Shillito *et al.* (1989)
	Plants	Prioli & Söndahl (1989)

Protoplasts isolated from grass species are essentially no different (apart from the previously mentioned regeneration problem) from protoplasts of any other species and thus can be used for fusion and genetic manipulation experiments. Some early reports have described grass/dicot heterokaryons (Kao & Michayluk, 1974; Dudits *et al.*, 1976) and also grass/grass heterokaryons (Brar *et al.*, 1980) but these fusion products either formed only microcalli, or one of the partners was eliminated after a few cell divisions. Recently, however, there have been reports of successful somatic interspecific hybridizations between grass species (Ozias-Akins, Ferl & Vasil, 1986; Tabaeizadeh, Ferl & Vasil, 1986). Ozias-Akins *et al.* (1986) fused amino-ethyl cysteine (AEC)-resistant *Pennisetum americanum* protoplasts, inactivated with iodoacetate, with *Panicum maximum* protoplasts, and selected for heterokaryons on AEC-containing medium. Somatic hybrids were verified by isozyme banding patterns and the use of a maize ribosomal DNA probe, which identified DNA sequences homologous to sequences from both parents in hybrid calli. These workers later analysed the mitochondrial genomes of three hybrid cell lines and found extensive rearrangements in all three. Restriction patterns showed both combinations of parental bands and also novel fragments not present in the parents (Ozias-Akins, Pring & Vasil, 1987). Tabaeizadeh *et al.* (1986) fused sugarcane protoplasts with the same AEC-resistant *P. americanum* line, using the same inactivation/selection scheme. Somatic hybrids were again verified by isozyme banding and the use of the maize rDNA probe. Subsequent analysis of the mitochondrial genomes of two *Saccharum/Pennisetum* hybrid cell lines showed one line to have mtDNA identical to that of the *Pennisetum* parent, while the other had a combination of parental mtDNAs, as well as novel bands suggesting mtDNA recombination (Tabaeizadeh, Pring & Vasil, 1987).

With the aim of crop improvement, protoplast fusion was used to obtain hybrid plants of rice (*Oryza sativa* L.) and four wild *Oryza* species, *O. officinalis*, *O. eichingeri*, *O. brachyantha* and *O. perrieri* to incorporate useful traits of the latter species into cultivated rice (Hayashi, Kyozuka & Shimamoto, 1989). The hybrid nature of the fused protoplast-derived plants was confirmed by karyotypic, morphological and isozyme analyses. The hybrids of *O. sativa* + *O. eichingeri*, *O. sativa* + *O. officinalis* and *O. sativa* + *O. perrieri* produced viable pollen and the *O. sativa* + *O. eichingeri* hybrid has produced a progeny plant. This material will be used now in the conventional breeding scheme for improvement of rice with agronomically useful traits such as disease resistance, insect tolerance and salt tolerance.

In a more distant combination rice protoplasts were fused with protoplasts of barnyard grass (*Echinochloa oryzicola*). Selection of hybrids was based on inactivation of rice protoplasts by iodoacetamide and the inability of barnyard grass protoplasts to divide (Terada *et al.*, 1987). Some shoots were obtained from hybrid calli and nine plantlets showed a morphology which was distinct from that of either parent. These studies demonstrate intergeneric nuclear genome combination in grasses and also indicate the opportunities for organellar genome fusion/recombination via protoplast fusion.

Asymmetric hybrids and limited gene transfer

In situations where the transfer of a few nuclear traits or of cytoplasmic traits is desired, asymmetric somatic hybridization (partial genome transfer) or cybridization (cytoplasm transfer) may be possible. In these techniques the nuclear genome of the donor species may be either inactivated or pulverized by irradiation, so that heterokaryons have only one fully functional nucleus, but two cytoplasms (Zelcer, Aviv & Galun, 1978), although fragments from the irradiated nucleus may be incorporated into the recipient genome. Recent studies, one involving tobacco and barley, and the other rice, illustrate partial genome transfer. Somers *et al.* (1986) complemented nitrate reductase-deficient tobacco mutants by the introduction of barley nitrate reductase genes from gamma-irradiated protoplasts. Yang *et al.* (1988, 1989) fused ^{60}Co-irradiated protoplasts of a cytoplasmic male-sterile rice line with iodoacetamide-treated protoplasts of the fertile (normal) rice cultivar 'Fujiminori'. The cybridity of the plants was confirmed by mitochondrial DNA analysis, plasmid-like DNA analysis, by isozyme analysis and by cytological and morphological investigations. The chromosome number of the cybrid plants was 24, the normal diploid chromosome number of rice.

Two potential methods for transferring only nuclear traits are the uptake of isolated nuclei or isolated chromosomes by protoplasts (Dudits & Praznovsky, 1985; Lörz, 1985). In dicot species the complementation of *Datura* auxotrophic mutants by the introduction of *Vicia* nuclei has been reported (Saxena *et al.*, 1986) but there have been few similar studies in grasses. Szabados, Hadlackzy & Dudits (1981) reported the uptake of wheat and parsley chromosomes by wheat and maize protoplasts, but their fate could be followed for only a few hours post uptake. The present availability of selectable marker genes and improving protoplast systems make gene transfer by the uptake of 'marked' nuclei or chromosomes increasingly feasible.

Many agronomically desired improvements, for example in yield, quality, disease resistance, stress tolerance or growth rate, involve several or many genes of presently unknown identity. Without detailed molecular knowledge of the genes involved, the transfer of such characteristics across sexual barriers can be approached with the methods mentioned above, namely non-recombinant DNA somatic cell fusion procedures (Cocking & Davey, 1987). It is therefore assumed that somatic cell genetics will, for quite some time, be an experimental tool for breeding of cereals and grasses, as the transfer of defined, isolated genes will be a part of future breeding programmes.

Genetic transformation

Different methods are presently under investigation to develop an efficient and widely applicable transformation technique for cereals and grasses and the topic has been reviewed extensively in recent years (Cocking & Davey, 1987; Göbel & Lörz, 1988; Lörz, Göbel & Brown, 1988; Ozias-Akins & Vasil, 1988).

The interest in this technology is for basic research, to study the function of genes, as well as for applied work, to develop the tools for plant breeders of the future to do breeding with defined, isolated genes. As in many other areas of plant biotechnology, transformation has been developed first with solanaceous species. Originally, *Agrobacterium tumefaciens* was the only vector available for plant transformation. As this natural vector system turned out to be limited in its host range and not suitable for transformation of grasses, numerous alternative transformation methods have been investigated over the last few years and will be discussed briefly in this section.

One major direction is the development of a protocol for plant regeneration from isolated protoplasts because protoplasts can be transformed efficiently, and they also allow the study of other aspects of somatic cell genetics. However, transfer of DNA to plants and cultured plant cells has been achieved now with several other methods, including microinjection, biolistic processes and pollen-mediated transformation among others, but the transformation efficiency and reproducibility when applied to different species are still limited. It can be foreseen, however, that the experimental difficulties will soon be overcome and the transformation technique will no longer be the limiting step for biotechnological modification of grasses.

Transformation of protoplasts

The incubation of freshly isolated protoplasts with naked DNA is a direct method of gene transfer. It was demonstrated first with tobacco protoplasts and subsequently also with protoplasts of different graminaceous species. DNA uptake is either stimulated by PEG-treatment, induced by electroporation, or brought about by treatments utilizing both stimuli in slightly modified techniques. Transformation of Gramineae was achieved first with protoplasts of *Triticum monoccocum* and *Lolium multiflorum*, and meanwhile also with protoplasts of *Zea mays*, *Oryza sativa*, *Pennisetum americanum*, *Panicum maximum* and *Hordeum vulgare* (Table 8.4). In all cases the plasmids used have contained a selectable marker gene (coding, for instance, for kanamycin or hygromycin resistance) and a constitutive promoter (*35S* gene of CaMV or nopaline synthase gene).

In respect of direct gene transfer cereal protoplasts are considered not to be different from protoplasts of other species. However, the limitation concerning grasses still remains the difficulty of regenerating plants from protoplasts. Whereas in maize and rice protoplast-derived and transformed plants have recently been obtained, in the other species mentioned DNA transfer to protoplasts has resulted in transformed callus

Table 8.4. *Transformation of grass protoplasts*

Species	Level of regeneration	Reference
Hordeum vulgare	Callus	Lazzeri *et al.* (1990)
Lolium multiflorum	Callus	Potrykus *et al.* (1985)
Oryza sativa	Plants	Zhang & Wu (1988)
	Plants	Shimamoto *et al.* (1989)
Panicum maximum	Callus	Hauptmann *et al.* (1988)
Panicum squamulatum	Callus	Hauptmann *et al.* (1988)
Triticum monococcum	Callus	Lörz *et al.* (1985)
Saccharum spp.	Callus	Chen *et al.* (1987)
Zea mays	Callus	Fromm *et al.* (1986)
	Plants	Rhodes *et al.* (1988*b*)

only. Further progress in plant regeneration from protoplasts should increase the number of transgenic cereal species in the future.

Direct gene transfer to intact cells and pollen

To circumvent the problems associated with the regeneration of protoplasts, more direct delivery systems for DNA transfer have been studied, primarily with cereals. These methods include micro- and macro-injection, microprojectiles, tissue imbibition and vector-mediated gene transfer.

Transformation via microinjection of DNA has been studied for several years and technical as well as applied aspects of this method have been reviewed recently (Schweiger *et al.*, 1987). The first successful transformation experiments applying microinjection of DNA were reported for tobacco and alfalfa protoplasts. Cereal protoplasts may also be injected using standard techniques, but very few cultures have sufficiently high division frequencies and they do not develop under the low density conditions used for post-injection culture.

A more attractive target than protoplasts for the injection of DNA are cells of small aggregates found in embryogenic suspensions. A problem with the injection of multicellular structures, however, is the fact that not all cells are accessible, resulting in a chimaeric tissue. This takes away one of the major attractions of microinjection, namely that individual cells can be injected and monitored without the need of applying a selection scheme.

Beside cells and aggregates from embryogenic suspensions, other competent cells such as pollen, microspores and microspore-derived embryoids are highly attractive for microinjection (Schweiger *et al.*, 1987). Anther culture of cereals has received much attention and for several species frequencies of callus and embryo formation have been improved significantly in the last few years. Wheat, barley and rice plants have been regenerated from isolated microspores, making them possible targets for microinjection. So far, however, no transformed plants have been obtained from this procedure.

Mature pollen grains are also being studied as recipient cells for microinjection (Kranz & Lörz, 1990). Pollen grains of maize are cultured *in vitro* to form a pollen tube or are placed on a stigma and thereafter are treated with a microcapillary for DNA delivery. It is anticipated that the foreign DNA will be 'transported' to the egg cell along with pollen tube growth and, in a so far unknown process, the DNA might be integrated during fertilization. This approach avoids the establishment of complex *in vitro* systems and attempts to make use of pollen as a natural vector system.

Another pollen-based transformation method is the incubation of pollen with total genomic DNA or recombinant DNA plasmids, followed by pollination and seed production. This procedure has been applied to *Petunia*, *Nicotiana*, maize (de Wet *et al.*, 1985; Ohta, 1986), wheat (Hess, 1987; Picard *et al.*, 1988) and rice (Luo & Wu, 1988). While in the earlier reports evidence for transformation was based mostly on phenotypic changes and formal genetic analyses, the recent publications have also included preliminary molecular evidence.

Biolistic transformation (Klein *et al.*, 1987) is a recently developed method which employs high-velocity microprojectiles to deliver substances into cells and tissues and is comparable to the injection methods. The new method has a great potential for transformation of intact cells and tissues, especially for species which are not transformable by other methods such as protoplast transformation or *Agrobacterium*-mediated gene transfer. Like the capillary microinjection, the particle gun method allows the delivery of DNA into cells with a cell wall. The number of possible target cells therefore is unlimited and theoretically all types of cells can be transformed. Several positive results for transformation applying the explosive-powered particle gun or a process of electric discharge to accelerate DNA-coated particles into cells have been published very recently (Table 8.5).

Suspension cells of different grass species and scutellum tissue of maize, wheat, barley and rice have been bombarded and expression of the introduced genes has been demonstrated by the activity of either neomycin phosphotransferase (NPT II), chloramphenicolacetyltrans-

Table 8.5. *Biolistic transformation experiments with grasses*

Species	Target tissue	Gene	Reference
Hordeum vulgare	Suspension cells	*GUS* *NPT*	Mendel *et al.* (1989)
Oryza sativa	Suspension cells	*GUS* *CAT*	Wang *et al.* (1988)
Triticum aestivum	Immature embryos Embryogenic callus	*GUS* *NPT*	Kiedrowski & Lörz (unpublished)
Triticum monoccocum	Suspension cells	*GUS* *CAT*	Wang *et al.* (1988)
Zea mays	Suspension cells	*NPT, GUS*	Klein *et al.* (1988)
	Zygotic embryos	*GUS*	Klein *et al.* (1988)

ferase (CAT) or glucoronidase (GUS). The major efforts now have to be directed to transform specific target cells, which are competent for plant regeneration and provide a high probability that the foreign genes are integrated also into the gametes and consequently are stably inherited. Such cells can be found in zygotic embryos, in young stages of floral tillers, in pollen grains, in haploid embryoids, in somatic embryos or any other tissue which can give rise to a fertile plant originating from a single, transformed cell.

Another transformation method which does not involve any tissue culture technique was developed for *Secale cereale* (de la Pena, Lörz & Schell, 1987). Plasmid DNA is injected into young floral tillers of rye at a specific stage during floral development. Although so far only a small number of transformants can be obtained and be confirmed by biochemical and molecular evidence, this approach provides clear-cut evidence that transformation of cereals is possible without using specific vectors, without using *in vitro* culture and independently of isolated single cells. Present limitations are seen mainly in the low efficiency of this method, and it remains to be seen whether this procedure can also be applied successfully to other species, such as wheat, barley, maize or rice.

A highly attractive and simple method of introducing genes directly into cereal cells has been demonstrated by incubating mechanically isolated wheat embryos in DNA solution (Töpfer *et al.*, 1989). Mature embryos isolated from dry seeds take up DNA rapidly by imbibition, and transient expression of chimaeric genes was demonstrated by assaying NPT- or CAT-activity in embryos of wheat, barley, rice, triticale, oats and maize. Embryos treated this way can efficiently and easily be regenerated to plants. Whether the foreign genes are stably integrated and transmitted to the progeny is presently under investigation. Instead of regenerating plants directly from the treated embryos, we have used the treated embryos of rice and durum wheat to establish embryogenic, scutellum-derived callus cultures. In a subsequent step these cultures were kept on a selective medium to select for transformed cells and eventually transformed regenerants. So far, however, this approach has not led to stably transformed *in vitro* cultures or plants.

Vector-mediated gene transfer

At present probably the most effective means of gene transfer into plants is infection of multicellular explants with *Agrobacterium*, followed by selection and regeneration of transformed plants, called 'leaf disc transformation' or modified versions thereof (Fraley, Rogers & Horsch, 1986). Most monocots, however, are considered to be outside the host range of agrobacteria and in a large-scale screening of monocotyledonous species only members of the order *Liliales* and *Arales* were found to be susceptible for *Agrobacterium* infection (De Cleene, 1985). Despite these indications of agrobacteria/monocot 'incompatibility', several reports have been published recently suggesting that this situation is not absolute. Treatments which activate infecting bacteria, i.e. which induce the virulence genes of the Ti-plasmid, or the identification of experimental conditions and specific target cells and tissues might increase artificially the sensitivity to infection. For several monocotyledonous species (Hooykaas & Schilperoort, 1987) *Agrobacterium tumefaciens*-mediated transformation has been described in the last few years. In grasses the only data so far reported are those on transformation of *Zea mays* with *A. tumefaciens*, and opine production in maize seedlings after incubation with *Agrobacterium* was taken as evidence (Graves & Goldman, 1986). This has subsequently been questioned as opines might be formed as a product of the arginine metabolism in non-transformed plant tissue (Christou, Platt & Ackerman, 1986).

Additional support for the usefulness of *Agrobacterium* as a vector for cereal transformation has been provided by so called 'agroinfection' experiments. In this work young maize seedlings were infected with an *A. tumefaciens* strain which carried a cloned dimer of maize streak virus (MSV) DNA in the T-DNA (Grimsley *et al.*, 1987). After a few weeks the treated plants showed typical symptoms of viral infection, whereas plants treated with viral DNA only or with wild-type *Agrobacterium* showed no symptoms and also were found 'negative' in subsequent DNA analysis for replication of MSV. Similar results have been obtained also after agroinfection of wheat with agrobacteria carrying recombinant plasmids with dimers of wheat dwarf virus (WDV) DNA (Woolston *et al.*, 1988).

Even after intensive research on *Agrobacterium*-mediated transformation of grasses in recent years, to date no convincing molecular evidence has been provided that T-DNA of *Agrobacterium* is stably integrated and in consequence stably transmitted to progeny, as has been shown in numerous dicotyledonous species.

Summary

In vitro culture and plant regeneration originating from multicellular explants has been achieved with numerous grass species and can be described as a routine procedure, even though genotypic differences in response are frequently observed. Anther and microspore culture are used intensively for applied purposes of haploid/homozygous plant production. Young inflorescences, immature and mature embryos, and to a lesser extent bases of leaves, nodes, shoot primordia and root tips are used as explants for the initiation of morphogenic callus and suspension cultures. Plant regeneration occurs via the pathways of shoot–root organogenesis or somatic embryogenesis. Special emphasis is put into the experimental development and biochemical and molecular analysis of somatic embryogenesis, as this pathway resembles closely the natural development of a zygote and evidence is available that somatic embryos, and therefore plants originate from a single somatic cell cultured *in vitro*.

Although considerable progress has been made recently with culture and regeneration of grass protoplasts, including those from such important crop species as rice, maize and barley, this procedure is far from being generally applicable to different species and numerous genotypes. The most promising approach so far is, first, the development of embryogenic suspension cultures, and thereafter the isolation of protoplasts from such morphogenic cultures. It has been shown in several instances that protoplasts express the same competence of capacity for regeneration as the cells from which they have been isolated.

Different routes have been followed to transfer genes to cereal crops. Protoplasts have been used for direct gene transfer experiments, giving rise to stably transformed plants and cell lines, and also for somatic cell genetic studies aiming at somatic hybrids or cybrids. Additional gene transfer methods applicable to cereals are designed primarily to circumvent the present difficulties of plant regeneration from protoplasts. These approaches include the microinjection of DNA into cells, pollen, microspore-derived embryoids or somatic embryos, the injection of genetic material directly into plants, the application of high-velocity projectiles for DNA transfer, the use of pollen or embryos for DNA uptake, and the use of vector systems based on *Agrobacterium* infection.

References

Abdullah, R., Cocking, E. C. & Thompson, J. A. (1986). Efficient plant regeneration from rice protoplasts through somatic embryogenesis. *Bio/ Technology*, **4**, 1087–90.

Abe, T. & Futsuhara, Y. (1986). Efficient plant regeneration by somatic embryogenesis from root callus tissue of rice. *J. Plant Physiol.* **121**, 111–18.

Armstrong, C. L. & Green, C. E. (1985). Establishment and maintenance of friable, embryogenic maize callus and the involvement of *L*-proline. *Planta*, **164**, 207–14.

Armstrong, C. L. & Phillips, R. C. (1988). Genetic and cytogenetic variation in plants regenerated from organogenic and friable, embryogenic tissue cultures of maize. *Crop Sci.* **28**, 363–9.

Benito Moreno, R. M., Macke, F., Alwen, A. & Heberle-Bors, E. (1988). *In-situ* seed production after pollination with *in vitro* matured, isolated pollen. *Planta*, **176**, 145–8.

Benzion, G., Phillips, R. L. & Rines, H. W. (1986). Case histories of genetic variability *in vitro*; oats and maize. In *Cell Culture and Somatic Cell Genetics of Plants, vol. 3*, ed. I. K. Vasil, pp. 435–48. Academic Press: New York.

Bhaskaran, S., Smith, R. H. & Schertz, K. (1983). Sodium chloride tolerant callus of *Sorghum bicolor* L. Moench. *Z. Pflanzenphysiol.* **112**, 459–63.

Boyes, C. J. & Vasil, I. K. (1987). *In vitro* selection for tolerance to S-(2-aminoethyl)-*L*-cysteine and overproduction of lysine by embryogenic calli and regenerated plants of *Pennisetum americanum* (L.) K. Schum. *Plant Sci.* **50**, 195–203.

Brar, D. S., Rambold, S., Constabel, F. & Gamborg, O. L. (1980). Isolation, fusion and culture of *Sorghum* and corn protoplasts. *Z. Pflanzenphysiol.* **96**, 269–75.

Brettell, R. I. S. & Thomas, E. (1980). Reversion of Texas male-sterile maize in culture to give fertile, T-toxin resistant plants. *Theor. Appl. Gen.* **58**, 55–8.

Bright, S. W. J. (1985). Selection *in vitro*. In *Cereal Tissue and Cell Culture*, ed. S. W. J. Bright & M. G. K. Jones, pp. 231–60. Martinus Nijhoff/Dr W. Junk Publishers: Dordrecht, The Netherlands.

Bright, S. W. J. & Jones, M. G. K. (ed.) (1985). *Cereal Tissue and Cell Culture.* Martinus Nijhoff/Dr W. Junk Publishers: Dordrecht, The Netherlands.

Cai, Q., Kuo, C., Qian, Y., Jiong, R. & Zhou, Y. (1988). Somatic embryogenesis and plant regeneration from protoplasts of maize (*Zea mays* L.). In *Progress in Protoplast Research*, ed. K. J. Puite *et al.*, p. 120. Kluwer: Dordrecht, The Netherlands.

Chaleff, R. S. (1983). Isolation of agronomically useful mutants from plant cell cultures. *Science*, **219**, 616–82.

Chaleff, R. S. & Ray, T. B. (1984). Herbicide-resistant mutants from tobacco cell cultures. *Science*, **233**, 1148–51.

Chandler, S. F. & Vasil, I. K. (1984). Optimisation of plant regeneration from longterm embryogenic callus cultures of *Pennisetum purpureum* Schum. (Napier grass). *J. Plant Physiol.* **117**, 147–56.

Chawla, H. S. & Wenzel, G. (1987). *In vitro* selection of barley and wheat for resistance against *Helminthosporium sativum*. *Theor. Appl. Genet.* **74**, 841–5.

Chen, D. & Xia, C.-A. (1987). Mature plant regeneration from cultured protoplasts of *Polypogon fugax*. *Scientica Sinica*, **30**, 698–703.

Chen, W. H., Gartland, K. M. A., Davey, M. R., Sotak, R., Gartland, J. S., Mulligan, B. J., Power, J. B. & Cocking, E. C. (1987). Transformation of sugarcane protoplasts by direct uptake of a selectable chimaeric gene. *Plant Cell Rep.* **6**, 297–301.

Cho, M. S. & Zapata, F. J. (1988). Callus formation and plant regeneration in isolated pollen culture of rice (*O. sativa* L. cv. Taipei 309). *Plant Sci.* **58**, 239–44.

Chourey, P. S. & Zuwarski, D. B. (1981). Callus formation from protoplasts of a maize cell culture. *Theor. Appl. Genet.* **59**, 341–4.

Christou, P., Platt, S. G. & Ackerman, M. C. (1986). Opine synthesis in wild-type plant tissue. *Plant Physiol.* **82**, 218–21.

Cocking, E. C. & Davey, M. R. (1987). Gene transfer in cereals. *Science*, **236**, 1259–62.

Coumans, M. P., Sohota, S. & Swanson, E. B. (1989). Plant development from isolated microspores of *Zea mays* L. *Plant Cell Rep.* **7**, 618–21.

Dale, P. J. & Dalton, S. J. (1983). Immature inflorescence culture in *Lolium*, *Festuca*, *Phleum* and *Dactylis*. *Z. Pflanzenphysiol.* **111**, 39–45.

Dale, P. J. & Webb, K. J. (1985). Germplasm storage and micropropagation. In *Cereal Tissue and Cell Culture*, ed. S. W. J. Bright & M. G. K. Jones, pp. 79–96. Martinus Nijhoff/Dr W. Junk Publishers: Dordrecht, The Netherlands.

Dalton, S. J. (1988*a*). Plant regeneration from cell suspension protoplasts of *Festuca arudinaceae*, *Lolium perenne* and *Lolium multiflorum*. *Plant Cell Tiss. Org. Cult.* **12**, 137–40.

(1988*b*). Plant regeneration from cell suspension protoplasts of *Festuca arudinaceae* (tall fescue) and *Lolium perenne* (perennial rye grass). *J. Plant Physiol.* **132**, 170–5.

Dalton, S. J. & Dale, P. J. (1985). The application of *in vitro* tiller induction in *Lolium multiflorum*. *Euphytica*, **34**, 897–904.

Datta, S. K. & Wenzel, G. (1987). Isolated microspore derived plant formation via embryogenesis in *Triticum aestivum*. *Plant Sci.* **48**, 49–54.

Daub, M. E. (1986). Tissue culture and the selection of resistance to pathogens. *Ann. Rev. Phytopathol.* **24**, 159–86.

Day, A. & Ellis, T. H. N. (1984). Chloroplast deletions associated with plants regenerated from pollen: possible basis for maternal inheritance of chloroplasts. *Cell*, **39**, 359–68.

Deaton, W. R., Metz, S. G., Armstrong, T. A. & Mascia, P. N. (1987). Genetic analysis of the anther culture response of three spring wheat crosses. *Theor. Appl. Genet.* **74**, 334–8.

de Cleene, M. (1985). The susceptibility of monocotyledons to *Agrobacterium tumefaciens*. *Phytopath. Z.* **113**, 81–9.

de Greef, W., Delon, R., de Block, M., Leemans, J. & Bottermans, J. (1989).

Evaluation of herbicide resistance in transgenic crops under field conditions. *Bio/Technology*, **1**, 61–4.

de la Pena, A., Lörz, H. & Schell, J. (1987). Transgenic rye plants obtained by injecting DNA into floral tillers. *Nature*, **325**, 274–6.

de Wet, J. M. J., Bergquist, R. R., Harland, J. R., Brink, D. E., Cohen, C. E., Newell, C. A. & de Wet, A. E. (1985). Exogenous gene transfer in maize (*Zea mays*) using DNA-treated pollen. In *Experimental Manipulation of Ovule Tissues*, ed. G. P. Chapman, S. H. Mantell & R. W. Daniels, pp. 197–209. Longman: New York, London. 256 pp.

Donn, G., Tischner, E., Smith, J. A. & Goodman, H. M. (1984). Herbicide-resistant alfalfa cells: an example of gene amplification in plants. *J. Mol. Appl. Genet.* **2**, 621–35.

Dudits, D. & Praznovsky, T. (1985). Intergenic gene transfer by protoplast fusion and uptake of isolated chromosomes. In *Biotechnology in Plant Science: relevance to agriculture in the eighties*, ed. M. Zaitlin, P. Day & A. Hollaender, pp. 115–27. Academic Press: New York.

Dudits, D., Kao, K. N., Constabel, F. & Gamborg, O. L. (1976). Fusion of carrot and barley protoplasts and division of heterokaryocytes. *Can. J. Genet. Cytol.* **18**, 263–9.

Duncan, D. R. & Widholm, J. M. (1986). Cell selection for crop improvement. *Plant Breeding Rev.* **4**, 153–73.

Duncan, D. R., Williams, M. E., Zehr, B. E. & Widholm, J. M. (1985). The production of callus capable of plant regeneration from immature embryos of numerous *Zea mays* genotypes. *Planta*, **165**, 322–32.

Dunstan, D. I., Short, K. C., Ohaliwal, H. & Thomas, E. (1979). Further studies on plantlet production from cultured tissues of *Sorghum bicolor*. *Protoplasma*, **101**, 355–61.

Dunwell, J. M. (1985). Anther and ovary culture. In *Cereal Tissue and Cell Culture*, ed. S. W. J. Bright & M. G. K. Jones, pp. 1–44. Martinus Nijhoff/Dr W. Junk Publishers: Dordrecht, The Netherlands.

(1986). Pollen, ovule and embryos culture as tools in plant breeding. In *Plant Tissue Culture and its Agricultural Applications*, ed. L. A. Withers & P. G. Alderson, pp. 375–404. Butterworth: London.

Epstein, E., Norlyn, J. D., Rush, D. W., Kingsbury, R. W., Kelley, D. B., Cunningham, G. A. & Wrona, A. F. (1980). Saline culture of crops: a genetic approach. *Science*, **210**, 399–404.

Foroughi-Wehr, B., Friedt, W. & Wenzel, G. (1982). On the genetic improvement of haploid formation in *Hordeum vulgare*. *Theor. Appl. Genet.* **62**, 233–9.

Fraley, R. T., Rogers, S. G. & Horsch, R. B. (1986). Genetic transformation in higher plants. *CRC Crit. Rev. Plant Sci.* **4**, 1–46.

Fromm, M. E., Taylor, L. P. & Walbot, V. (1986). Stable transformation of maize after gene transfer by electroporation. *Nature*, **319**, 791–3.

Fujimura, T., Sakurai, M., Akagi, H., Negishi, T. & Hirose, A. (1985). Regeneration of rice plants from protoplasts. *Plant Tiss. Cult. Lett.* **2**, 74–5.

Fukui, K. (1983). Sequential occurrence of mutations in a growing rice callus. *Theor. Appl. Genet.* **65**, 225–30.

Gamborg, O. L., Constabel, F. & Miller, R. A. (1970). Embryogenesis and production of albino plants from cell cultures of *Bromus inermis*. *Planta*, **95**, 355–8.

Gendloff, E. H., Scheffler, R. P. & Somerville, S. C. (1987). An improved bioassay for victorin based on the use of oat protoplasts. *Physiol. Molec. Plant Path.* **31**, 421–7.

Gengenbach, B. D. & Green, C. E. (1975). Selection of T-cytoplasm maize callus cultures resistant to *Helminthosporium maydis* race T. Pathotoxin. *Crop Sci.* **15**, 645–9.

Gengenbach, B. G., Green, C. E. & Donovan, C. M. (1977). Inheritance of selected pathotoxin resistance in maize plants regenerated from callus cultures. *Proc. Nat. Acad. Sci.* **74**, 5113–17.

Göbel, E. & Lörz, H. (1988). Genetic manipulation of cereals. *Oxford Surv. Plant Mol. Cell Biol.* **5**, 1–22.

Graves, A. C. F. & Goldman, S. L. (1986). The transformation of *Zea mays* seedlings with *Agrobacterium tumefaciens*. Detection of T-DNA specific enzyme activities. *Plant Mol. Biol.* **7**, 43–50.

Gray, D. J., Conger, B. V. & Hanning, G. E. (1984). Somatic embryogenesis in suspension and suspension-derived callus cultures of *Dactylis glomerata*. *Protoplasma*, **122**, 196–202.

Green, C. E. & Rhodes, C. A. (1982). Plant regeneration in tissue cultures of maize. In *Maize for Biological Research*, ed. W. F. Sheridan, pp. 367–72. Plant Molecular Biology Association: Charlottesville, VA.

Grimsley, N., Hohn, T., Davies, J. W. & Hohn, B. (1987). Agrobacterium-mediated delivery of infectious maize streak virus into maize plants. *Nature*, **325**, 177–9.

Harris, R., Wright, M., Byrne, M., Varnum, J., Brightwell, B. & Schubert, K. (1988). Callus formation and plantlet regeneration from protoplasts derived from suspension cultures of wheat (*Triticum aestivum*). *Plant Cell Rep.* **7**, 337–40.

Hauptmann, R. M., Vasil, V., Ozias-Akins, P., Tabaeizadeh, Z., Rogers, S. G., Fraley, R. T., Horsch, R. B. & Vasil, I. K. (1988). Evaluation of selectable markers for obtaining stable transformants in the Gramineae. *Plant Physiol.* **86**, 602–6.

Hayashi, Y., Kyozuka, J. & Shimamoto, K. (1989). Hybrids of rice (*Oryza sativa* L.) and wild *Oryza* species obtained by cell fusion. *Mol. Gen. Genet.* **214**, 6–10.

Hess, D. (1987). Pollen based techniques in genetic manipulation. In *Pollen – cytology and development*, ed. K. L. Giles & J. Prakash, pp. 367–95. Academic Press: Orlando.

Heyser, J. W. (1984). Callus and shoot regeneration from protoplasts of proso millet (*Panicum miliaceum* L.). *Z. Pflanzenphysiol.* **113**, 293–9.

Ho, W. & Vasil, I. K. (1983). Somatic embryogenesis in sugarcane (*Saccharum officinarum* L.) II. The growth of and plant regeneration from embryogenic cell suspension cultures. *Ann. Bot.* **51**, 719–26.

Hodges, T. K., Kamo, K. K., Imbrie, C. W. & Becwar, M. R. (1986). Genotype specifity of somatic embryogenesis and regeneration in maize. *Bio/Technology*, **4**, 219–23.

Hodgkin, T. (1988). In vitro pollen selection in *Brassica napus* L. In *Sexual*

Reproduction in Higher Plants, ed. M. Cresti, P. Gori & E. Pacini, pp. 57–62. Springer-Verlag: New York, London, Paris, Tokyo.

Hooykaas, P. J. J. & Schilperoort, R. A. (1987). Detection of monocot transformation via *Agrobacterium tumefaciens*. *Methods in Enzymology*, **153**, 305–13.

Horn, M. E., Conger, B. V., & Harms, C. T. (1988). Plant regeneration from protoplasts of embryogenic suspension cultures of orchard grass (*Dactylis glomerata* L.). *Plant Cell Rep.* **7**, 371–4.

Hunter, C. P. (1987). 'Plant regeneration method'. European patent application, Number EP 0245898 A2.

Imbrie-Milligan, C. W. & Hodges, T. K. (1986). Microcallus formation from maize protoplasts prepared from embryogenic callus. *Planta*, **168**, 395–401.

Johnson, B. B. & Worthington, M. (1987). Established suspension cultures from seeds of Plains Bluestem and regeneration plants via somatic embryos. In vitro *Cell Dev. Biol.* **23**, 783–8.

Jones, M. G. K. & Dale, P. J. (1982). Reproducible regeneration of callus from suspension culture protoplasts of the grass *Lolium multiflorum*. *Z. Pflanzenphysiol.* **105**, 267–74.

Kamo, K. K. & Hodges, T. K. (1986). Establishment and characterisation of long-term embryogenic maize callus and cell suspension cultures. *Plant Sci.* **45**, 111–17.

Kao, K. N. & Michayluk, M. R. (1974). A method for high-frequency intergeneric fusion of plant protoplasts. *Planta*, **115**, 355–67.

Kao, K. N., Gamborg, O. L., Michayluk, M. R., Keller, W. A. & Miller, R. A. (1973). The effects of sugars and inorganic salts on cell regeneration and sustained division in plant protoplasts. *Colleg. Intern. CNRS*, **212**, 207–13.

Kartha, K. K. (1985). Elimination of viruses. In *Cell Cultures and Somatic Cell Genetics of Plants*, ed. I. K. Vasil, pp. 577–83. Academic Press: Orlando.

Kavi Kishor, P. B. & Reddy, G. M. (1986). Retention and revival of regeneration ability by osmotic adjustment in long-term cultures of four rice varieties. *J. Plant Physiol.* **126**, 49–54.

King, P. J. (1980). Cell proliferation and growth in suspension cultures. In *Perspectives in Plant Cell and Tissue Culture*, ed. I. K. Vasil. *Int. Rev. Cytol. Suppl.* **11A**, 25–53.

Klein, T. M., Gradziel, T., Fromm, M. E. & Sanford, J. C. (1988). Factors influencing gene delivery into *Zea mays* cells by high-velocity microprojectiles. *Bio/Technology*, **6**, 559–63.

Klein, T. M., Wolf, E. D., Wu, R. & Sanford, J. C. (1987). High-velocity microprojectiles for delivering nucleic acids into living cells. *Nature*, **327**, 70–3.

Koblitz, H. (1976). Isolierung und Kultivierung von Protoplasten aus Kalluskulturen der Gerste. *Biochem. Physiol. Pflanzen*, **170**, 287–93.

Köhler, F. & Wenzel, G. (1985). Regeneration of isolated barley microspores in conditioned media and trials to characterise the responsible factor. *J. Plant Physiol.* **121**, 181–91.

Kott, L. S. & Kasha, K. J. (1984). Initiation and morphological development of somatic embryoids from barley cell cultures. *Can. J. Bot.* **62**, 1245–9.

Kranz, E. & Lörz, H. (1990). In vitro germination, micromanipulation and pollination of single pollen grains of maize (*Zea mays* L.). *Sexual Plant Reprod.* **3** (in press).

Kueh, J. S. H. & Bright, S. W. S. (1981). Proline accumulation in a barley mutant resistant to trans-4-hydroxy-*L*-proline. *Planta*, **153**, 166–71.

Kuhlmann, K. & Foroughi-Wehr, B. (1989). Production of doubled haploid lines in frequencies sufficient for barley breeding programs. *Plant Cell Rep.* **8**, 78–81.

Kyozuka, J., Mayasmi, Y. & Shimamoto, K. (1987). High frequency of plant regeneration from rice protoplasts by novel nurse culture methods. *Mol. Gen. Genet.* **206**, 408–13.

Kyozuka, J., Otoo, E. & Shimamoto, K. (1988). Plant regeneration from protoplasts of indica rice: genotypic differences in culture response. *Theor. Appl. Genet.* **76**, 887–90.

Larkin, P. J. & Scowcroft, W. R. (1983). Somaclonal variation and eyespot toxin tolerance in sugarcane. *Plant Cell Tiss. & Org. Cult.* **2**, 111–21.

Lazzeri, P. A. & Lörz, H. (1988). *In vitro* genetic manipulation of cereals and grasses. *Adv. Cell Cult.* **6**, 291–325.

(1989). Regenerable suspension and protoplast cultures of barley, and stable transformation via DNA uptake into protoplasts. In *Genetic Engineering of Crop Plants* (49th Nottingham Easter School), ed. G. W. Lycett & D. Grierson, pp. 231–8. Butterworth Scientific: London (in press).

Lazzeri P., Lührs, R. & Lörz, H. (1990). Stable transformation of barley via PEG-induced DNA uptake into protoplasts (submitted).

Leu, L. S. (1972). Freeing sugarcane from Mosaic Virus by apical meristem culture and tissue culture. *Taiwan Sugar Experiment Station Report*, **57**, 57–63.

Lihua, S., Jianming, S. & Xuefeng, L. (1986). Selection of mutants of *Xanthomonas oryzae* by tissue culture in rice. 1. *In vitro* induction and screening of mutants resistant to *Xanthomonas oryzae* Dowson from callus culture in rice. *Acta Genetica Sinica*, **13**, 188–93.

Ling, D. H., Vidhyaseharan, P., Borromeo, E. S., Zapata, F. J. & Mew, T. W. (1985). In vitro screening of rice germplasm for resistance to brown spot disease using phytotoxin. *Theor. Appl. Genet.* **71**, 133–5.

Lo, P. F., Chen, C. H. & Ross, J. G. (1980). Vegetative propagation of temperate forage grasses through tissue culture. *Crop Sci.* **20**, 363–7.

Lörz, H. (1985). Isolated cell organelles and subprotoplasts: their roles in somatic cell genetics. In *Plant Genetic Engineering*, ed. J. H. Dodds, pp. 27–59. Cambridge University Press.

Lörz, H. & Brown, P. T. H. (1986). Variability in tissue culture derived plants – possible origins, drawbacks and advantages. In *Genetic Manipulation in Plant Breeding*, ed. W. Horn, C. J. Jensen, W. Odenbach & O. Schieder, pp. 513–34. Walter de Gruyter: Berlin.

Lörz, H., Baker, B. & Schell, J. (1985). Gene transfer to cereal cells mediated by protoplast transformation. *Mol. Gen. Genet.* **199**, 178–82.

Lörz, H., Göbel, E. & Brown, P. T. H. (1988). Advances in culture and progress towards genetic transformation of cereals. *Plant Breed.* **100**, 1–25.

Lu, C. & Vasil, I. K. (1981). Somatic embryogenesis and plant regeneration from suspended cells and cell groups of *Panicum maximum in vitro*. *Ann. Bot.* **47**, 543–8.

Lu, C. Y., Vasil, V. & Vasil, I. K. (1981). Isolation and culture of protoplasts of *Panicum maximum* (Guinea Grass): somatic embryogenesis and plant formation. *Z. Pflanzenphysiol.* **104**, 311–18.

Lührs, R. & Lörz, H. (1988). Initiation of morphogenic cell suspension and protoplast cultures of barley (*Hordeum vulgare* L.). *Planta*, **175**, 71–81.

Luo, Z. X. & Wu, R. (1988). A simple method for the transformation of rice via the pollen-tube pathway. *Plant Mol. Biol. Rep.* **6**, 165–74.

Lupotto, E., Mongodi, M. & Lusardi, M. C. (1988). Salt tolerance: in vitro selection with regard to the regenerative potential. *Maize Genetics Corporation Newsletter*, **62**, 30–1.

Maddock, S. E. (1987). Suspension and protoplast culture of hexaploid wheat (*Triticum aestivum*). *Plant Cell Rep.* **6**, 23–6.

Maliga, P. (1984). Isolation and characterisation of mutants in plant cell culture. *Ann. Rev. Plant Physiol.* **35**, 519–42.

Mayo, O. (1988). Conventional plant breeding and the new genetics. In *Plant Breeding and Genetic Engineering*, ed. A. H. Zakri, pp. 1–22. SABRAO (Malaysia).

Mendel, R. R., Müller, B., Schulze, J., Kolesniker, V. & Zelenin, A. (1989). Delivery of foreign DNA to intact barley cells by high-velocity microprojectiles. *Theor. Appl. Genet.* **78**, 31–4.

Miao, S., Duncan, D. R. & Widholm, J. (1988). Selection of regenerable maize callus cultures resistant to 5-methyl-DL-tryptophan, S-2-aminoethyl-*L*-cysteine and high levels of *L*-lysine plus *L*-threonine. *Plant Cell Tiss. & Org. Cult.* **14**, 3–14.

Morrison, R. A. & Evans, D. A. (1987). Gametoclonal variation. In *Plant Breeding Reviews*, Vol. 5, ed. J. Janick, pp. 359–91. AVI Press: Westport, Connecticut.

(1988). Haploid plants from tissue culture: new plant varieties in a shortened time frame. *Bio/Technology*, **6**, 684–90.

Müller, E., Brown, P. T. H., Hartke, S. & Lörz, H. (1990). DNA variation in tissue culture-derived rice plants. *Theor. Appl. Genet.* (in press).

Nabors, M. W., Kroskey, C. S. & McHugh, D. M. (1982). Greenspots are predictors of high callus growth rates and shoot formation in normal and in salt stressed tissue cultures of oat (*Avena sativa* L.). *Z. Pflanzenphysiol.* **105**, 341–9.

Nelson, O. E. (1969). Genetic modification of protein quality in plants. *Advan. Agron.* **21**, 171–94.

Ohta, Y. (1986). High-efficiency genetic transformation of maize by a mixture of pollen and exogenous DNA. *Proc. Nat. Acad. Sci.* **83**, 715–19.

Oswald, T. H., Smith, A. E. & Phillips, D. V. (1977). Herbicide-tolerance developed in cell suspension cultures of perennial white clover. *Can. J. Bot.* **55**, 1351–8.

Ozias-Akins, P., Ferl, R. J. & Vasil, I. K. (1986). Somatic hybridisation in the graminae: *Pennisetum americanum* L. K. Schum. (pearl millet) + *Panicum maximum* Jacq. (guinea grass). *Mol. Gen. Genet.* **203**, 365–70.

Ozias-Akins, P., Pring, D. R. & Vasil, I. K. (1987). Rearrangements in the mitochondrial genome of somatic hybrid cell lines of *Pennisetum americanum* L. K. Schum. + *Panicum maximum* Jacq. *Theor. Appl. Genet.* **74**, 15–20.

Ozias-Akins, P. & Vasil, I. K. (1988). *In vitro* regeneration and genetic manipulation of grasses. *Physiol. Plantarum*, 73, 565–9.

Pauly, M. H., Shane, W. W. & Gengenbach, B. G. (1987). Selection for bacterial blight phytotoxin resistance in wheat tissue culture. *Crop Sci.* **27**, 340–4.

Petolino, J. F. & Thompson, S. A. (1987). Genetic analysis of anther culture response in maize. *Theor. Appl. Genet.* **74**, 284–6.

Picard, E., Jacquemin, J. M., Granier, F., Bobin, M. & Forgeois, P. (1988). Genetic transformation of wheat (*Triticum aestivum*) by plasmid DNA uptake during pollen tube germination. *Proc. 7th Int. Wheat Genetics Symp.*, ed. T. E. Miller & R. M. D. Koebner, pp. 779–81. Bath Press: Bath.

Potrykus, I., Saul, M. W., Petruska, J. & Shillito, R. (1985). Direct gene transfer to cells of a graminaceous monocot. *Mol. Gen. Genet.* **199**, 181–8.

Prioli, L. M. & Söndahl, M. R. (1989). Plant regeneration and recovery of fertile plants from protoplasts of maize (*Zea mays* L.). *Bio/Technology*, **7**, 589–94.

Racchi, M. L., Forlani, G., Pelanda, R. & Nielsen, E. (1989). Glyphosate effects on growth and EPSP synthase isozymes activity in cultured maize cells. *Proceedings 12th EUCARPIA Congress*, Poster Abstracts, Abstract 26-11. 12th EUCARPIA Congress, Göttingen, February 1989.

Rajasekaran, K., Schank, S. C. & Vasil, I. K. (1986). Characterisation of biomass production and phenotypes of plants regenerated from embryogenic callus cultures of *Pennisetum americanum* × *P. purpureum* (hybrid napier grass). *Theor. Appl. Genet.* **73**, 4–10.

Rangan, T. S. & Vasil, I. K. (1983). Sodium chloride tolerant embryogenic cell lines of *Pennisetum americanum* L. K. Schum. *Ann. Bot.* **52**, 59–64.

Reddy, P. J. & Vaidyanath, K. (1986). In vitro characterisation of salt stress effects and the selection of salt tolerance plants in rice (*Oryza sativa* L.). *Theor. Appl. Genet.* **71**, 757–60.

Rhodes, C. A., Lowe, K. I. S. & Ruby, K. L. (1988*a*). Plant regeneration from protoplasts isolated from embryogenic maize cell cultures. *Bio/Technology*, **6**, 65–70.

Rhodes, C. A., Pierce, D. A., Mettler, I. J., Mascarenhas, D. & Detmar, J. J. (1988*b*). Genetically transformed maize plants from protoplasts. *Science*, **240**, 204–7.

Rines, H. W. & Luke, H. H. (1985). Selection and regeneration of toxin-insensitive plants from tissue cultures of oats (*Avena sativa*) susceptible to *Helminthosporium victoriae*. *Theor. Appl. Genet.* **71**, 16–21.

Saxena, P. K., Mii, M., Crosby, W. L., Fowke, L. C. & King, J. (1986). Transplantation of isolated nuclei into plant protoplasts. *Planta*, **168**, 29–35.

Schweiger, H.-G., Dirk, J., Koop, H.-U., Kranz, E., Neuhaus, G., Sangenberg, G. & Wolff, D. (1987). Individual selection, culture and manipulation of higher plant cells. *Theor. Appl. Genet.* **73**, 769–83.

Shaner, D. L. & Anderson, P. C. (1985). Mechanism of action of imidazolinones and cell culture selection of tolerant maize. In *Biotechnology in Plant Science: relevance to agriculture in the eighties*, ed. M. Zaitlin, P. Day & A. Hollaender, pp. 287–99. Academic Press: Orlando.

Shillito, R. D., Carswell, G. K., Johnson, C. M., Dimaio, J. S. & Harms, C. T. (1989). Regeneration of fertile plants from plant protoplasts of elite inbred maize. *Bio/Technology*, **7**, 581–7.

Shimamoto, K., Terada, R., Izawa, T. & Fujimoto, H. (1989). Fertile transgenic rice plants regenerated from transformed protoplasts. *Nature*, **388**, 274–6.

Snape, J. N., Simpson, E., Parker, B. B., Friedt, W. & Foroughi-Wehr, B. (1986). Criteria for the selection and use of doubled haploid systems in cereal breeding programmes. In *Genetic Manipulation in Plant Breeding*, ed. W. Horn, C. J. Jensen, W. Odenbach & O. Schieder, pp. 513–34. Walter de Gruyter: Berlin.

Somers, D. A., Narayanan, K. R., Kleinhofs, A., Cooper-Bland, S. & Cocking, E. C. (1986). Immunological evidence for transfer of the barley nitrate reductase structural gene to *Nicotiana tabacum* by protoplast fusion. *Mol. Gen. Genet.* **204**, 296–301.

Srinivasan, C. & Vasil, I. K. (1986). Plant regeneration from protoplasts of sugarcane (*Saccharum officinarum* L.). *J. Plant Physiol.* **126**, 41–8.

Stolarz, A. & Lörz, H. (1986). Somatic embryogenesis, cell and protoplast culture of *Triticale* (× *Triticosecale* Wittmack). In *Genetic Manipulation in Plant Breeding*, ed. W. Horn, C. J. Jensen, W. Odenbach & O. Schieder, pp. 499–501. Walter de Gruyter: Berlin.

Swanson, E. B., Coumans, M. P., Brown, G. L., Jayanti, D. P. & Beversdorf, W. D. (1988). The characterisation of herbicide tolerant plants in *Brassica napus* L. after *in vitro* selection of microspores and protoplasts. *Plant Cell Rep.* **7**, 83–7.

Szabados, L., Hadlackzy, G. & Dudits, D. (1981). Uptake of protoplasts by plant protoplasts. *Planta*, **151**, 141–5.

Tabaeizadeh, Z., Ferl, R. J. & Vasil, I. K. (1986). Somatic hybridisation in the Graminae: *Saccharum officinarum* L. (sugarcane) + *Pennisetum americanum* (L.) K. Schum. (pearl millet). *Proc. Nat. Acad. Sci.* **83**, 5616–19.

Tabaeizadeh, Z., Pring, D. R. & Vasil, I. K. (1987). Analysis of mitochondrial DNA from somatic cell lines of *Saccharum officinarum* (sugarcane) + *Pennisetum americanum* (pearl millet). *Plant Mol. Biol.* **8**, 509–13.

Terada, R., Kyozuka, J., Nishibayashi, S. & Shimamoto, K. (1987). Plant regeneration from somatic hybrids of rice (*Oryza sativa* L.) and barnyard grass (*Echinochloa oryzicola* Vasing). *Mol. Gen. Genet.* **210**, 39–43.

Tomes, D. T. & Smith, G. S. (1985). The effect of parental genotype on initiation of embryogenic callus from elite maize (*Zea mays*) germplasm. *Theor. Appl. Genet.* **70**, 505–9.

Töpfer, R., Gronenborn, B., Schell, J. & Steinbiss, H.-H. (1989). Uptake and transient expression of chimaeric genes in seed-derived embryos. *The Plant Cell*, **1**, 133–9.

Toriyama, K., Hinata, K. & Sasaki, T. (1986). Haploid and diploid plant regeneration from protoplasts of anther callus in rice. *Theor. Appl. Genet.* **73**, 16–19.

van der Valk, P., Zaal, M. A. C. M. & Creemers-Molenaar, J. (1988). Regeneration of albino plantlets from suspension culture derived protoplasts of *Poa pratensis* L. (Kentucky bluegrass). *Euphytica*, **S**, 169–73.

Vasil, I. K. (1986). Regeneration in cereal and other grass species. In *Cell Culture and Somatic Cell Genetics of Plants, vol. 3*, ed. I. K. Vasil, pp. 121–50. Academic Press: New York.

(1987). Developing cell and tissue culture systems for the improvement of

cereal and grass crops. *J. Plant Physiol.* **128**, 193–218.
Vasil, V. & Vasil, I. K. (1980). Isolation and culture of cereal protoplasts. II: Embryogenesis and plantlet formation from protoplasts of *Pennisetum americanum*. *Theor. Appl. Genet.* **56**, 97–9.
(1981*a*). Somatic embryogenesis and plant regeneration from suspension cultures of pearl millet (*Pennisetum americanum*). *Ann. Bot.* **47**, 669–78.
(1981*b*). Somatic embryogenesis and plant regeneration from tissue cultures of *Pennisetum americanum* and *P. americanum* × *P. purpureum* hybrid. *Amer. J. Bot.* **68**, 864–72.
(1986). Plant regeneration from friable embryogenic callus and cell suspension cultures of *Zea mays*. *J. Plant Physiol.* **124**, 399–408.
Vasil, V., Wang, D. & Vasil, I. K. (1983). Plant regeneration from protoplasts of *Pennisetum purpureum* Schum. (Napier grass). *Z. Pflanzenphysiol.* **111**, 319–25.
Wakasa, K. & Widholm, J. M. (1987). A 5-methyltryptophan resistant rice mutant, MTR1 selected in tissue culture. *Theor. Appl. Genet.* **74**, 49–54.
Wang, Y. C., Klein, T. M., Fromm, M., Cao, J., Sanford, J. C. & Wu, R. (1988). Transient expression of foreign genes in rice, wheat, and soybean cells following particle bombardment. *Plant Mol. Biol.* **11**, 433–9.
Wenzel, G. (1985). Strategies in unconventional breeding for disease resistance. *Ann. Rev. Phytopathol.* **23**, 149–72.
Wenzel, G. & Foroughi-Wehr, B. (1984). Anther culture of cereals and grasses. In *Cell Culture and Somatic Cell Genetics of Plants, vol. 1*, ed. I. K. Vasil, pp. 311–27. Academic Press: New York.
Wernicke, W. & Milkovits, L. (1987*a*). Rates of uptake and metabolism of IAA and 2,4-D by cultured leaf segments at different stages of development in wheat. *Physiol. Plant*, **69**, 23–8.
(1987*b*). Effect of auxin on the mitotic cycle in cultured leaf segments at different stages of development in wheat. *Physiol. Plant*, **69**, 16–22.
Wernicke, W., Gorst, J. & Milkovits, L. (1986). The ambiguous role of 2,4-D in wheat tissue culture. *Physiol. Plant*, **68**, 597–602.
Withers, L. A. (1985). Cryopresentation of cultured cells and meristems. In *Cell Cultures and Somatic Cell Genetics of Plants, vol. 2*, ed. I. K. Vasil, pp. 253–316. Academic Press: Orlando.
Woolston, C. J., Barker, R., Gunn, H., Boulton, M. I. & Mullineaux, P. M. (1988). Agroinfection and nucleotide sequence of cloned wheat dwarf virus DNA. *Plant Mol. Biol.* **11**, 35–43.
Yamada, Y., Yang, Z. Q. & Tang, D. T. (1986). Plant regeneration from protoplast derived callus of rice (*Oryza sativa*). *Plant Cell Rep.* **5**, 85–8.
Yang, M. Y. & Zhou, C. (1982). *In vitro* induction of haploid plants from unpollinated ovaries and ovules. *Theor. Appl. Genet.* **63**, 97–104.
Yang, Z. Q., Shikanai, T. & Yamada, Y. (1988). Asymmetric hybridisation between cytoplasmic male-sterile (cms) and fertile rice (*Oryza sativa* L.) protoplasts. *Theor. Appl. Genet.* **76**, 801–8.
Yang, Z. Q., Shikanai, T., Mori, K. & Yamada, Y. (1989). Plant regeneration from cytoplasmic hybrids of rice (*Oryza sativa* L.). *Theor. Appl. Genet.* **77**, 305–10.

Ye, J. M., Kao, K. N., Harvey, B. L. & Rossnagel, B. G. (1987). Screening salt-tolerant barley genotypes via F1 anther culture in salt stress media. *Theor. Appl. Genet.* **74**, 426–9.

Zelcer, A., Aviv, D. & Galun, E. (1978). Interspecific transfer of cytoplasmic male sterility by fusion between protoplasts of normal *Nicotiana sylvestris* and X-ray irradiated protoplasts of male sterile *N. tabacum*. *Z. Pflanzenphysiol.* **90**, 397–407.

Zhang, W. & Wu, R. (1988). Efficient regeneration of transgenic plants from rice protoplasts and correctly regulated expression of the foreign gene in the plants. *Theor. Appl. Genet.* **76**, 835–40.

Zimny, J. & Lörz, H. (1986). Plant regeneration and initiation of cell suspensions from root tip derived callus of *Oryza sativa* L. (rice). *Plant Cell Rep.* **5**, 89–92.

9

Reproduction and recognition phenomena in the Poaceae

R. Bruce Knox and Mohan B. Singh

Introduction

In recent years, several reproductive technologies have become available for application to the grasses, including refinement of electron microscopy, and access to monoclonal antibodies and to recombinant DNA technology. Grass biology can be seen in a new perspective. This chapter examines the data and the exciting new concepts that are emerging.

First, there is the 100-year-old conundrum of *double fertilization*. It is nearly 100 years since this was first interpreted cytologically in flowering plants, and textbook explanations have changed little in the intervening years. In the last decade, however, with the application of new quantitative computer-based techniques, some surprising findings have emerged. The grasses are taking their place in the case histories that are being painstakingly assembled. Fertilization ultimately represents sperm–egg recognition. New hypotheses are being formulated on the regulation of this process.

These processes are the end-point of pollen–pistil recognition, which commences within certain species of Poaceae with the interaction of the *S*,*Z* alleles of self-incompatibility. The question may be asked as to what progress has been made in understanding the mechanisms involved.

Finally, grass pollen shows an unexpected form of recognition, i.e. specific interactions with human defence cells of the immune system in triggering allergic asthma in susceptible patients. Grass pollen possesses the most important molecules, *allergens*, of the outdoor environment in temperate climates each spring and early summer. The immediate (type 1) hypersensitive response occurs in about one in five humans (hay fever, asthma, eczema). Molecular biological techniques have been applied to understanding structure/function relationships of allergens. The prospect is that we may soon understand why these molecules are present in airborne pollen, and the grasses are the leading experimental system for this research. An unexpected bonus is that the technology can lead not only to the development of low allergen rye-grass strains. The development of male sterile lines of cereals for hybrid seed production is an application in plant breeding.

Double fertilization reappraised

There is increasingly cytological evidence from transmission EM studies combined with three-dimensional reconstruction techniques that the products of the generative cell division, the pair of sperm cells traditionally referred to as 'male gametes', are not identical: for example, *Plumbago zeylanica* (Russell & Cass, 1983; Russell, 1984); *Brassica campestris* and *B. oleracea* (McConchie *et al.*, 1985, 1987) and *Hordeum vulgare* (Mogensen & Wagner, 1987; Mogensen, this volume). The hypothesis has been proposed that it is not necessarily a matter of chance which of the pair of sperm cells in each pollen tube fuses with the egg cell at fertilization, and which fuses with the central cell (Knox & Singh, 1987; Knox, Southworth & Singh, 1988). The very definition of a *gamete* puts the situation in perspective: a *gamete* is a reproductive cell involved in embryo formation. Accordingly, the *egg* is the female gamete, but which of the *pair of sperm cells* in each angiosperm pollen tube is the male gamete?

If these findings are generalized, it becomes necessary to redescribe 'double fertilization' along, perhaps, the following lines. The pollen grain arises at the first pollen mitosis when the microspore becomes bicellular, undergoing division to yield the vegetative and generative cells. During later development, either these cells then pass into the pollen tube at its germination, or there is a further division of the generative cell within the grain to yield a *male gamete* to fertilize the egg cell and an *associate cell* destined to fuse with the polar nuclei to form the primary endosperm cell.

In more than 20 genera of both dicots and monocots, there is evidence for the organization of the reproductive cells and vegetative nucleus to form male germ units. These are single units which result in the transfer of the sperm cells and vegetative nucleus, i.e. all the DNA of male heredity, within the pollen tube (Dumas *et al.*, 1984*a*,*b*; see Mogensen, this volume).

The present, very limited evidence suggests that each species either has its own version of a male germ unit, or has none that has been detected (see Knox & Singh, 1987). Between the male gamete and the associate cell, there is scope for differences in cellular organization and size, nuclear control, and the number and proportions of plastids to mitochondria.

If this is indeed the case, such differences may explain the difficulties of interspecies crossing and even why in some cases a cross, but not its reciprocal, might succeed. A further and important consequence is that any genetic modification induced in the generative cell *before* it divides would have unpredictable consequences. Any genetic modification applied after this division would *directly* affect the zygote only if it were applied to the male gamete. If the associate cell were affected, possible modifications would be to the endosperm and this could *indirectly* affect the zygote.

Assuming for the moment that this distinction between male gamete and the associate cell proves to be justified, there is the fascinating prospect that the surface recognition molecules that develop and redistribute in animal sperms during their maturation might have different counterparts on both the male gamete and the associate cell surfaces (Hough *et al.*, 1986; Knox *et al.*, 1988). Any emerging reproductive technology that sought to influence events at or around fertilization would have to take account of these new concepts.

Evidence for or against this hypothesis is being obtained in several ways. In *Plumbago*, Russell (1985) demonstrated the existence of preferential fertilization by TEM studies of fertilized zygotes. Several laboratories have isolated sperm cells and are exploring their cell biology. Sperm cells from mature pollen of maize and wheat have been isolated (Dupuis *et al.*, 1987; see Mogensen, this volume). A characteristic feature is that the elongate sperm cells round off when isolated, appearing as spherical protoplasts or gametoplasts.

Any differences in surface antigens of the pairs of sperm cells can be detected using monoclonal antibody technology to detect regional heterogeneity and surface domains as in animal sperm (Myles & Primakoff, 1985). This has been initiated for the sperm of two dicots: *Brassica* (Hough *et al.*, 1986) and *Plumbago* (Pennell *et al.*, 1988).

The stage is set for another approach to investigate double fertilization. Living sperms have been isolated from pollen of wheat and corn, and techniques have been established to isolate protoplasts from egg cells, and even whole embryo sacs in several systems (Zhou & Yang, 1985). It is now technically feasible to set up *in vitro* culture conditions for single fertilization, and cereal systems are a leading candidate for these experiments. Such approaches have the added benefit of determining whether both sperms can fuse with the egg, and whether sperms of plants have fusibility factors, like the sperms of animals (Ashida & Scofield, 1987). Given further improvements in recombinant DNA technology, it should be possible in future to detect sperm-specific gene expression associated with fertilization.

Pollen–pistil recognition

Pollen grains, including those of grasses, carry the male gamete or its progenitor in flowering plants. They are envelopes for the male reproductive cells. To serve this function, the pollen wall has been shown to possess numerous adaptations (see Knox, 1984*a*,*b*). The outer exine is made of sporopollenin. This polymer has remarkable strength and elasticity which are displayed when the grain hydrates at the commencement of fertilization. The exine is patterned with species or group-specific features. In some cases there is a fishnet pattern, with the pollencoat materials filling the holes or cavities on the surface. In the grasses, there are surface micropores that lead into crypt-like cavities filled with pollencoat materials. The inner cellulosic intine layer contains tubular or ribbon-like evaginations of the plasma membrane, filled with proteins. These proteins at the surface of pollen grains include a range of hydrolytic enzymes, as well as allergens, antigens and stigma recognition factors.

Recognition between pollen and stigma has been explored in grasses. Knox & Heslop-Harrison (1970) showed that grass pollen contains a massively thickened intine at the single aperture or pore, and the wall polymer contains hydrolytic enzymes, which are secreted to the stigma surface at the pore. In later studies of the early events of pollen–stigma interactions in *Phalaris aquatica*, a self-incompatible species, pollen-wall proteins were shown by immunocytochemistry to bind to surface components of the stigma. This occurred in both compatible and self-incompatible matings (Knox & Heslop-Harrison, 1971). However, in the latter, the germinating pollen did not adhere to the stigmas. After ultrastructural studies of fertilization in other species of grasses, the sequence of events have been described (Heslop-Harrison, 1979; Heslop-Harrison & Heslop-Harrison, 1982, 1988). While these studies are important for understanding the physiology and cytology of fertilization, the early events of pollen–stigma recognition remain unknown. Recent molecular biological studies of one of the major pollen components offer this prospect.

The molecular biology of allergens

Many grass pollen species are highly allergenic. The pollen present in the atmosphere during pollination can provoke the immediate hypersensitive response of allergic asthma (see review by Howlett &

Fig. 9.1. How the symptoms of allergic asthma are induced, involving production of specific types of human defence molecules of the immune system, Immunoglobulin E (IgE) (from Howlett & Knox, 1984).

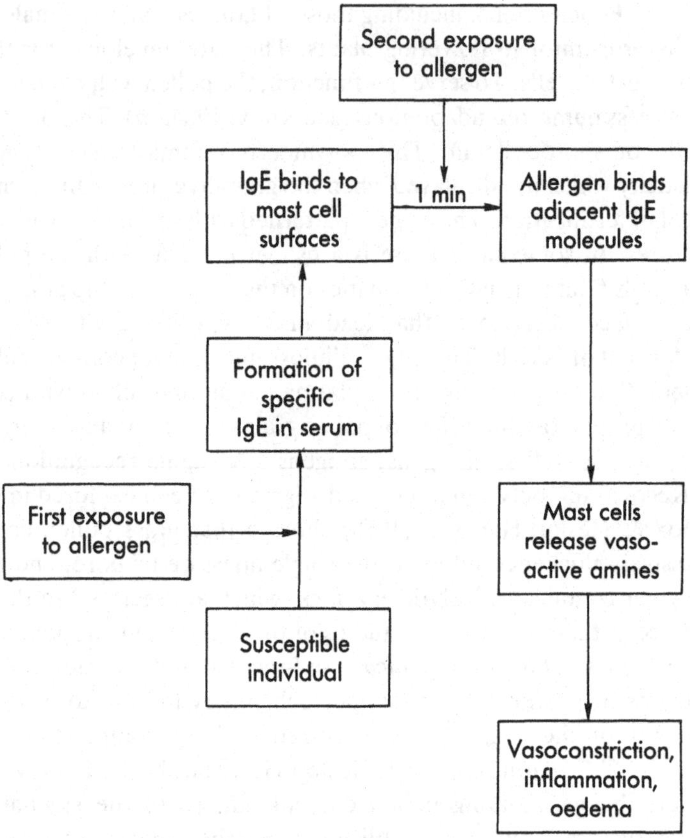

Knox, 1984). This occurs in about one in five of the human population. Fig. 9.1 shows how the symptoms are induced, involving production of specific types of human defence molecules of the immune system, Immunoglobulin G (IgG) and Immunoglubulin E (IgE). The pollen allergen can bind specifically to these human molecules. The allergic response is then evoked in susceptible subjects. The main sources are certain temperate grasses of the Poideae: rye-grass *Lolium perenne*; cock's foot *Dactylis glomerata*; timothy *Phleum pratense*; sweet vernal grass *Anthoxanthum odoratum*; Yorkshire fog or velvet grass *Holcus lanatus*; Kentucky blue grass *Poa pratensis*. In addition, Bermuda or couch grass, *Cynodon dactylon*, of the Chlorideae is important in sub-tropical climates.

Table 9.1. *Some properties of defined rye-grass pollen allergens*

Allergen	MR (kD)	Positive reaction (%)[a]	References[b]
Immunodominant allergen			
*Lol p*I	34	84–95	1, 2, 4, 6
Other defined allergens			
*Lol p*II	11–16	43–60	3, 4
*Lol p*III	11–14	70	2, 4
*Lol p*IV	50–67	20–48	2, 4
HMBA	57	74	4, 5

[a] Percentage of rye-grass-sensitive patients showing positive reaction by skin test or RAST.
[b] *References:* 1: Johnson & Marsh (1965); 2: Marsh (1975); 3: Smart *et al.* (1983); 4: Ford & Baldo (1986); 5: Ekramoddoulah *et al.* (1983); 6: Freidhoff *et al.* (1986).

Rye-grass has been studied more than other grasses because the pollen of this species is the major outdoor allergen in many temperate climates. Rye-grass produces very high levels of pollen output (Smart, Tuddenham & Knox, 1979). Rye-grass pollen output per anther is high, around 5000 grains per anther, which multiplies to over 2 million grains per spike, and finally to estimates of 2.1×10^{13} per hectare. Assuming that the mass of a rye-grass pollen grain is 22×10^{-9} g, this gives the figure of nearly 0.5 tonnes of pollen output per hectare from a pure rye-grass pasture if not mown or grazed in a single season in Melbourne.

The allergens of rye-grass pollen were first isolated using biochemical techniques at Cambridge in 1965 (see Table 9.1). There are at least four groups of allergens, separated according to their molecular weight. The principal allergen is a glycoprotein of 27–35 kD molecular weight. The molecule contains 5% carbohydrate that does not contribute to its allergenicity (Johnson & Marsh, 1966). The allergen, known originally as Group I allergen, but now by convention as *Lol p*I, is the immunodominant component, i.e. more than 95% of patients possess specific IgE that binds to this pollen protein. Subsequently, the Group I allergen has been found to be separable into four components by electrophoretic techniques. These components have molecular weights of 28, 30, 32 and 35 kD.

The structure of the rye-grass allergens has been the subject of many immunochemical and biochemical studies (Lynch & Turner, 1974; Marsh, 1975; Cottam, Moran & Standring, 1986; Ford & Baldo, 1986; Olson & Klapper, 1986; Ansari, Kihara & Marsh, 1987; Esch, 1989; Esch & Klapper, 1989; Marsh, Zwollo & Ansari, 1989). Furthermore, immunological relatives of similar molecular weight are present in other related clinically important grass genera. These allergens share antigenic and allergenic determinants with *Lolp*I, as determined by antibody-binding sites with monoclonal antibodies (Fig. 9.2), and specific IgE antibodies (Singh & Knox, 1985*a*; Singh *et al.*, 1989). Allergenic determinants immunochemically similar to rye-grass can even be shared by pollen of dicots, for example, *Acacia* pollen (Howlett, Hill & Knox, 1982).

These other grasses, mainly of the Pooidae, contain allergenic proteins: for example, *Dactylis glomerata* (Peltre, Lapeyre & David, 1982; Ford, Tovey & Baldo, 1985); *Festuca elatior* (Esch & Klapper, 1989); *Poa pratensis* (Ekramoddoullah, Kisil & Sehon, 1986). Substantial allergenic cross-reactivity between grass pollens has been demonstrated using an IgE-binding assay, the radioallergo-sorbent test (RAST) (see Marsh, Haddad & Campbell, 1970; Leiferman & Gleich, 1976; Lowenstein, 1978). However, there is no cross-reactivity with distantly related grasses such as Bermuda grass, *Cynodon dactylon* (Ford & Baldo, 1987).

*Lol p*I amounts to 5% of the soluble pollen protein (Marsh, 1975), and is the major protein component of the grains. The allergen is located at the surface of the grain. This has been determined by immunocytochemistry (Vithanage *et al.*, 1982; Staff *et al.*, 1989) and by a new membrane print technique (O'Neill, Singh & Knox, 1986, 1989). Monoclonal antibodies or specific IgE are the probes (Fig. 9.2). Yet the function of this allergen has only just begun to be understood through molecular analysis. The amino acid sequence of a highly conserved

Fig. 9.2. Diversity of proteins, antigens and allergens in rye-grass pollen compared with a panel of 16 other grasses; Lane 1: *Bromus inermis*; 2: *Agropyron cristatum*; 3: rye, *Secale cereale*; 4: cock's foot or orchard grass, *Dactylis glomerata*; 5: *Festuca elatior*; 6: rye-grass, *Lolium perenne*; 7: *L. multiflorum*; 8: *Poa compressa*; 9: oat, *Avena sativa*; 10: *Holcus lanatus*; 11: sweet vernal, *Anthoxanthum odoratum*; 12: *Agrostis alba*; 13: timothy, *Phleum pratense*; 14: canary, *Phalaris arundinacea*; 15: Bermuda, *Cynodon dactylon*; 16: *Sorghum halepensis*; 17: maize, *Zea mays*.
(*a*) Coomassie blue stained protein profiles of pollen extracts after SDS-PAGE under reducing conditions. The major allergen *Lol p*I is evident as a series of bands in the region 28 to 35 kD. (*b*) Western blot of same gel probed with monoclonal antibody 40.1 specific for the major allergen which is detected as one or more bands in rye-grass (arrow) and in other genera of the panel of grass pollens. (*c*) Western blot of same gel probed with specific IgE antibodies from pooled grass pollen-allergic human sera showing that *Lol p*I (28 to 35 kD) is the major allergen (from Singh *et al.*, 1989).

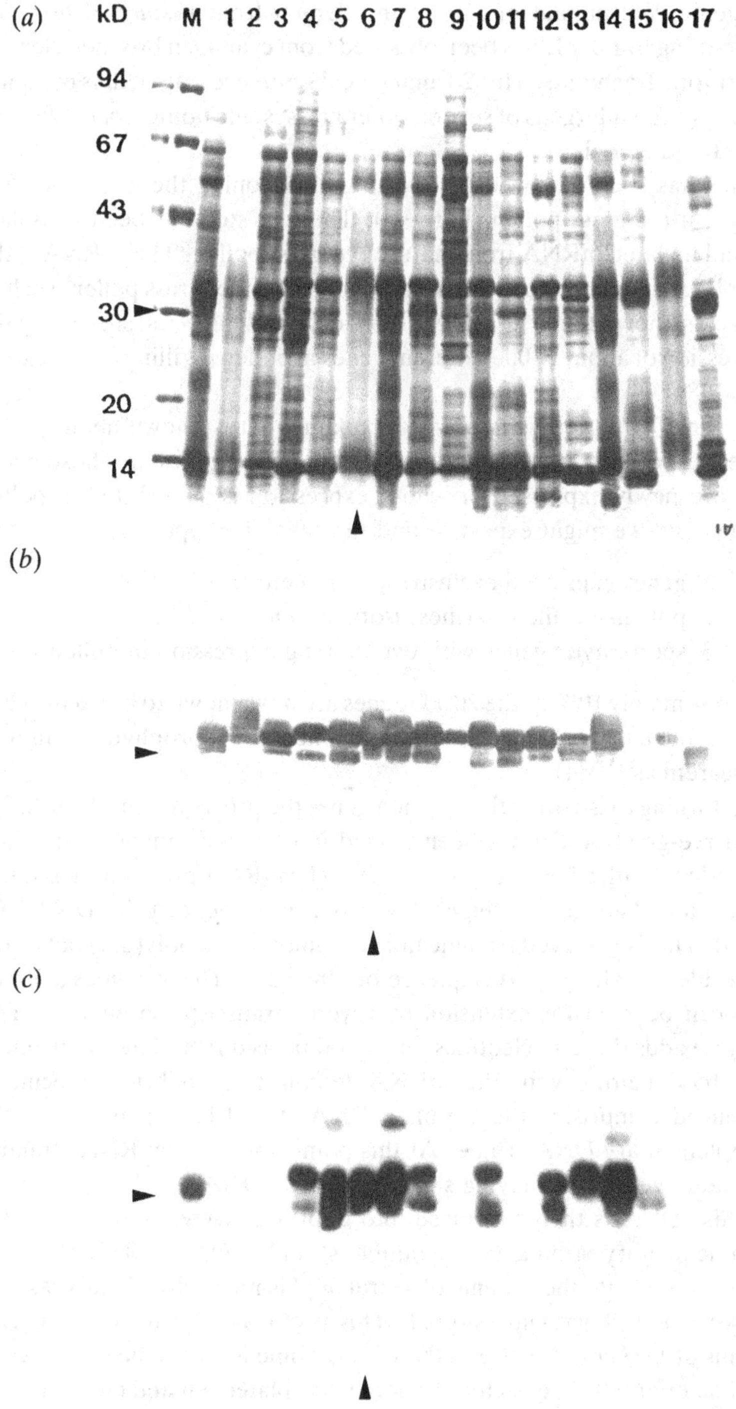
(a)
kD
M 1 2 3 4 5 6 7 8 9 10 11 12 13 14 15 16 17
94
67
43
30
20
14
(b)
(c)

antigenic determinant of *Fes e*I, an allergen from *Festuca elatior* corresponding to *Lol p*I, has been obtained from cyanogen bromide cleaved and tryptic fragments. This 28 amino acid sequence fragment is common to the group I allergens of several other grasses, including *Lol p*I, *Poa p*I, *Ant o*I and *Agr a*I.

This has been achieved by isolating and cloning the gene encoding *Lol p*I in rye-grass pollen (Fig. 9.3). The first step has been to isolate polyadenylated mRNA from mature rye-grass pollen. This mRNA is the template used to produce the protein *in vivo*. Other grass pollen, such as maize, is known to possess up to 20 000 different mRNAs, suggesting the expression of about 20 000 different genes in pollen (Willing & Mascarenhas, 1984).

A short digression is needed to explain what is known about pollen genes, especially in the pollen of grasses. Pollen genes are those genes that are newly expressed or whose expression is augmented in pollen grains, and we might expect to find the following types:

1 genes expressed exclusively in pollen;
2 pollen-specific isozymes, isotypes and variants;
3 sporophytic genes with overlapping expression in pollen.

Approximately 10% of the 20 000 genes are now known to be exclusive to pollen, the balance being expressed in pollen and sporophyte (Willing & Mascarenhas, 1984).

Returning to isolating the allergen gene, the procedure in which *Lol p*I from rye-grass pollen has been cloned into bacteria involves standard procedures with a few interesting twists. The mRNA preparation is taken as the template, and a single-stranded complementary (c)-DNA prepared. This is achieved by annealing a primer to the poly(dA) tail of the molecule, usually a short sequence of oligo(dT). This provides a 3′ end that can be used for extension by reverse transcriptase enzyme. This enzyme adds deoxynucleotides one at a time as directed by complementary base pairing with the mRNA template. A hybrid molecule is produced comprising the template RNA strand base paired with the complementary DNA strand. At this point, the original RNA strand is degraded with alkali, giving single-stranded cDNA.

This cDNA is then converted into duplex DNA via synthesis of the complementary strand, giving double-stranded cDNA. This cDNA is incorporated into the genome of a viral or plasmid vector. In this case the vector is the phage lambda-gt 11. This vector is used to infect specific strains of *Escherichia coli* and the viral genome is able to be expressed in the bacterial cell. The bacterial colonies are plated out and multiplied. In

Fig. 9.3. Localization of the major allergen of rye-grass pollen. (*a*) Scanning electron micrograph of mature rye-grass pollen grains (after aldehyde fixation and critical point drying). Preparation and photograph by Philip Taylor. Bar = 10 μm. (*b*) Thin section of part of rye-grass pollen grain after anhydrous processing to prevent leaching of allergens from grains, and treatment for immunocytochemistry using specific monoclonal antibodies and colloidal gold as probe. Transmission electron micrograph in which the allergen sites are indicated by the gold labelling (×26 000; from Staff *et al.*, 1989).

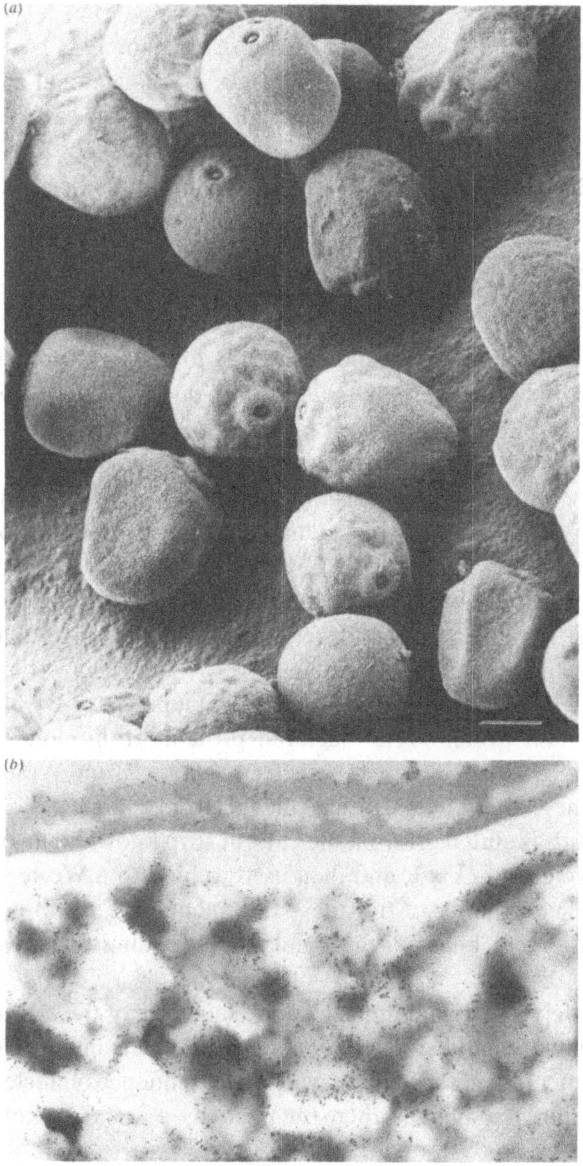

this way, the genes have been cloned, forming a cDNA library. This library should contain copies of all the mRNA species existing in the original pollen grains.

At this stage, screening for the individual gene products, now produced by the protein synthetic machinery of the bacteria, is carried out. Each clump of bacteria now make the gene product of the particular one of the 20 000 mRNAs in the cDNA library that has been produced. To obtain a particular cDNA, it is a matter of recognizing the clone using a specific cellular probe.

The plaque lifts of the bacterial lawn of recombinant bacteria are treated with the probe. For detection of *Lol p*I, monoclonal antibodies known to be specific to the allergen, and IgE in the sera of patients known to be allergic to grass pollen (Knox *et al.*, 1989), are used. The selected clones are themselves multiplied, and re-screened. Identical colonies of recombinant bacteria bind both probes (Fig. 9.4). These colonies are selected and individually multiplied and numbered. Each will contain cDNA clones representing different lengths of the gene encoding *Lol p*I. Clone 12R is 1200 bp in length and binds to both MAbs and IgE. A second clone, 6R, is 600 bp in length and binds only to MAbs. Apparently, 6R does not encode the region of the molecule-binding IgE (Knox *et al.*, 1989).

How is the identity of the cDNA clones with *Lol p*I verified? This can be done in several ways. First, several different MAbs are used to bind dot blots of native protein and compare this with binding to similar recombinant preparations. Clone 12R binds to both antibodies used to isolate the clone, and to those obtained from other workers in the field. Second, IgE obtained from different patients is employed to check specificity of the recombinant protein.

A particularly elegant method has been developed to prove identity of the recombinant protein. The bacterial proteins from clone 12R are separated by SDS-polyacrylamide gel electrophoresis, and Western blots prepared. These are treated with specific IgE from allergic patients' sera. The bound IgE is eluted off from the cut-out strip known to be of similar molecular weight to *Lol p*I, and then used to bind to a Western blot of rye-grass pollen proteins. After these treatments, the result is that the recombinant *Lol p*I binds to the band corresponding to native *Lol p*I (Singh *et al.*, 1989).

The next step is to obtain the nucleotide sequence of the cDNA clone. This is done by preparing a restriction map, after cleaving the clone into smaller fragments using specific enzymes. The sequence of nucleotides in each restriction fragment is then obtained by a standard sequencing

Fig. 9.4. Isolation of cDNA clones specific for the group I allergens, as shown by recognition of positive clones by three different MAbs (1) 40.1, (2) 21.3, (3) 12.1, and (4) IgE from allergic patients' sera, (5) control for 1, 2 and 3 in which primary MAb omitted. Poly(A^+) mRNA isolated from mature rye-grass pollen was used to construct a cDNA library in the vector lambda-gt 11. The library was then screened with antibody probes to detect sequences expressing group I proteins. *E. coli* Y1090 transfected with 3×10^4 recombinant phages were plated and incubated then the plates were overlaid with a dry nitrocellulose filter previously soaked in 10 mm IPTG. After incubation for 3 h, the filters were carefully peeled off and washed. A second set of filters was placed on phage plates and incubated and treated as above. Both sets of NC filters were tested for binding of MAb 40.1 to plaques. The antibody positive plaques were picked off, purified then replated and tested for binding to probes (from Singh *et al.*, 1989).

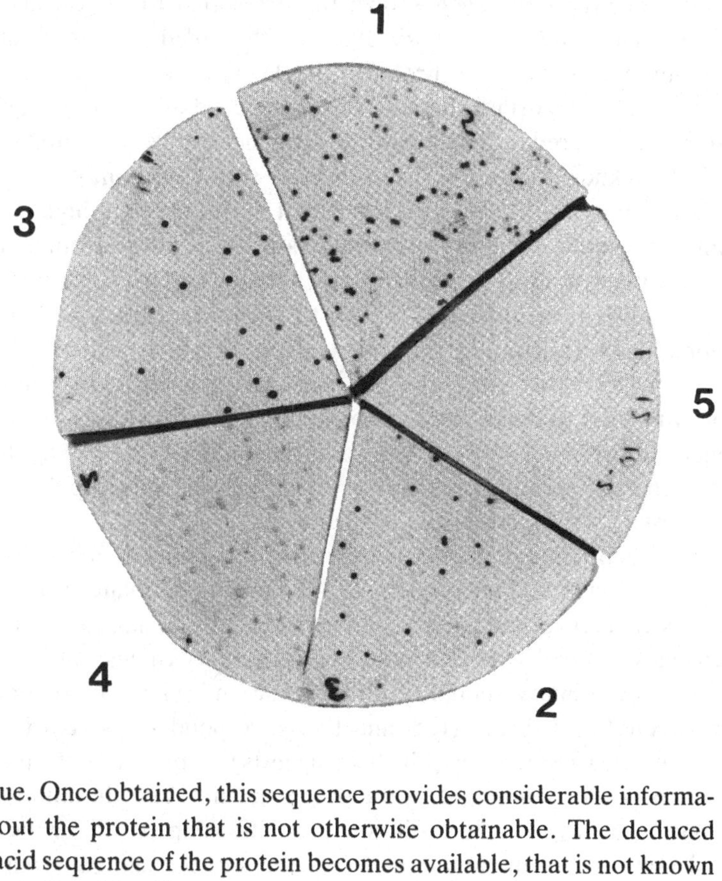

technique. Once obtained, this sequence provides considerable information about the protein that is not otherwise obtainable. The deduced amino acid sequence of the protein becomes available, that is not known for *Lol*pI. For this allergen, only total amino acid anaylses have been published. A hydrophilicity profile can be plotted that indicates the hydrophilic, surface amino acid sites which are likely to be the antigenic determinants. Each restriction fragment can be expressed in bacteria and produce part of the protein molecule. When this is done for *Lol p*I, IgE binding is restricted to a large fragment representing only one-half of the molecule, while the other half does not bind to specific IgE. Thus, the IgE-binding determinant/s occur in discrete regions of the molecule, and are expected to comprise characteristic sequences of amino acids.

The cloned allergen from rye-grass proved to be a proline-rich protein (Singh *et al.*, 1989), and in this property it is like extensin and the hydroxyproline-rich glycoprotein, HPRG, from plant cell walls, and collagen from animal cells. This amino acid composition is quite different from that reported previously for *Lol p*I (Johnson & Marsh, 1966; Klapper, Goodfriend & Capra, 1980; Howlett & Clarke, 1981). The cloned allergen has been shown to correspond to the 30 and 32 kD proteins of *Lol p*I (Singh *et al.*, 1989), so the *Lol p*I previously subjected to amino acid analysis is probably the 35 kD protein.

A second important function of the deduced amino acid sequence is that once entered on a data base, its sequence can be compared with that of other known proteins. The sequence shows no homology with any known protein. Already, we know that *Lol p*I shows a high degree of immunochemical identity with bromelain, the protease from pineapple. This is interesting, as one other major allergen that has been cloned, that from house dust mite, has proved to show more than 70% sequence homology with actinidin, the protease from kiwi-fruit (Chua *et al.*, 1988). This suggests that the major allergens may exhibit hydrolytic enzyme activity, and perhaps explain one of their functions to alter the permeability of human cells. At the same time, identification of the function of such an important human recognition molecule opens up the potential to explore their natural role in plants.

Of potentially even more interest is the possibility of being able to use the *Lol p*I cDNA clones as heterologous probes to isolate corresponding genes in other related grasses, and so study DNA or amino acid sequence changes as a molecular indicator of evolutionary changes in the grasses. The extraordinary similarity in antigenic and allergenic determinants between *Lol p*I of rye-grass and the corresponding protein in related grasses shown previously (Fig. 9.2) suggests that regions of the molecules may be highly conserved, while others are not, and the differences in homology may indicate evolutionary relationships at the genus level.

Knowledge of the amino acid sequences of the allergens will enable precise mapping of the antibody-binding regions for both MAbs and IgE, and we are refining these domains by analysis of overlapping synthetic peptides. This approach will be valuable in determining why allergenic responses are characterized by IgE interactions, and why only certain subjects develop clinical symptoms of allergy.

A striking feature of our results is that specific IgE antibody-binding sites are restricted to the region of the molecule adjacent to the COOH terminus, while MAb-binding sites are more commonly distributed along the length of the allergen molecule. This finding agrees with recent studies of *Fes* *e*I (Esch, 1989; Esch & Klapper, 1989). MAbs hyperimmunized with group I allergens recognize a more diverse array of epitopes than those that stimulate an allergic response. When we consider that the entire protein surface is potentially antigenic, the apparent restriction of IgE antibody specificity may be due to several regulatory factors. Marsh (1989) has recently suggested that IgE antibody specificity to rye-grass pollen allergens is an immunogenetic response, and has a primary association with the DR3 region of HLA, the human major histocompatibility complex (MHC). The possible tight genetic linkage in the immune responsiveness to different proteins in the 27–35 kD region may explain why the majority of patients with grass pollen hypersensitivity have IgE antibodies to these rye-grass components.

Applications in biotechnology

The gene *Lol* *p*I is known to be expressed exclusively in pollen. This has been determined by slot blots and Northern analyses. In these experiments, samples of mRNA from other organs of the rye-grass plant are hybridized with *Lol* *p*I cDNA, either directly on nitrocellulose membrane or after separation by electrophoresis. The results are unequivocal: the only hybridization is with pollen (Fig. 9.5). *Lol* *p*I is the only gene of known function to possess this property.

What determines that *Lol* *p*I is expressed exclusively in pollen? According to classic recombinant DNA dogma, expression of the encoding region of a gene is regulated by a controlling region upstream, the promoter. It is likely that *Lol* *p*I has a promoter sequence that can be sequenced and isolated from genomic DNA, and this is in progress. Promoters have been isolated and characterized which regulate tissue, organ and developmental gene expression. In plants these include promoters for heat-shock proteins, leaf- and seed-specific promoters (Shaw *et al.*, 1986; Stockhaus *et al.*, 1987).

The regulatory role of promoters is being investigated using transgenic plants, and promoters and putative gene regulatory sequences are analysed by making fusions with reporter genes that express a known

Fig. 9.5. Organ specificity of allergen gene expression. Samples of mRNA from other organs of the rye-grass plant are hybridized with *Lol p*I cDNA, directly on nitrocellulose membrane as slot blots. The lambda-12R cDNA shows hybridization of ^{32}P-labelled cDNA to mRNA isolated from mature pollen but not from vegetative organs of rye-grass. Lanes: 1, pollen; 2, leaf; 3, root; 4, imbibed seed.

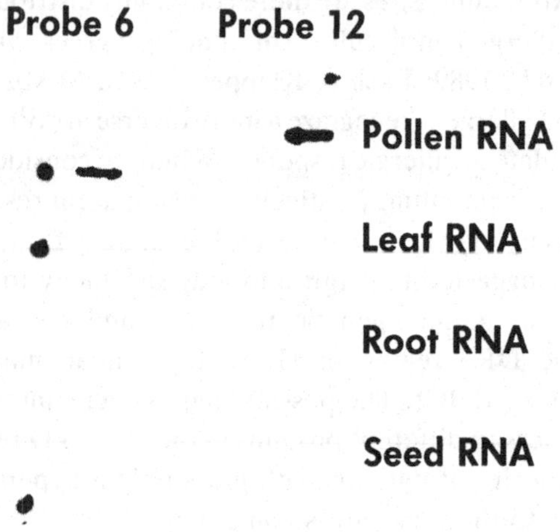

exotic gene product (Willmitzer, 1988), for example, octopine and nopaline synthase (from *Agrobacterium tumefaciens* T-DNA), CAT (chloramphenicol acetyltransferase), NPT-II (neomycin phosphotransferase II) or GUS (beta-glucuronidase).

Pollen gene expression can be modified using *Lol p*I. The strategy is to isolate and characterize the pollen-specific promoter, and splice this to the encoding region of a gene that is desired to be expressed in pollen. Since *Lol p*I or a very similar molecule is present in the pollen of a wide range of cereals as well as grasses, the probability is that its promoter will operate in the cereals. This means that strategies can be developed to manipulate pollen of cereals in a way not previously possible.

One useful development would be the production of lines of rye-grass that produce fertile pollen that is low in the major allergens. Because the number of allergenic determinants is small, and the allergens of rye-grass appear to share common determinants, this is not such a daunting task. New techniques of *in situ* mutagenesis can be applied that can block the expression of specific genes. A problem here is that the allergens may be essential for fertilization, hence their non-expression would be lethal. This is unlikely. In a parallel case in oilseed rape, a gene has been identified that controls the expression of the enzyme beta-galactosidase (Singh & Knox, 1985*b*), and mutant pollen grains defective in enzyme activity are still effective in fertilization. Differences in activity of a peptidase have been detected in pollen from the same anthers of rye-grass (Bhalla, Singh & Knox, 1987), providing a possible precedent in the grasses.

Probably the most useful strategy is to develop a system for controlling male sterility for hybrid seed production in cereals. Assuming that the *Lol p*I gene is expressed early in microsporogenesis, techniques are now available to splice the promoter sequence to a lethal construct and, by transformation, insert this construct into the genome of a cereal crop. Such plants should exhibit nuclear male sterility, and could be used as female parents for hybrid seed production by standard plant breeding techniques. If either the antisense or ribozyme ('gene shears') technique (Haseloff & Gerlach, 1988) is used to create the lethal construct, then fertility restoration is feasible, offering the possibility of backcrossing. Hybrid seed production is currently very difficult in the cereals. Genetic engineering strategies like this could enable hybrid cereals to become a reality.

References

Ansari, A. A., Kihara, T. K. & Marsh, D. G. (1987). Immunochemical studies of *Lolium perenne* (ryegrass) pollen allergen, *Lol p*I, II and III. *J. Immunol.* **139**, 4034–41.

Ashida, E. R. & Scofield, V. L. (1987). Lymphocyte major histocompatibility complex-encoded class II structures may act as sperm receptors. *Proc. Nat. Acad. Sci. USA*, **84**, 3395–9.

Bhalla, P. L., Singh, M. B. & Knox, R. B. (1987). Application of cytochemical methods for the detection and localization of plant proteolytic enzymes. In *Plant Proteolytic Enzymes*, ed. M. J. Dalling. CRC Press: Boca Raton, Florida.

Chua, K. Y., Stewart, G. A., Thomas, W. R., Simpson, R. J., Dilworth, R. J., Plozza, T. M. & Turner, K. J. (1988). Sequence analysis of cDNA coding for a major housedust mite allergen *Der p*I homology with cysteine proteases. *J. Exp. Med.* **167**, 175–82.

Cottam, G. P., Moran, D. M. & Standring, R. (1986). Physicochemical and immunochemical characterization of allergenic proteins from rye-grass (*Lolium perenne*) pollen prepared by a rapid and efficient purification method. *Biochem. J.* **234**, 305–10.

Dumas, C., Knox, R. B. & Gaude, T. (1984*a*). Pollen–pistil recognition: new concepts from electron microscopy and cytochemistry. *Int. Rev. Cyt.* **90**, 239–72.

Dumas, C., Knox, R. B., McConchie, C. A. & Russell, S. D. (1984*b*). Emerging physiological concepts in fertilization. *What's New in Plant Physiology*, **15**, 177–20.

Dupuis, I., Roeckel, P., Matthys-Rochon, E. & Dumas, C. (1987). Procedure to isolate viable sperm cells from corn (*Zea mays* L.) pollen grains. *Plant Physiol.* **85**, 876–8.

Ekramoddoullah, A. K. M., Kisil, F. T. & Sehon, A. H. (1983). Immunochemical characterization of a high molecular weight basic allergen (HMBA) of rye-grass (*Lolium perenne*) pollen. *Molec. Immunol.* **20**, 465–73.

(1986). Isolation of a Kentucky blue grass pollen allergen using a murine monoclonal antibody immunosorbent. *Int. Arch. Allergy appl. Immunol.* **80**, 100–6.

Esch, R. E. (1989). Allergenic extracts. In *Allergy and Molecular Biology*, ed. A. Said El Shami & T. G. Merrett, pp. 3–11. Pergamon Press: Oxford.

Esch, R. E. & Klapper, D. G. (1989). Identification and localisation of allergenic determinants on grass group I antigens using monoclonal antibodies. *J. Immunol.* **142**, 179–84.

Ford, S. A. & Baldo, B. A. (1986). A re-examination of ryegrass (*Lolium perenne*) pollen allergens. *Int. Arch. Allergy appl. Immunol.* **81**, 193–202.

(1987). Identification of Bermuda grass (*Cynodon dactylon*)-pollen allergens by electroblotting. *J. Allergy Clin. Immunol.* **79**, 711–20.

Ford, S. A., Tovey, E. R. & Baldo, B. A. (1985). Identification of orchard grass (*Dactylis glomerata*) pollen allergens following electrophoretic transfer to nitrocellulose. *Int. Arch. Allergy appl. Immunol.* **78**, 15–21.

Freidhoff, L. R., Ehrlich-Kantzky, E., Grant, J. H., Meyers, D. A. & Marsh, D. G. (1986). A study of the human immune response to *Lolium perenne* (rye) pollen and its components *Lol p*I and *Lol p*II (rye-I and rye-II). I. Prevalence of reactivity to the allergens among skin test, IgE antibody and IgG antibody data. *J. Allergy Clin. Immunol.* **78**, 1190–201.

Haseloff, J. & Gerlach, W. L. (1988). Simple RNA enzymes with new and highly specific endoribonucleotide activities. *Nature*, **334**, 585–91.

Heslop-Harrison, J. (1979). Aspects of the structure, cytochemistry and germination of the pollen of rye (*Secale cereale*). *Ann. Bot. Suppl. 1*, **44**, 1–47.

Heslop-Harrison, J. & Heslop-Harrison, Y. (1982). The growth of the pollen tube 1. Characteristics of the polysaccharide particle (P particles) associated with apical growth. *Protoplasma*, **112**, 71–80.

(1988). The pollen tube: motility and cytoskeleton. In *Sexual Reproduction in Higher Plants*, ed. M. Cresti, P. Gori & E. Pacini, pp. 195–203. Springer-Verlag: Berlin.

Hough, T., Singh, M. B., Smart, I. J. & Knox, R. B. (1986). Immunofluorescent screening of monoclonal antibodies to surface antigens of animal and plant cells bound to polycarbonate membranes. *J. immunol. Methods*, **92**, 103–7.

Howlett, B. J. & Clarke, A. E. (1981). Isolation and partial characterization of two antigenic glycoproteins from rye-grass (*Lolium perenne*) pollen. *Biochem. J.* **197**, 707–14.

Howlett, B. J. & Knox, R. B. (1984). Allergic interactions. *Encyclopaedia of Plant Physiology, Vol. 17*, ed. H. F. Linskens & J. Heslop-Harrison, pp. 655–74. Springer-Verlag: Berlin, New York.

Howlett, B. J., Hill, D. J. & Knox, R. B. (1982). Cross-reactivity between *Acacia* (wattle) and ryegrass pollen allergens. Detection of allergens in *Acacia* (wattle) pollen. *Clin. Allergy*, **12**, 259–68.

Johnson, P. & Marsh, D. G. (1965). 'Isoallergens' from rye-grass pollen. *Nature*, **206**, 935–7.

(1966). Allergens from common rye-grass pollen (*Lolium perenne*). I. Chemical composition and structure. *Immunochemistry*, **3**, 91–100.

Klapper, D., Goodfriend, L. & Capra, J. D. (1980). Amino acid sequences of ragweed allergen Ra3. *Biochemistry*, **19**, 5729–34.

Knox, R. B. (1984*a*). The pollen grain. In *Embryology of Plants*, ed. M. B. Johri, pp. 197–271. Springer-Verlag: Berlin, New York.

(1984*b*). Pollen–pistil interactions. In *Encyclopaedia of Plant Physiology, Vol. 17*, ed. H. F. Linskens & J. Heslop-Harrison, pp. 508–608. Springer-Verlag: Berlin, New York.

Knox, R. B. & Heslop-Harrison, J. (1970). Pollen-wall proteins: cytochemical, localization and enzymatic activity. *J. Cell Sci.* **6**, 1–27.

(1971). Pollen-wall proteins: the fate of intine-held antigens on the stigma in compatible and incompatible pollinations of *Phalaris tuberosa* L. *J. Cell Sci.* **9**, 239–51.

Knox, R. B. & Singh, M. B. (1987). New perspectives in pollen biology and fertilization. *Ann. Bot.* (Centenary volume), **60**, 15–37.

Knox, R., Southworth, D. & Singh, M. (1988). Sperm cell determinants and control of fertilization in plants. In *Eukaryote Cell Recognition*, ed. G. P. Chapman, pp. 175–93. Cambridge University Press.

Knox, R. B., Singh, M. B., Hough, T. & Theerakulpisut, P. (1989). The rye-grass pollen allergen, *Lol p*I. In *Allergy and Molecular Biology*, ed. T. Merrett. Proc. 1st International Symposium, Laguna Niguel, California, USA. *Adv. Biosci.* **74**, 161–71. Pergamon Press: Oxford.

Leiferman, K. M. & Gleich, G. J. (1976). The cross-reactivity of IgE antibodies with pollen allergens. I. Analyses of various species of grass pollens. *J. Allergy Clin. Immunol.* **58**, 129–39.

Lowenstein, H. (1978). Quantitative immunoelectrophoretic methods as a tool for the analysis and isolation of allergens. *Prog. Allergy*, **25**, 1–62.

Lynch, N. R. & Turner, K. J. (1974). Application of *in vitro* and *in vivo* assay techniques in the isolation of rye grass (*Lolium perenne*) pollen allergens. *Int. Arch. Allergy*, **47**, 818–28.

McConchie, A. C., Jobson, S. & Knox, R. B. (1985). Computer-assisted reconstruction of the male germ unit in pollen of *Brassica campestris*. *Protoplasma*, **127**, 57–63.

McConchie, C. A., Russell, S. D., Dumas, C., Tuohy, M. & Knox, R. B. (1987). Quantitative cytology of the sperm cells of *Brassica campestris* and *B. oleracea*. *Planta*, **170**, 446–52.

Marsh, D. G. (1975). Allergens and the genetics of allergy. In *The Antigens, Vol. 3*, ed. M. Sela, pp. 271–359. Academic Press: London, New York.

Marsh, D. G., Haddad, Z. H. & Campbell, D. H. (1970). A new method for determining the distribution of allergenic fractions in biological materials: its applications to grass pollen extracts. *J. Allergy*, **46**, 107–12.

Marsh, D. G., Zwollo, P. & Ansari, A. A. (1989). Toward a total human immune response fingerprint: the allergy model. In *Allergy and Molecular Biology*, ed. A. Said El Shami & T. G. Merrett, pp. 65–82. Pergamon Press: Oxford.

Mogensen, H. L. & Wagner, V. T. (1987). Associations among components of the male germ unit following *in vivo* pollination in barley. *Protoplasma*, **138**, 161–72.

Myles, D. G. & Primakoff, P. (1985). Sperm surface domains. In *Hybridoma Technology in the Biosciences and Medicine*, ed. T. A. Springer, pp. 239–50. Plenum Press: New York.

Olson, J. & Klapper, D. G. (1986). Two major human allergenic sites on ragweed pollen allergen antigen E identified by using monoclonal antibodies. *J. Immunol.* **136**, 2109–15.

O'Neill, P., Singh, M. B. & Knox, R. B. (1986). Applications of a new membrane print technique in biotechnology. In *Pollen Biotechnology and Ecology*, ed. D. G. Mulcahy, G. B. Mulcahy & E. Ottaviano, pp. 203–8. Springer-Verlag: New York.

(1989). Grass pollen allergens: detection on surface of living pollen grains using membrane print technique. *Int. Arch. Allergy appl. Immunol.* (in press).

Peltre, G. Lapeyre, J. & David, B. (1982). Heterogeneity of grass pollen allergens (*Dactylis glomerata*) recognized by antibodies in human patients sera by a new nitrocellulose immunoprint technique. *Immunol. Lett.* **5**, 127–31.

Pennell, R. I., Geltz, N. R., Russell, S. D. & Koren, E. (1988). Production and partial characterization of hybridoma antibodies elicited to the sperm of *Plumbago zeylanica. Bot. Gaz.* **148**, 401–6.

Russell, S. D. (1984). Ultrastructure of the sperm of *Plumbago zeylanica* II. Quantitative cytology and three-dimensional organization. *Planta*, **162**, 385–91.

(1985). Preferential fertilization in *Plumbago*: ultrastructural evidence for gamete level recognition in an angiosperm. *Proc. Nat. Acad. Sci. USA*, **82**, 6129–32.

Russell, S. D. & Cass, D. D. (1983). Unequal distribution of plastids and mitochondria during sperm cell formation in *Plumbago zeylanica.* In *Pollen: biology and implications for plant breeding*, ed. D. L. Mulcahy & E. Ottaviano, pp. 135–40. Elsevier: Amsterdam.

Shaw, C. H., Sanders, D. M., Bates, M. R. & Shaw, G. H. (1986). Light regulation of an *ss rubisco-nos* chaemeric gene – photoregulatory control sequences from a C-3 plant function in cells of a CAM plant. *Nucl. Acids Res.* **14**, 6603–13.

Singh, M. B. & Knox, R. B. (1985*a*). Grass pollen allergens: antigenic relationships detected using monoclonal antibodies and dot blotting immunoassay. *Int. Arch. Allergy appl. Immunol.* **78**, 300–4.

(1985*b*). A gene controlling beta-galactosidase deficiency in pollen of oilseed rape *Brassica campestris. J. Hered.* **76**, 199–201.

Singh, M. B., Hough, T., Theerakulpisut, P., Avjioglu, A., Smith, P., Davies, S. & Knox, R. B. (1989). Molecular cloning and characterization of major grass pollen allergen. *Nature* (in press).

Smart, I. J., Tuddenham, W. G. & Knox, R. B. (1979). Aerobiology of grass pollen in the city atmosphere of Melbourne: effects of weather parameters and pollen sources. *Aust. J. Bot.* **27**, 333–42.

Smart, I. J., Heddle, R. J., Zola, M. & Bradley, J. (1983). Development of monoclonal antibodies specific for allergenic components in rye grass, *Lolium perenne. Int. Arch. Allergy appl. Immunol.* **72**, 243–8.

Staff, I. A., Taylor, P. E., Singh, M. B. & Knox, R. B. (1989). Cellular localization of water-soluble, pollen-specific allergenic proteins in rye-grass (*Lolium perenne*) pollen using monoclonal antibodies and specific IgE antibodies with immuno-gold probes. *Histochem. J.* (in press).

Stockhaus, J., Eckes, P., Blau, A., Schell, J. & Willmitzer, L. (1987). Organ-specific and dosage-dependent expression of a leaf-stem specific gene from potato after tagging and transfer into potato and tobacco plants. *Nucl. Acids Res.* **15**, 3479–90.

Vithanage, H. I. M. V., Howlett, B. J., Jobson, S. & Knox, R. B. (1982). Immunocytochemical localization of water soluble glycoproteins, including Group 1 Allergen, in pollen of ryegrass, *Lolium perenne* using ferritin-labelled antibody. *Histochem. J.* **14**, 949–66.

Willing, R. P. & Mascarenhas, J. P. (1984). Analysis of the complexity and diversity of mRNAs from pollen and shoots of *Tradescantia*. *Plant Physiol.* **75**, 865–8.

Willmitzer, L. (1988). The use of transgenic plants to study plant gene expression. *Trends in Genetics*, **4**, 13–18.

Zhou, C. & Yang, Y. H. (1985). Observations on enzymatically isolated living and fixed embryo sacs in several angiosperm species. *Planta*, **165**, 225–31.

10

The widening perspective: reproductive biology of bamboos, some dryland grasses and cereals

G. P. Chapman

When Arber wrote *The Gramineae: a study of cereals, bamboo and grass*, she sought a synthesis among obviously different but representative taxa. Today, a modernized version would have to take account not only of the immense literature generated since 1934 but also its very considerable bias towards cereals. Even if attention were confined only to reproductive biology, this difficulty remains. Watson's remark (this volume) that, from a taxonomist's viewpoint, most economically orientated grass research seems too introverted and that the family as a whole should be seen as a magnificent resource is one that should be far more widely appreciated. Of 39 species of grass listed as forage plants for arid and semi-arid Africa (Anon., 1984), 33 were described as having no known seed collections, although descriptive botany in many cases recognizes morphological and ecological variants that suggest prospects for selection and improvement. The situation is unlikely to have changed radically for the better with such grasses. What follows here is an attempt to supply a wider perspective emphasizing some contrasts with the cereals and northern pasture grasses that provide familiar teaching and research models. It is, too, partly a response to the preceding chapters.

Bamboos

While attempts to find links with groups ancestral to grasses remain tentative, within the Poaceae primitive, intermediate and advanced taxa can be convincingly defined. Bamboos are recognizably primitive in many but not all respects, while C_4 grasses, for example, seem comparatively highly evolved. Clifford (1961) demonstrated that the bamboo *Arundinaria* has a floral formula where, by reduction, the range of bilaterally symmetrical flowers could theoretically be derived that are represented among existing grasses. A problem with bamboos is their massive indurated stems. Is this a later, specialized development peculiar to this group or was it a mainstream evolutionary feature that modern, 'slimmed down' grasses have discarded?

Many bamboos have a reproductive phase at long intervals varying in duration from one genus to another. Although often holocarpic, this is not invariably so and *Swallenochloa*, for example, blooms frequently without adverse consequences (Pohl, 1974). Synchronous flowering was discussed by Arber (1934) and has been reviewed in detail by Janzen (1976).

Self-incompatibility in bamboos is problematic since virtually all evidence is circumstantial, an exception being provided by Bokareva (1974), who was unable to self-fertilize *Phyllostachys*. Janzen (1976) infers that isolated specimens introduced as pieces of rhizome or seeds gathered from one point in a mast crop would explain the little or no seeding involved. Bamboos possibly share an *S,Z* incompatibility system with other grasses and, if instances of self-fertilization occur, it is this system that has been modified presumably.

Grass pollen is commonly believed to be entirely windborne and informed beekeepers in England, for example, assume grass pollen irrelevant to the honey bee. In the tropics a modified view is necessary. Among Hymenoptera the subfamily Apoidea (bees) contains about 12 000 species. Within *Apis* not only *A. mellifera* but also *A. dorsata* and *A. florea* occur. Beyond this genus are bees either much bigger or smaller than the familiar honey bee and among the smaller are many quite unrecognizable at first as bees. Some bees are stingless (but can bite) and some tropical bee species have pollinating males as well as females. In addition to bees there is a wealth of other pollen-feeding insects.

Davis & Richards (1933) recorded *Pariana* visited by flies and stingless bees (*Melipona*). Janzen (1976) implicates *Bambusa polymorpha* and *Chusquea abietifolia* as insect pollinated. Jackson & Woodbury (1976) report carpenter bees (*Xylocopa*) collecting pollen from male flowers of *Dendrocalamus stricta*. Wong (1981) records flowers of *Schizostachum zollingeri* heavily visited by bees (*Apis*, and *Trigona* which is a stingless bee). A detailed review of insect pollination of *Pariana* and *Olyra* is that of Söderstrom & Calderon (1971).

A simple assumption is that infrequent seeding impedes bamboo evolution relative to other grasses. As with size, however, it is unclear whether some bamboos secondarily adopted this or modern grasses have discarded infrequent flowering. On balance it seems likely that bamboos as 'living fossils' perpetuate earlier types of flowering structures although greatly increased stamen number in *Ochlandra* and baccate fruits in *Melocanna*, for example, might also be secondary acquisitions.

What is known of population genetics based on temperate grasses cannot yet readily be extended to bamboos and one possibility, the study of distribution and diversity of iso-enzymes, coupled with computer modelling, remains apparently unexplored.

A pattern of grass evolution?

Clayton & Renvoize (1986) in a stimulating reworking of the grass family presented an interpretation of relationships among various taxa.

Genera with larger numbers of species – for example, *Digitaria* (230), *Eragrostis* (350), *Festuca* (450), *Panicum* (470) and *Poa* (500) – are not among the most derived grasses. Beyond these large genera it is possible to recognize increasingly 'peripheral' groups. For example, *Saccharum* or *Sorghum* are more typical andropogonoids but *Tripsacum* with its separated sexes seems a rather extreme derivative. Even more extreme is *Zea* among whose species is that curious and atypical grass *Zea mays*. Though of quite different lineage, *Triticum* and *Hordeum* are, in this sense, peripheral.

Among the bamboos the likely basic number is $x = 12$ although most are multiples of this. For panicoids $x = 10$ and for pooids $x = 7$ typically, although in both cases many variants are known and throughout the grasses polyploidy is commonplace. Very low basic numbers occur and could reasonably be regarded as specialized offshoots of larger genera, indicated here in parentheses. These include *Airopsis* $x = 4$, *Peribalia* $x = 4$ (*Deschampsia*), *Zingeria* $x = 2$ (*Agrostis*), *Anthoxanthum* $x = 5$ (*Hierochloë*) and *Iseilema* $x = 3$ (*Andropogon*). Such low basic numbers might represent end points unless there is an escape via allopolyploidy.

Was there then a burst of evolutionary activity in, say, the 'protobamboos' leaving a now static bamboo remnant but with the remainder diversifying away from forests and forest margins into more open habitats, diminishing reliance on insect pollination, 'streamlining' their morphology and diversifying their genetics? Is it accidental that the major cereals such as *Hordeum*, *Triticum* and *Zea* are so responsive to the breeders' efforts or is this genetic turbulence at the margins of a still expanding family? Are these currently the 'inventive' genera while those towards the older, more central parts of the Poaceae have settled down apart from some locally persistent or temporary hotspots such as *Oryza*, *Pennisetum* and *Sorghum*?

Alternatively, is the underlying pattern that any part of the Poaceae, whether young or old, primitive or derived can, given ecological opportunity, find the appropriate response? Again, are merely 'large' genera such as *Digitaria*, *Eragrostis*, *Festuca*, *Panicum*, *Poa* and *Stipa* the crucial genetic repositories with the smaller genera being derived from them as clusters of more or less successful expedients?

Finally, in addition to the speciating effects of conventional polyploidy well known in the Poaceae, are there more unusual mechanisms at work induced, for example, by wide hybridization?

Hybridization and polyploidy

A simplifying assumption is that, radiating from the origin of the Poaceae, is an overall diminution of basic chromosome number from which periodically have arisen polyploid episodes. Where a decline in basic number genuinely represents loss of DNA it can imply specialization. Forms of allopolyploidy in combining these specialized types reverse the trend, thus giving more widely adapted entities.

Several qualifications are, of course, required. At any level of ploidy DNA amplification and mutation can occur, as is well known in wheat and its relatives (Flavell *et al.*, 1987). Polyploidy that subsequently involves various forms of chromosome rearrangement and further hybridization enlarges the evolutionary arena. A latent tendency to perennialism, for example, could be intensified by partial sterility, allowing the plant to explore recombinant options not only over seasons but over years. Among the progeny the outcome might be reviving seediness and a return to annualism (as with some cereals), or that their sexuality is diverted towards apomixis, as seems likely among many semi-desert and desert perennial grasses.

An important model is that for *Aegilops* proposed by Zohary & Feldman (1962) and Zohary (1965), of which the following are the essential details. Among about 20 species are diploids each taxonomically a 'good' species and with a distinct seed dispersal type. The six genomic groups form sterile intergroup hybrids only. Each diploid (with three significant exceptions) has a relatively restricted distribution, namely *Ae. squarrosa* (DD), *Ae. umbellulata* (C^uC^u) and *Triticum boeticum* (AA). (Nomenclature is difficult, Stebbins (1956), for example, advocating pooling *Triticum* with *Aegilops*. Miller (1987) provides a detailed discussion of this problem and a synonomy.)

By contrast, the tetraploid *Aegilops* species are variable and widely distributed and comparatively 'weedy' (ecologically aggressive). Three of these tetraploids, for example, *Ae. variabilis* ($C^yC^yS^vS^v$), *Ae. ovata* ($C^uC^uM^oM^o$) and *Ae. biuncialis* ($C^uC^uM^bM^b$) share the C^u genome and will hybridize and segregate. Backcrossing and selfing also occur. The C^u genome is thus seen as 'pivotal', exercising a buffering effect permitting advantageous reorganization of the other genome. Since clusters of other tetraploids involve either AA or DD, these too are seen as pivotal genomes. Zohary (1965) regards the mixture of isolated diploids and hybridizing tetraploids as a flexible and dynamic system utilizing a common gene pool that accounts for the spread of weedy *Aegilops* species.

A quite different system occurs in *Saccharum*, where a series of high polyploids based supposedly on $n = 10$ is known. In the cross *S. officinarum* × *S. spontaneum* the former contributes an unreduced chromosome number. The hybrid as female parent if pollinated only once more with *S. spontaneum* will again contribute its unreduced chromosome number (Bremer, 1961). This curious process forms the basis of 'nobilization' in sugarcane and it is noteworthy that reciprocals and other crosses within *Saccharum* or with its related genera give the expected (reduced) chromosome numbers in their gametes.

A further contrast is the concept of the compilo-species. *Dichanthium intermedium* by a combination of sexual fertilization (with or without unreduced female gametes) and apoximis can both incorporate and exploit related germ plasm (Harlan & de Wet, 1963). Bashaw & Hanna (this volume) expand on this and regard a combination of apomixis and sexuality as a potent evolutionary mechanism.

Among grasses at low latitudes, as with those elsewhere, variations in the basic number for the genus are known. If lower basic numbers imply specialization, polyploids based on higher and lower numbers might represent 'conservative' and 'advanced' assemblages respectively. Conceivably, conservatism and apomixis could be mutually reinforcing on occasion.

Particularly as attention is turned towards semi-desert perennials and where apomixis is a significant factor, the models derived from *Aegilops*, *Saccharum* and *Dichanthium*, for example, may contribute but with modification to an emerging awareness of 'eco-apomixis'. What is clearly required is for more genera to be examined from this viewpoint.

Wide hybridization

C_4 pathways occur largely, though not exclusively, towards the more peripheral grasses. *Neurachne*, a three-styled, seemingly rather primitive genus, is an exception having C_3 and C_4 types and curiously in *N. minor* C_3–C_4 (Hattersley *et al.*, 1986).

Three alternative explanations for C_4 distributions among grasses have been suggested, namely that C_4 types have arisen independently on several occasions (Brown, 1977; Clayton & Renvoize, 1986; Watson, this volume) or the interesting suggestions that C_4 potential arose early in the history of the group but was concealed or that genetic information is transmitted 'clandestinely' by a natural version of genetic engineering (Watson, Clifford & Dallwitz, 1985). These possibilities prompt two questions: first, do such mechanisms need to be confined to transfer of photosynthetic pathways, and second, what is the true extent of *wide* hybridization in the Poaceae? Watson (this volume) lists known hybrids but brings a quite proper scepticism to some of the more surprising claims. Data on wide crosses normally are available only from crop genera and even if in Nature rare cases of similar sorts did occur we would probably be ill-placed to detect them assuming they survived. An interesting exception is *Tarigidia*, a genus based on one species *T. aequiglumis*. Loxton (1974) considers this a natural though rarely occurring hybrid of *Anthephora pubescens* and a *Digitaria* species. Within subfamily Panicoideae and tribe Paniceae this would imply a cross between the subtribes Cenchrinae and Digitariinae. Table 10.1 sets out five examples in approximate increasing order of taxonomic distance as a basis against which future claims about wide hybridization might be made, supported, one would suggest, preferably by voucher herbarium specimens, photographs, mitotic and meiotic analyses and some evidence of transfer of DNA.

Other examples of wide cross-pollination are known where, although some sexual interaction occurs, the effect is transitory and not passed on to the subsequent generations. These include *Triticum* × *Avena* giving pseudogamous seeds (Kruse, 1969) and *Triticum* × *Zea* giving haploids of the former (Laurie & Bennett, 1988).

For the breeder, the rationale of hybridization is that separate advantages of different plants might be combined in one individual and subsequently multiplied to a variety. Nature interposes an array of recognition mechanisms upon attempts at wide hybridization which the breeder seeks to overcome, moderate or bypass. Fusing somatic cells is straightforward but differentiation thereafter is problematic (see Lazzeri, Kollmorgen & Lörz, this volume). The search therefore has increasingly turned towards the zygote, a cell *par excellence* programmed for differentiation to a whole plant, to which, via fertilization or some modification of it, defined portions of DNA might be added in ways that are direct, precise and reliable, a matter to which we return when considering syngamy and recognition.

Table 10.1. *A framework for assessing wide hybridization in the* Poaceae

Hybrid or quasi-hybrid			Comments	Reference
Group 1. Within subtribe (Tripsacinae)			Describes 'counterfeit hybrids'	de Wet *et al.* (1984)
	Zea × *Tripsacum*			
Group 2. Within subfamily (Panicoideae)			Requires embryo rescue. Some chromosomes lost in the hybrid	Nitsch *et al.* (1986)
Tribes:	Paniceae,	Andropogoneae		
Subtribes:	Cenchrinae,	Tripsacinae		
	Pennisetum	× *Zea*		
Group 3. Between subfamilies				
(*a*) **Subfamilies:**			Contains expected no. of chromosomes.	Farooq & Naqvi (1987)
	Bambusoideae,	Chloridoideae	$(12 + 10 = 22)$	
Tribes:	Oryzeae,	Eragrostideae	$C_3 \times C_4$ (NAD-Me)	
Subtribes	—	Eleusiniae		
	Oryza	× *Leptochloa*		
(*b*) **Subfamilies:**			$C_3 \times C_4$ (NAD-Me)	Wu & Tsai (1963)
	Bambusoideae,	Panicoideae	'Hybrid' was sterile and resembled rice	
Tribes:	Oryzeae,	Paniceae		
Subtribe:	—	Cenchrinae		
	Oryza	× *Pennisetum*		
(*c*) **Same subfamilies as** (*b*)			$C_3 \times C_4$ (NADP-Me)	Zhou *et al.* (1981)
Tribes:	Oryzeae,	Andropogoneae	Hybrid resembled rice but acquired *Sorghum* grain colour and a band of esterases	
Subtribe:	—	Sorghinae		
	Oryza	× *Sorghum*		

Based on Clayton & Renvoize (1986) nomenclature.

Departures from sexuality

Cleistogamy and apomixis taken to extremes negate or avoid the consequences of sexuality since, by different means but leading to the same end, genotypes are perpetuated unchanged. In such extreme cases the obvious conclusion is that such genotypes might represent near-limits of adaptive refinement and recombination would be mostly deleterious. In practice both cleistogamous and apomictic biotypes undergo occasional hybridizations with perhaps consequent heterozygosis, and the ability to respond to selection is retained.

Some grasses oscillate between different modes on a seasonal basis: for example, *Bothriochloa decipiens* is cleistogamous in short and chasmogamous in long days (Heslop-Harrison, 1961); *Dichanthium aristatum* varies between apomixis and sexuality in short and long days respectively (Knox & Heslop-Harrison, 1963). Quarin (1986) reported increased frequency of apomixis at peak flowering in polyploids of *Paspalum cromyorrhizon*. *Rottboellia exaltata* not only shows a cleistogamy/chasmogamy switch from short to long days but extends the response into the vegetative behaviour of the resulting seeds, those arising from chasmogamy manifesting both dormancy and a propensity to tillering (Heslop-Harrison, 1961). As Heslop-Harrison (1983) remarked, the genetic system itself exhibits phenotypic plasticity. This contrasts with the system in those bamboos where time-to-flowering seems to be driven by an internal clock largely independent of external conditions.

Some grasses of arid and semi-arid regions

Among the grasses examined by Kernick (this volume) botanists have made numerous collections and various ecotypes have been recognized. Cutting regimes and digestibility studies have been begun, as have those on saline tolerance and photosynthetic efficiency in some species (see Anon., 1984). For genetics, however, work has hardly gone beyond scattered chromosome counts, although *Dichanthium* provides an early and interesting exception (see Harlan & de Wet, 1963). Again, *Panicum turgidum* is known to exist in three modes, a sand binder, a forage and a primitive cereal (Williams & Farias, 1972). It will persist in areas with 20–30 mm of rainfall per year, will colonize unstable dunes and can root beyond a depth of 1 m. Genetically, it remains unexplored. It might be that as a sand binder it is of greatest value but if the facility for seediness were increased it might assist the establishment of new populations on unstable dunes.

Few institutes in the arid regions are equipped to sustain breeding programmes although presumably the potentially valuable benefits will come to be recognized eventually and the dearth of viable seed collections has already been referred to. What, then, is an appropriately tentative assessment?

1 Diversity

One can distinguish between diploid and polyploid grasses, the latter often but not invariably tending towards perennialism. Where the basic number for the genus can vary, diploidy or polyploidy based on higher rather than lower basic numbers would make their possessors genetically more 'conservative'. Examples include *Hyparrhenia* $x = 10$, 15 *H. hirta* 30, 44 *Oryzopsis* $x = 11$, 12 *O. holciformis* 24, *Panicum* $x = 7$, 9, 10 *P. antidotale* 18 *Pennisetum* $x = 7$, 9 *P. squamulatum* 54 ($6x$). Polyploid complexes in stress situations can include apomicts. The best prospects at low latitudes must be with C_4 grasses and it is noteworthy that of the 33 genera showing apomixis listed (e.g. Watson, this volume) many are C_4 grasses. One could envisage a breeding strategy based logically on genera with high species numbers where access to genetic diversity might be facilitated although it should be recalled that several small genera here have contributed useful species including *Cynodon* (8), *Eleusine* (9), *Lasiurus* (1) and *Shoenefeldia* (2). (Species per genus are in parentheses.)

2 Longevity

Harberd (1961, 1962) estimated clone age for *Festuca rubra* at up to 400 years and for *F. ovina* up to 1000 years. *Phragmites australis* quoted by Richards (this volume) having an estimated 6000 years puts it well beyond the conifers *Sequoia* and *Taxodium* believed to be about 4000 years old and thus among the older living things on the planet.

For desert and semi-desert grasses we appear to have no comparable estimates. Herbarium specimens of xeromorphic desert grasses when set beside known rainfall patterns suggest, sometimes, tenacious long-term occupancy with periods of mostly minimal physiological activity. In the field, this author has seen growth circles of some Australian *Eragrostis* and *Triodia* species which, especially in open habitats, might be convenient for longevity estimates. The existence of 'resurrection' plants in southern Africa such as *Eragrostis nindenensis*, *Oropetium capense* and *Tripogon minimus* (Gaff & Ellis, 1974) and in Australia *Eragrostiella bifaria*, *Tripogon loliiformis* and various species of *Micraria* and *Sporobolus portobolus* (Gaff & Latz, 1978) adds a further dimension of interest. Therefore, although individual plants of desert and semi-desert grasses might, by inspection, appear to be of great age, no convincing quantitative data are known to the author. Obviously, too, it is unsurprising that the age of genotypes perpetuated unchanged through apomictic seed generations has attracted little scientific attention so far.

3 Pollination and seed dispersal

Although typically anemophilous, the insect pollination known to occur in bamboos has parallels in other grasses, and in Jamaica it is not unusual, for example, to see honey bees collecting pollen of *Cynodon dactylon*. Bogdan (1962) lists for Kenya 20 genera representing 52 grass species visited for pollen honey bees or *Nomia* or both. It is evident that activity is not perfunctory. Bogdan reports 'sometimes thousands of bees were working in the nursery plots'. Of *Nomia* visiting *Panicum coloratum* he mentions 'astonishing numbers . . . the hissing noise made by these small insects can be heard 3–4 yards away'. Bogdan's list includes *Chloris gayana* (*Nomia*), *Cynodon dactylon*, *Hyparrhenia hirta* (*Apis*) and *H. dissoluta* (*Nomia* and *Apis*) and also lists earlier references to honey bees collecting pollen of pearl millet and maize.

Grass seeds of species beyond the temperate regions frequently depend in part on animals for their distribution, as indicated by Clayton (this volume). A curious instance is that involving the dryland grass *Yakirra*, an Australian grass formerly included in *Panicum* (Lazarides & Webster, 1984). The rachillar stipe is prolonged and extended below the terminal floret. It appears to function as an elaisome by which means ants gather the caryopses (Berg, 1985). Latz (personal communication) notes that bandicoots collect the caryopses from ants' nests. There is, however, a human involvement: Aborigines also collect such caryopses and then put them in a 'coolamon', a shallow wooden bowl, and by 'yandying' (a kind of winnowing) separate out unwanted matter.

Kernick (this volume) has argued for emphasis on indigenous species for use in environmentally stressed situations. If this policy were adopted, it would mean, in practice, that we would need to know much more about grasses that so far have been among the least studied, but the prospect is one that surely combines great need and great interest.

Cereals

The reproductive systems of cereals, especially those crops of major importance, are well understood and Mogensen (this volume) has set out a detailed review. Looking to the future one can attempt to identify areas of increasing interest, but before doing so, it is worth asking what makes for 'domesticability'. Harlan (personal communication) points out that more cereals are self- than open-pollinated but that maize, rye and pearl millet, all open-pollinated, are very successful. The annual habit is common to all cereals, perhaps since the most metabolizable energy is put into seeds, and cereals can be found among triticoids, panicoids, andropogonoids, oryzoids and Avenae. Harlan further remarks that hunter-gatherers might first have been attracted to large-seeded annuals that occurred in massive stands and were, of course, unaware that they spanned the range of ploidy. His conclusion is that 'the pattern is no pattern'. Earlier, Harlan, de Wet & Price (1973) had shown that, regardless of species, domestication imposes similar selection pressures, among them free threshing and non-shattering. One might add 'suitability', to include diminished dormancy, productivity, preferred flavour and texture and the ability to travel with humankind moving from one ecology to another. There are degrees of 'domesticatedness', it could be argued, that have gone to completion only when a cereal is totally human-dependent for perpetuation, examples of which would be bread-wheat and maize.

An instructive genus here is *Pennisetum*. Some of its perennials are polyploid on $x = 9$. Morphologically, it is close to *Cenchrus* (where curiously $x = 17$), the latter genus perhaps deriving in whole or in part from *Pennisetum*. A separate lineage within *Pennisetum* is where $x = 7$ with diploids *P. mollisimum* and *P. violaceum*, the closely related domesticate *P. americanum* (pearl millet) and a related tetraploid *P. purpureum* (napier grass). Pearl millet is an impressive cereal in semi-arid situations with substantial genetic diversity, though apparently not including apomixis. It is noteworthy that many farmers would prefer to grow sorghum to pearl millet and maize to sorghum given the choice. Is it that a 'modernizing' trend in *Pennisetum* from $x = 9$ to $x = 7$ opened the way for eventual domestication? Even so, it is unable to compete on equal terms with a far more 'modernized' grass, namely maize, and is considered only for areas where maize will not grow. In temperate regions wheat is the preferred cereal, a fact that has long interested scholars.

In 1704 John Ray commented:

> It is worthy the noting, That *Wheat*, which is the beſt Sort of Grain, of which the pureſt, moſt ſavory and wholeſome Bread is made, is patient of both Extreams, Heat and Cold, growing and bringing its Seed to Maturity, not only in temperate Countries, but alſo on one Hand in the Cold and Northern, *viz. Scotland, Denmark*, &c. on the other, in the hotteſt and moſt Southerly, as *Egypt, Barbary, Mauritania*, the *Eaſt-Indies, Guinea, Madagaſcar*, &c. ſcarce refuſing any Climate.

but in so doing was echoing Pliny, from Roman times, that nothing is more fruitful than wheat.

Clayton & Renvoize (1986) in considering the taxonomic problem of the Triticeae commented on its hybridity. Watson (this volume) regards this as demonstrating relatively recent reticulate evolution. One is reminded of Vilmorin as quoted by Darwin (1868).

> The most celebrated horticulturist in France, namely Vilmorin, even maintains that, when any particular variation is desired, the first step is to get the plant to vary in any manner whatever, and to go on selecting the most variable individuals, even though they vary in the wrong direction; for the fixed character of the species being once broken, the desired variation will sooner or later appear.

The Triticeae thus seem to be already primed by natural hybridization (a possible expression of the 'genetic turbulence' at the margin of the family mentioned earlier) and this may help to explain the success of both wheat domestication and modern breeding.

For whatever reason, maize, wheat and rice have shown themselves responsive to the breeders' efforts. This in turn has drawn in further scientific resources and around these in particular there has gathered an impressive technology that serves increasingly to isolate them from other cereals. Scientifically, they set the pace. What remains unexplained is whether our plant science here is wholly directed by our food preferences or whether there is some underlying flexibility or responsiveness in these cereals that makes them especially suitable both for domestication and for science.

Syphonogamy and recognition

Events from pollination to fertilization involve sequential recognition at the stigma surface, in the style, synergid, at the egg cell boundary and at nuclear fusion and what fertilization normally achieves is a zygote highly conserved genetically. That is to say, each gamete contributes matching chromosome homologues and the resulting offspring is recognizably of the same species and adapted to the same ecology. Angiosperm diversity depends for its continuance partly on the precision of syphonogamous reproduction. The egg cell is programmed to double, by addition from outside, the amount of DNA it contains initially and is expressly a *receptive* structure. Syphonogamy effectively protects it from entry by DNA greatly dissimilar to what is there already. From there it is a short step to compromise the precision of syphonogamy and seek to introduce dissimilar DNA by interspecies or even intergeneric hybridization. A modification of this approach is to work *within* a species thus operating all the recognition events efficiently but incorporating foreign DNA at some point in the process. The object is to achieve a modified zygote which, as the foundation cell of a new individual, would, theoretically, convey the change to every descendent cell including those in the germ line, an approach that has been applied successfully to animals (see Hammer *et al.* (1985) and Clarke (1988). What then are the prospects for cereals?

The pollen pathway

The 'counterfeit hybrids' that comprise 'tripsacoid' maize suggested that alien DNA could be added to an otherwise intact maize genome. De Wet *et al.* (1985, 1986) presented evidence that germinated maize pollen treated with *Tripsacum* DNA could in some cases impart tripsacoid characters to the resulting maize plant. Again, a maize susceptible to *Puccinia sorghi* self-pollinated with its pollen treated with DNA from a resistant maize line could impart rust resistance to the F_1 and thereafter be inherited as a simple Mendelian dominant. In much the same way red grain colour could sometimes be added to hitherto white-grained plants. A substantially similar approach was that of Ohta (1986) who also reported transformation.

It was far from clear how DNA molecules, quite apart from integrating themselves into the male genome, could even enter the pollen tube and the matter was examined critically by Heslop-Harrison (1988). A different approach was taken by Roeckel *et al.* (1988) which accepted that, if DNA were to enter the pollen tube by some means, it would need to be protected from the host nucleases.

An alternative method used with cotton for the incorporation of alien DNA was that after fertilization the style was decapitated and such DNA was applied to the severed pollen tubes which it was assumed acted as 'micro pipettes' providing ready access to the newly formed zygote (Zhou *et al.*, 1983; Zhou, 1985). Again, transformation was claimed. Similar claims were made by Duan & Chen (1985) for transformation in rice. Using this approach, Luo & Wu (1988) presented molecular evidence for the incorporation of neomycin phosphotransferase (npt II) but, significantly, refer to the 'pollen tube pathway'.

Two other papers are perhaps significant in seeking to understand what was apparently happening. Gong *et al.* (1988) utilized H^3 labelled DNA applied to pollen tubes in cotton styles. It reached the embryo sac but was found not inside the pollen tubes but on the *outside*. In another study using *Hemerocallis*, *Raphanus* and *Vicia*, Sanders & Lord (1989) presented the surprising finding that inert latex particles introduced into cut styles travelled towards the ovules at rates comparable to those of pollen tubes and for *Vicia* and *Raphanus* in a few cases reached the micropyle. The mechanism remained unexplained.

From a variety of sources evidence, seemingly, has accumulated that alien DNA might travel down the pollen tube pathway, enter the zygote, be incorporated, expressed and subsequently inherited and thus to have compromised the precision of syphonogamy. If eventually it were shown that the approach outlined here were reliable, it would have considerable attractions for the breeder.

The pollen tube pathway leads to the embryo sac. If the embryo sac could be isolated in viable condition, the pathway itself could be discarded and alien DNA could be applied directly. Enzymic isolation of the *Zea mays* embryo sac has been described by Wagner *et al.* (1989). True *in vitro* fertilization (i.e. direct fusion of male and female gametes) requires a corresponding isolation of viable male gametoplasts and this has been studied by Dupuis *et al.* (1987). Knox, Southworth & Singh (1988) suggested that DNA might be used to transform the male gamete for *in vitro* fertilization, but an alternative is that of adding DNA directly to the egg cell along with rather than through the male gamete (Wagner *et al.*, 1989).

Perhaps therefore an emerging technology will centre on the cereal zygote for several reasons. Cereals monopolize our plant genetics. *Agrobacterium* systems applied to cereals have had, so far, relatively little success, somatic cells show restricted embryonic potential and animal scientists have pioneered the transgenic possibilities of zygote manipulation. It is, however, premature to abandon or advocate any one approach (Lazzeri *et al.*, this volume). What emerges is that cereal breeders have the advantages of classical breeding techniques combined with the first option on whatever new technology seems potentially useful, as, for example, with the ingenious approach to F_1 hybrid production in wheat indicated by Knox & Singh (this volume).

Plant breeding is a science and one with which molecular biology interacts. It remains true that plant breeding is also an art in which long time scales, a cumulative shared memory among breeders and subtle shifts of emphasis are important. For a variety to sell it must combine both productivity and wide ecological tolerance. Even so, the author recalls being shown a barley variety trial where, ignoring the plot labels, the guide identified the individual breeders responsible from the appearance of their varieties. Amid preoccupation with performance, the breeder had created varieties upon which it was still possible for him to impose his own distinctive image of the barley plant.

Graminum mysteria

Arber asked the question quoted at the beginning of this book about the intrinsic character of grasses. She subsequently concluded that we are confronted with an 'abiding mystery'.

Fashions change and nowadays many scientists would consider such speculation either a luxury or an impediment. In Arber's case it is evident, especially from her later writings (1950, 1954), that a penetrating mind was inseparable from a sense of wonder about the plants with which she worked. Perhaps her eloquence can be offered in place of our silence on these matters.

The context as well as the fashions of science change. It is not only botanists who are now aware that the earth's green mantle is a fragile and inceasingly threadbare garment, and the chapter by Kernick here is essential to a modern perspective. The message is unequivocal. We have to abandon exploitation and embrace stewardship.

As we confront the problems of a planet under pressure ecologically due to the combined effects of pollution, desertification and salinization set in train by population growth, we need many kinds of grass. Let us have a proper respect for those Neolithic farmers who, in response to their own situation, taught us the valuable lesson that some grasses are better than others and a few are very good indeed.

Our problem, though, is different both in kind and in scale. We have made havoc in a way that does not readily correct itself. In retrieving the situation, grasses have an immense importance. This is not merely the chosen few that are our crops but the many to whose existence until now we have paid small heed but which, if we bestir ourselves towards them, harbour untold benefit.

Why this should be so, and it is, is today's part of the *graminum mysteria*.

References

Anon. (1984). *Forage and Browse Plants for Arid and Semi-arid Africa.* IBPGR/RBG Kew. 293 pp.

Arber, A. (1934). *The Gramineae: a study of cereal, bamboo and grass.* Cambridge University Press. 480 pp.

(1950). *The Natural Philosophy of Plant Form.* Cambridge University Press. 247 pp.

(1954). *The Mind and the Eye. A Study of the Biologist's Standpoint.* Cambridge University Press. 146 pp.

Berg, R. Y. (1985). Spikelet structure in *Panicum australiense* (Poaceae): taxonomic and ecological implications. *Aust. J. Bot.* **33**, 579–83.

Bogdan, A. V. (1962). Grass pollination by bees in Kenya. *Proc. Linn. Soc.* **173**, 57–60.

Bokareva, L. I. (1974). Features of the flowering and fruiting of bamboo in the Abkhazian ASSR. *Referativnyi Zhurnal*, **10**, 55. 623. 232–43. (Eng. abstract.)

Bremer, G. (1961). Problems in breeding and cytology of sugarcane. IV. The origin of the increase of chromosome number in species hybrids of *Saccharum. Euphytica*, **10**, 325–42.

Brown, W. V. (1977). The Kranz Syndrome and its subtypes in grass systematics. *Mem. Torr. Bot. Club*, **23** VI and **97**, 6, 25, 26, 28.

Clarke, A. J. (1988). Transgenic biology. *AFRC Ann. Rep. Edinburgh*, pp. 1–9. Research Station, Roslin.

Clayton, W. D. & Renvoize, S. A. (1986). *Genera Graminum: grasses of the world.* Kew Bull. Addit. Ser. 13. Royal Botanic Gardens, Kew: London.

Clifford, H. T. (1961). Floral evolution in the family Gramineae. *Evolution*, **15**, 455–60.

Darwin, C. (1868). *Animals and Plants under Domestication* (1st ed.), 2 vols. John Murray: London.

Davis, T. A. W. & Richards, P. W. (1933). The vegetation of Moraballi Creek, British Guiana: an ecological study of a limited area of tropical rain forest. Part I. *J. Ecol.* **21**, 350–84.

de Wet, J. M. J., Newell, C. A. & Brink, D. E. (1984). Counterfeit hybrids between *Tripsacum* and *Zea* (Gramineae). *Amer. J. Bot.* **71**, 245–51.

de Wet, J. M. J., de Wet, A. E., Brink, D. E., Hepburn, A. G. & Woods, J. A. (1986). Gametophyte transformation in maize (*Zea mays*), Gramineae. In *Biotechnology and Ecology of Pollen*, ed. D. L. Mulcahy, G. B. Mulcahy & E. Ottaviani, pp. 59–64. Springer-Verlag: New York. 530 pp.

de Wet, J.M.J., Bergquist, R.R., Harland, J.R., Brink, D.E., Cohen, C.E., Newell, C. A. & de Wet, A. E. (1985). Exogenous gene transfer in maize (*Zea mays*) using DNA-treated pollen. In *Experimental Manipulation of Ovule Tissues*, ed. G. P. Chapman, S. H. Mantell & R. W. Daniels, pp. 197–209. Longman: New York, London. 256 pp.

Duan, X. & Chen, S. (1985). Variation of the characters in rice (*Oryza sativa*) induced by foreign DNA uptake. *China Agric. Sci.* **3**, 6–9. (Chinese. Referred to by Luo & Wu, 1988.)

Dupuis, I., Roeckel, P., Matthys-Rochon, E. & Dumas, C. (1987). Procedure to isolate viable sperm cells from corn (*Zea mays* L.) pollen grains. *Plant Physiol.* **85**, 876–8.

Farooq, S. & Naqvi, S. H. M. (1987). Problems and prospects of rice × kalla grass by hybridisation. *Pakistan J. Sci. Res.* **30**, 660–3.

Flavell, R. B., Bennett, M. D., Seal, A. G. & Hutchinson, J. (1987). Chromosome structure and organisation. In *Wheat Breeding. Its Scientific Basis*, ed. F. G. H. Lupton, pp. 211–68. Chapman & Hall: London. 566 pp.

Gaff, D. R. & Ellis, R. P. (1974). Southern African grasses with foliage that revives after dehydration. *Bothalia*, **11**, 305–8.

Gaff, D. F. & Latz, P. K. (1978). The occurrence of resurrection plants in the Australian flora. *Aust. J. Bot.* **26**, 485–92.

Gong, Z., Shen, W., Zhou, G., Huang, J. & Qian, S. (1988). Introducing exogenous DNA into plants after pollination. *Scientia Sinica (Ser. B)* **31**, 1080–7.

Hammer, R. E., Pursel, V. G., Rexroad, C. G. Jr., Wall, R. J., Bolt, D. J., Ebert, K. M., Palmiter, R. D. & Brinster, R. L. (1985). Production of transgenic rabbits, sheep and pigs by micro-injection. *Nature*, **315**, 680–3.
Harberd, D. J. (1961). Observations on population structure and longevity of *Festuca rubra*. *New Phytol.* **60**, 184–206.
(1962). Some observations of natural clones in *Festuca ovina*. *New Phytol.* **61**, 85–100.
Harlan, J. R. & de Wet, J. M. J. (1963). The compilo-species concept. *Evolution*, **17**, 497–501.
Harlan, J. R., de Wet, J. M. J. & Price, E. G. (1973). Comparative evolution of cereals. *Evolution*, **27**, 311–25.
Hattersley, P. W., Wong, S.-C., Perry, S. & Roksandie, Z. (1986). Comparative ultrastructure and gas exchange characteristics of the 3-carbon photosynthetic pathway, 4-carbon photosynthetic pathway intermediate *Neurachne minor* Poaceae. *Plant Cell Env.* **9**, 217–34.
Heslop-Harrison, J. (1961). Photoperiodic affects on sexuality, breeding system and seed germination in *Rottboellia exaltata*. *Apomixis, environment and adaptation. Proc. 11th Int. Bot. Congr. Montreal*, pp. 891–5.
(1983). The reproductive versatility of plants: an overview. In *Strategies of Plant Reproduction*, ed. W. I. Mendt, pp. 3–18. Beltsville Symposia in Agricultural Research. No. 6. Allenheld Osmun: London. 386 pp.
(1988). Some permeability properties of angiosperm pollen grains, pollen tubes and generative cells. *Sex. Plant Reprod.* **1**, 65–73.
Jackson, G. C. & Woodbury, R. O. (1976). Host plants of the carpenter bee (*Xylocopa brasilianorum* L.) (Hymenoptera: Apoidea) in Puerto Rico. *J. Agr. Univ. Puerto Rico*, **60**, 639–60.
Janzen, D. H. (1976). Why bamboos wait so long to flower. *Ann. Rev. Ecol. and Syst.* **7**, 347–91.
Knox, R. B. & Heslop-Harrison, J. (1963). Experimental control of apomixis in a grass of the Andropogoneae. *Bot. Notis.* **116**, 127–41.
Knox, R. B., Southworth, D. & Singh, M. B. (1988). Sperm cell determinants and control of fertilisation in plants. In *Eukaryote Cell Recognition: concepts and model systems*, ed. G. P. Chapman, C. C. Ainsworth & C. J. Chatham, pp. 175–93. Cambridge University Press. 315 pp.
Kruse, A. (1969). Inter generic hybrids between *Triticum aestivum* L. (V. Kogall, $2n = 42$) and *Avena sativa* L. (v. Stal. $2n = 42$) with pseudogamous seed formation. *Prelim. Rept. in Royal Vet. & Agric. Coll. Yearbook*, p. 188. Copenhagen.
Laurie, D. A. & Bennett, M. D. (1988). The production of haploid wheat plants from wheat × maize crosses. *Theor. Appl. Gen.* **76**, 393–7.
Lazarides, M. & Webster, R. D. (1984). Yakirra (Paniceae, Poaceae), a new genus for Australia. *Brunonia*, **7**, 289–96.
Loxton, A. E. (1974). A note on a possible bigeneric hybrid between *Digitaria* and *Anthephora*. *Bothalia*, **11**, 285–6.
Luo, Z.-X. & Wu, R. (1988). A simple method for the transformation of rice via the pollen tube pathway. *Plant Mol. Biol. Rep.* **6**, 165–74.
Miller, T. E. (1987). Systematics and evolution. In *Wheat Breeding. Its Scientific Basis*, ed. F. G. H. Lupton, pp. 1–30. Chapman & Hall: London. 566 pp.

Nitsch, C., Mornan, K. & Godard, M. (1986). Intergeneric crosses between *Zea* and *Pennisetum* reciprocally by *in vitro* methods. In *Biotechnology and Ecology of Pollen*, ed. D. L. Mulcahy, G. B. Mulcahy & E. Ottaviani, pp. 53–8. Springer-Verlag: New York. 530 pp.

Ohta, Y. (1986). High efficiency genetic transformation of maize by a mixture of pollen and exogenous DNA. *Proc. Nat. Acad. Sci. USA*, **83**, 715–19.

Pohl, R. W. (1974). Blooming behaviour of bamboos in Costa Rica. *Amer. J. Bot.* **61** (5, suppl.), 48–9.

Quarin, C. L. (1986). Seasonal changes in the incidence of diploid, triploid and tetraploid plants of *Paspalum cromyorrhizon*. *Euphytica*, **35**, 515–22.

Ray, J. (1704). *The Wisdom of God Manifest in the Works of Creation*, 4th edn. (For a detailed review, see Raven, C. E. (1950). *John Ray, Naturalist. His life and works.* Cambridge University Press. 506 pp.)

Roeckel, P., Heizmann, P., Dubois, M. & Dumas, C. (1988). Attempts to transform *Zea mays* via pollen grains. *Sex. Plant Reprod.* **1**, 156–63.

Sanders, L. C. & Lord, E. M. (1989). Directed movement of latex particles in the gynoecia of three species of flowering plants. *Science*, **243**, 1606–8.

Söderstrom, T. R. & Calderon, C. E. (1971). Insect pollination in tropical rain forest grasses. *Biotropica*, **3**, 1–16.

Stebbins, C. L. (1956). Taxonomy and evolution of genera with special reference to the family Gramineae. *Evolution*, **10**, 235–45.

Wagner, V. T., Song, Y. C., Maltby, S., Rochon, E. & Dumas, C. (1989). Observations of the isolated embryosacs of *Zea mays* L. *Plant Sci.* **59**, 127–32.

Watson, L., Clifford, H. T. & Dallwitz, M. J. (1985). The classification of Poaceae, subfamilies and supertribes. *Aust. J. Bot.* **33**, 433–84.

Williams, J. T. & Farias, R. M. (1972). Utilisation and taxonomy of the desert grass *Panicum turgidum*. *Econ. Bot.* **26**, 13–20.

Wong, K. M. (1981). Flowering, fruiting and germination of the bamboo *Schizostachyum sollingeri* in Perlis. *Malaysian Forester*, **44**, 453–63.

Wu, S. H. & Tsai, L.-K. (1963). Cytological observations on the F_1 hybrid (*Oryza sativa* L. × *Pennisetum* sp). *Acta Bot. Sin.* **11**, 293–307. (Chinese.)

Zhou, G.-Y. (1985). Genetic manipulation of the ovule after pollination. In *Experimental Manipulaton of Ovule Tissues*, ed. G. P. Chapman, S. H. Mantell & R. W. Daniels. Longman: New York, London. 256 pp.

Zhou, G.-Y., Weng, J., Zeng, Y., Huang, J., Qian, S. & Liu, G. (1983). Introduction of exogenous DNA into cotton embryos. *Method Enzymol.* **101**, 433–81.

Zhou, G.-Y., Zen, Y. & Yang, W. (1981). The molecular basis of remote hybridisation. An evidence for the possible integration of sorghum DNA into the rice genome. *Sci. Sin.* **24**, 701–9.

Zohary, D. (1965). Colonizer species in the wheat group. In *The Genetics of Colonizing Species*, ed. H. G. Baker & G. L. Stebbins, pp. 403–19. Academic Press: New York, London. 588 pp.

Zohary, D. & Feldman, M. (1962). Hybridisation between amphidiploids and the evolution of polyploids in the wheat (*Aegilops–Triticum*) group. *Evolution*, **16**, 44–61.

See Note added in proof on page 266.

Appendix

World grass genera

L. Watson

Classification into subfamilies, supertribes, tribes and (for Andropogoneae only) subtribes. Genera with 40 or more species indicated by †; species numbers for tribes, and for genera with 100 or more species, in parentheses.

POOIDEAE

Annual or perennial, herbaceous; culms unbranched above, usually with hollow internodes. Leaf blades not pseudopetiolate, without transverse veins. Ligule an unfringed membrane. Inflorescences various, but not comprising spikelike main branches; espatheate, espatheolate. Spikelets nearly always laterally compressed or terete, with 1–many hermaphrodite florets, rarely with incomplete proximal florets. Lemma without a germination flap. Palea usually 2-keeled and apically notched. Lodicules 2, usually membranous. Stigmas white. Hilum long-linear or short. Embryo small, usually with an epiblast, with neither mesocotyl internode nor scutellar tail, the embryonic leaf margins meeting. *Abaxial leaf blade epidermis*. Microhairs and papillae absent; costal silica bodies various, but hardly ever 'panicoid-type' or saddle-shaped; costal short-cells not in long rows. Stomatal guard-cells overlapped by the interstomatals. *Physiology, transverse section of leaf blade*. C_3. Adaxial surface usually ribbed. Mesophyll without fusoids and arm cells, without colourless columns. Midrib usually with a single vascular bundle. All vascular bundles accompanied by sclerenchyma. *Cytology*. Basic chromosome number usually $x = 7$. Group mean diploid 2*c* DNA value 8.9 pg. Rusts: *Puccinia*. Smuts: *Entyloma*, *Tilletia*, *Urocystis*, *Ustilago*.

North and south temperate, tropical mountains (with a Laurasian diversification).

Triticodae

Inflorescence usually spicate, commonly disarticulating at the joints. Glumes often lateral or 'displaced'. Lemma awns non-geniculate, usually entered by several veins. Ovary apex hairy, lodicules often ciliate. Endosperm containing only simple starch grains. *Abaxial leaf blade epidermis*. Crown cells sometimes present; stomata often very large. *Cytology*. Group mean diploid 2*c* DNA value 10.6 pg.

Triticeae (358): *Aegilops*, *Agropyron*, *Amblyopyrum*, *Australopyrum*, *Cockaynea*, *Crithopsis*, *Daspyrum*, †*Elymus* (150), *Elytrigia*, *Eremopyrum*, *Festucopsis*, *Henrardia*, *Heteranthelium*, *Hordelymus*, †*Hordeum*, *Hystrix*, *Leymus*,

Lophopyrum, *Malacurus*, *Pascopyrum*, *Psathyrostachys*, *Pseudoroegneria*, *Secale*, *Sitanion*, *Taeniatherum*, *Thinopyrum*, *Triticum*.
Brachypodieae (16): *Brachypodium*.
Bromeae (150): *Boissiera*, †*Bromus*.

Poodae

Inflorescence usually paniculate with persistent axes, or a persistent spike. Spikelets usually disarticulating above the glumes. Lemma awn geniculate or straight, entered by only one vein. Ovary apex usually glabrous; lodicules usually glabrous. Endosperm usually containing compound starch grains. Group mean 2*c* DNA value 7.9 pg.
Aveneae (including Agrostideae, Phalarideae – 1050): †*Agrostis* (220), *Aira*, *Airopsis*, *Alopecurus*, *Ammochloa*, *Ammophila*, *Amphibromus*, *Ancistragrostis*, *Aniselytron*, *Anthoxanthum*, *Antinoria*, *Apera*, *Arctagrostis*, *Arrhenatherum*, *Avellinia*, *Avena*, *Beckmannia*, †*Calamagrostis* (230), *Chaetopogon*, *Cinna*, *Cornucopiae*, *Corynephorus*, *Cyathopus*, *Dasypoa*, †*Deschampsia*, *Deyeuxia*, *Dichelachne*, *Dielsiochloa*, *Echinopogon*, *Euthryptochloa*, *Gastridium*, *Gaudinia*, *Gaudiniopsis*, †*Helictotrichon*, *Hierochloë*, *Holcus*, *Hyalopoa*, *Hypseochloa*, †*Koeleria*, *Lagurus*, *Leptagrostis*, *Libyella*, *Limnas*, *Limnodea*, *Maillea*, *Mibora*, *Milium*, *Nephelochloa*, *Pentapogon*, *Periballia*, *Peyritschia*, *Phalaris*, *Phleum*, *Pilgerochloa*, *Polypogon*, *Pseudarrhenatherum*, *Pseudophleum*, *Relchela*, *Rhizocephalus*, *Scribneria*, *Sinochasea*, *Sphenopholis*, *Stephanachne*, *Stilpnophleum*, *Tovarochloa*, *Triplachne*, †*Trisetum*, *Vahlodea*, *Ventenata*, *Zingeria*.

Upper glume nearly always long relative to the adjacent lemma; female-fertile florets 1–2(–7).
Meliceae (136): *Brylkinia*, *Catabrosa*, †*Glyceria*, *Lycochloa*, †*Melica*, *Pleuropogon*, *Schizachne*, *Streblochaete*, *Triniochloa*.

Leaf sheath margins joined. Basic chromosome number $x = 9$ or 10.
Seslerieae (33): *Echinaria*, *Oreochloa*, *Psilathera*, *Sesleria*, *Sesleriella*.
Poeae (including Hainardieae, Monermeae – 1124): *Agropyropsis*, *Anthochloa*, *Aphanelytrum*, *Arctophila*, *Austrofestuca*, *Bellardiochloa*, *Briza*, *Calosteca*, *Castellia*, *Catabrosella*, *Catapodium*, *Coleanthus*, *Colpodium*, *Ctenopsis*, *Cutandia*, *Cynosurus*, *Dactylis*, *Desmazeria*, *Dissanthelium*, *Dryopoa*, *Dupontia*, *Eremopoa*, *Erianthecium*, †*Festuca* (360), *Festucella*, *Gymnachne*, *Hainardia*, *Helleria*, *Hookerochloa*, *Lamarckia*, *Leucopoa*, *Lindbergella*, *Littledalea*, *Loliolum*, *Lolium*, *Lombardochloa*, *Megalachne*, *Microbriza*, *Micropyropsis*, *Micropyrum*, *Narduroides*, *Parafestuca*, *Parapholis*, *Phippsia*, *Pholiurus*, †*Poa* (500), *Podophorus*, *Poidium*, *Pseudobromus*, *Psilurus*, †*Puccinellia*, *Rhomboelytrum*, *Sclerochloa*, *Scolochloa*, *Simplicia*, *Sphenopus*, *Torreyochloa*, *Tsvelevia*, *Vulpia*, *Vulpiella*, *Wangenheimia*.

Glumes usually short relative to adjacent lemma; female-fertile florets usually (1–)2–many.

BAMBUSOIDEAE

Mostly perennial, culms woody or herbaceous. Leaf blades often pseudopetiolate, often with transverse veins, often disarticulating. Inflorescence usually paniculate, often spatheate. Lemmas with non-geniculate awns or awnless. Palea keel-less, 1- or 2-keeled, notched or entire. Lodicules 1–10 (often 3). Hilium usually long-linear. Embryo usually small, with an epiblast, usually with a scutellar tail and overlapping leaf margins. *Abaxial leaf blade epidermis.* Microhairs present, panicoid-type. Papillae often present. Costal silica bodies often 'panicoid-type', 'oryzoid-type' or saddle-shaped. Stomatal subsidiaries usually triangular or dome-shaped. *Physiology, transverse section of leaf blade.* C_3. Adaxial surface often flat. Mesophyll commonly with arm-cells and/or fusoids. Midrib usually with more than one vascular bundle, often with complex vasularization. All vascular bundles accompanied by sclerenchyma. *Cytology.* Chromosome base number usually $x = 10$, 11 or 12. Rusts: *Dasturella, Physopella, Stereostratum, Puccinia.* Smuts: *Entyloma, Tilletia, Sorosporium, Tolyposporium, Ustilago.*

Tropical/warm temperate, mostly forest/woodland and wet places.

Oryzodae

'Grasses', or to varying degrees more or less bambusoid in appearance; mostly herbaceous, culms commonly unbranched above. Lodicules commonly 2, rarely ciliate. Stigmas usually 2. Mesophyll with or without arm-cells and/or fusoids. Mostly diploids. Group mean diploid 2*c* DNA value 3.0 pg.

Oryzeae (73): *Chikusichloa, Hydrochloa, Hygroryza, Leersia, Luziola, Maltebrunia, Oryza, Porteresia, Potamophilia, Prosphytochloa, Rhynchoryza, Zizania, Zizaniopsis.*

Spikelets laterally compressed, with one hermaphrodite floret and sometimes with a proximal sterile lemma. Glumes absent or reduced to a minute cupule. Hilum long-linear. Mostly hydrophytic or helophytic.

Olyreae (119); *Arberella, Buergersiochloa, Cryptochloa, Diandrolyra, Ekmanochloa, Froesiochloa, Lithachne, Maclurolyra, Mniochloa, Olyra, Pariana, Parodiolyra, Piresia, Raddia, Raddiella, Rehia, Reitzia, Sucrea.*

Monoecious, the spikelets mixed or in separate inflorescences. Female spikelets usually dorsally compressed, with well-developed glumes and one floret; lemma hard, usually with a germination flap.

Centotheceae (33): *Bromuniola, Calderonella, Centotheca, Chasmanthium, Chevalierella, Gouldochloa, Lophatherum, Megastachya, Orthoclada, Pohlidium, Zeugites.*

Anomochloeae (1): *Anomochloa.*

Brachyelytreae (1): *Brachyelytrum.*

Diarrheneae (4–5): *Diarrhena.*

Ehrharteae (44): *Ehrharta, Microlaena, Petriella, Tetrarrhena.*

Spikelets compressed laterally or terete; with two proximal sterile florets and one hermaphrodite floret.

Phaenospermateae (1): *Phaenosperma.*

Phyllorhachideae (3): *Humbertochloa, Phyllorhachis.*

Phareae (14): *Leptaspis, Pharus, Scrotochloa, Suddia.*

Streptochaeteae (2): *Streptochaeta.*

Streptogyneae (2): *Streptogyna.*

Bambusodae

Woody bamboos with branching culms. Lodicules usually 3 or more, ciliate. Stigmas usually 3 or more. Usually tetra- or hexaploid.

Guaduelleae (8): *Guaduella.*
Puelieae (6): *Puelia.*
Bambuseae (820): *Acidosasa, Actinocladum, Alvimia, Apoclada, Arthrostylidium, †Arundinaria, Athroostachys, Atractantha, Aulonemia, †Bambusa* (120), *Cephalostachyum, Chimonobambusa, †Chusquea* (100), *Colanthelia, Decaryochloa, Dendrocalamus, Dendrochloa, Dinochloa, Elytrostachys, Fargesia, Gigantochloa, Glaziophyton, Greslania, Hickelia, Hitchcockella, Indocalamus, Indosasa, Melocalamus, Melocanna, †Merostachys, Metasasa, Myriocladus, Nastus, Neohouzeaua, Neurolepis, Ochlandra, Olmeca, Oreobambos, Otatea, Oxytenanthera, Perrierbambus, †Phyllostachys, Pseudocoix, Pseudosasa, Pseudostachyum, Racemobambos, Rhipidocladum, †Sasa, Schizostachyum, Semiarundinaria, Shibataea, †Sinarundinaria, Sinobambusa, Sphaerobambos, Swallenochloa, Teinostachyum, Thamnocalamus, Thyrsostachys, Yushania.*

ARUNDINOIDEAE

Perennial herbs, often caespitose (sometimes 'bambusoid'). Leaf blades sometimes disarticulating. Ligule usually a fringed membrane or a fringe of hairs. Inflorescence usually paniculate, espatheate, axes persistent. Spikelets usually laterally compressed or terete and disarticulating above the glumes; with 1–many hermaphrodite florets, occasionally with proximal incomplete florets. Lemmas usually hairy, incised, usually awned, awns straight or geniculate. Palea usually 2-nerved and 2-keeled. Lodicules fleshy or membranous, ciliate or glabrous. Stigmas 2. Endosperm usually with compound starch grains. *Abaxial leaf blade epidermis.* Microhairs present or absent (then usually present somewhere on the plant). Non-papillate. Costal silica bodies of various forms, but hardly ever 'pooid-type'. Stomatal guard-cells usually not overlapped by the interstomatals; subsidiaries usually triangular to dome-shaped. *Physiology, transverse section of leaf blade.* C_3 or C_4 type NADP-ME. Adaxial surface usually ribbed. Mesophyll without fusoids, very rarely with arm-cells. Midrib usually with a single bundle. Smallest vascular bundles accompanied by sclerenchyma. *Cytology.* Chromosome base numbers variable. Rusts: *Dasturella, Puccinia.* Smuts: *Neovossia, Tilletia, Urocystis, Sorosporium, Sphacelotheca, Tolyposporium, Ustilago.*

Cosmopolitan, but with a marked Gondwanan bias.

Stipeae (419): *Aciachne, Ampelodesmos, Anemanthele, Anisopogon, Danthoniastrum, Lorenzochloa, Nassella, Orthachne, Oryzopsis, Piptatherum, Piptochaetium, Psammochloa, †Stipa* (300), *Trikeraia.*

Spikelets with one hermaphrodite floret only; lemma indurated, without a germination flap; with a geniculate apical awn entered by several veins. Palea keel-less. Lodicules often 3, glabrous. Hilum long-linear. *Abaxial leaf blade epidermis.* Microhairs absent (but a peculiar form sometimes found elsewhere on the plant). Costal silica bodies often 'panicoid-type', crescentic or rounded. *Physiology.* C_3.

Steyermarkochloeae (2): *Arundoclaytonia, Steyemarkochloa.*
Nardeae (1): *Nardus.*
Lygeae (1): *Lygeum.*
Arundineae (7): *Arundo, Phragmites, Thysanolaena.*

Mostly tall reeds. Spikelets with 2–many florets. Hilum short.

Danthionieae (324): *Afrachneria, Alloeochaete, Amphipogon, Centropodia, Chaetobromus, Chionochloa, Cortaderia, Crinipes, Danthonia, Danthonidium,*

Dichaetaria, *Diplopogon*, *Dregeochloa*, *Duthiea*, *Elytrophorus*, *Erythranthera*, *Gynerium*, *Habrochloa*, *Hakonechloa*, *Karroochloa*, *Lamprothyrsus*, *Merxmuellera*, *Metcalfia*, *Molinia*, *Monachather*, *Monostachya*, *Nematopoa*, *Notochloë*, *Pentameris*, †*Pentaschistis*, *Phaenanthoecium*, *Plinthanthesis*, *Poagrostis*, *Prionanthium*, *Pseudodanthonia*, *Pseudopentameris*, *Pyrrhanthera*, †*Rytidosperma*, *Schismus*, *Sieglingia*, *Styppeiochloa*, *Tribolium*, *Urochlaena*, *Zenkeria*.

Mostly less than 250 cm high. Spikelets with 2–many hermaphrodite florets, without proximal incomplete florets. Nearly all C_3 (exception *Centropodia*).

Spartochloeae (1): *Spartochloa*.

Cyperochloeae (1): *Cyperochloa*.

Micrairieae (13): *Micraira*.

Aristideae (344): †*Aristida* (290), *Sartidia*, †*Stipagrostis*.

Spikelets with one hermaphrodite floret, no incomplete florets. Lemma hardened, with or without a germination flap; with a characteristic trifid awn (or derivative of this). Palea reduced. C_3 (*Sartidia*) or C_4. Chromosome base number usually 11.

Eriachneae (42): †*Eriachne*, *Pheidochloa*.

CHLORIDOIDEAE

Culms herbaceous, branched or unbranched above. Leaf blades not pseudopetiolate. Ligule nearly always a fringed membrane or of hairs. Inflorescences various, commonly of dorsiventral, spikelike main branches or paniculate, espatheate; axes usually persistent. Sometimes dioecious or monoecious with unisexual spikelets, usually with hermaphrodite spikelets and hermaphrodite florets. *Female-fertile spikelets* usually disarticulating above the glumes, compression lateral to dorsiventral; lower glume usually 1-nerved; very rarely with proximal incomplete florets; hermaphrodite (or female) florets 1–many. Lemmas without a germination flap, awn if present non-geniculate. Palea 2-nerved and 2-keeled. Lodicules 2, fleshy, glabrous. Pericarp often free or loose. Hilum short. Embryo large; usually with an epiblast; scutellar tail and mesocotyl internode present, embryonic leaf margins meeting. *Abaxial leaf blade epidermis*. Microhairs present, chloridoid-type or (less often) panicoid-type. Costal silica-bodies mostly panicoid-type or saddle shaped, the costal short-cells in long rows. Stomatal guard-cells not overlapped by the interstomatals; subsidiaries triangular to dome-shaped. *Physiology, transverse section of leaf blade*. C_4 (one known exception in *Eragrostis*), types PCK and NAD-ME (all XyMS+). Adaxial surface often flat. Mesophyll often traversed by colourless columns; fusoids absent, arm-cells (nearly always) absent. Bulliforms commonly combined with colourless cells to form deep-penetrating fans. Smallest vascular bundles accompanied by sclerenchyma. *Cytology*. Chromosome base number usually $x = 10$. Group mean 2c DNA value 1.1 pg. Rusts: *Physopella*, *Puccinia*. Smuts: *Entyloma*, *Melanotaenium*, *Tilletia*, *Sorosporium*, *Sphacelotheca*, *Tolyposporella*, *Ustilago*.

Mostly tropical and subtropical (especially Old World and Gondwanan), especially dry climates.

Triodieae (54): *Monodia*, *Plectrachne*, *Symplectrodia*, *Triodia*.

Pappophoreae (41): *Cottea*, *Enneapogon*, *Kaokochloa*, *Pappophorum*, *Schmidtia*.

Orcuttieae (9): *Neostapfia*, *Orcuttia*, *Tuctoria*.

Chlorideae (including Cynodonteae, Eragrosteae, Sporoboleae, Aeluropodeae, Jouveae, Unioleae, Leptureae, Lappagineae, Spartineae, Trageae, Perotideae, Pommereulleae – 1303): *Acamptoclados, Acrachne, Aegopogon, Aeluropus, Afrotrichloris, Allolepis, Apochiton, Astrebla, Austrochloris, Bealia, Bewsia, Blepharidachne, Blepharoneuron,* †*Bouteloua, Brachyachne, Brachychloa, Buchloë, Buchlomimus, Calamovilfa, Catalepis, Cathestechum, Chaetostichium,* †*Chloris, Chrysochloa, Cladoraphis, Coelachyropsis, Coelachyrum, Craspedorhachis, Crypsis, Ctenium, Cyclostachya, Cynodon, Cypholepis, Dactyloctenium, Daknopholis, Dasychloa, Decaryella, Desmostachya, Diandrochloa, Dignathia, Dinebra, Diplachne, Distichlis, Drake-Brockmania, Ectrosia, Ectrosiopsis, Eleusine, Enteropogon, Entoplocamia, Eragrostiella,* †*Eragrostis* (350), *Erioneuron, Eustachys, Farrago, Fingerhuthia, Gouinia, Griffithsochloa, Gymnopogon, Halopyrum, Harpachne, Harpochloa, Heterachne, Heterocarpha, Hilaria, Hubbardochloa, Indopoa, Ischnurus, Jouvea, Kampochloa, Kengia, Leptocarydion, Leptochloa, Leptochloöpsis, Leptothrium, Lepturella, Lepturidium, Lepturopetium, Lepturus, Lintonia, Lophachme, Lopholepis, Lycurus, Melanocenchris, Microchloa, Monanthochloë, Monelytrum, Mosdenia,* †*Muhlenbergia* (160), *Munroa, Myriostachya, Neeragrostis, Neesiochloa, Neobouteloua, Neostapfiella, Neyraudia, Ochthochloa, Odyssea, Opizia, Orinus, Oropetium, Oxychloris, Pentarrhaphis, Pereilema, Perotis, Piptophyllum, Planichloa, Pogonarthria, Pogoneura, Pogonochloa, Polevansia, Pommereulla, Pringleochloa, Psammagrostis, Pseudozoysia, Psilolemma, Pterochloris, Redfieldia, Reederochloa, Rendlia, Richardsiella, Saugetia, Schaffnerella, Schedonnardus, Schoenefeldia, Sclerodactylon, Scleropogon, Silentvalleya, Soderstromia, Sohnsia?, Spartina,* †*Sporobolus* (160), *Steirachne, Stiburus, Swallenia, Tetrachaete, Tetrachne, Tetrapogon, Thellungia, Tragus, Trichoneura, Tridens, Triplasis, Tripogon, Triraphis, Uniola, Urochondra, Vaseyochloa, Viguierella, Willkommia, Zoysia.*

PANICOIDEAE

Culms mostly herbaceous, usually branching above, internodes more often solid than hollow. Ligule a fringed or unfringed membrane, or a fringe of hairs. Inflorescence commonly paniculate, equally commonly of spikelike main branches (these often dorsiventral), spatheate or not; the axes disarticulating or persistent. Sometimes dioecious or monoecious, or with hermaphrodite spikelets. Female-fertile spikelets usually compressed dorsiventrally and falling with the glumes; rachilla not prolonged apically; nearly always with one proximal incomplete (male or sterile) floret and one hermaphrodite (rarely female-only) floret. Lodicules fleshy, usually glabrous. Stigmas usually red-pigmented. Hilum usually short, Endosperm starch grains usually simple. Embryo usually large; without an epiblast; scutellar tail and mesocotyl internode present; embryonic leaf margins overlapping. *Abaxial leaf blade epidermis.* Microhairs present, panicoid-type. Costal silica-bodies nearly always 'panicoid-type' (horizontally-elongated, cross-shaped, dumb-bell-shaped or nodular); costal short-cells in long rows. Sometimes papillate. Stomatal guard-cells not overlapped by the subsidiaries; subsidiaries triangular or dome-shaped. *Physiology, transverse section of leaf blade.* C_3 or C_4. Adaxial surface often flat. Mesophyll without arm-cells, very rarely with fusoids. Midrib usually with an arc of bundles, often with adaxial colourless tissue. Smallest bundles commonly unaccompanied by sclerenchyma. *Cytology*. Chromosome base numbers mostly $x = 5, 9$ or 10. Group mean diploid 2*c* DNA value 3.0 pg. Rusts: *Dasturella*, *Phakopsora*, *Physopella*, *Puccinia*. Smuts: *Entyloma*, *Melanotaenium*, *Tilletia*, *Sorosporium*, *Sphacelotheca*, *Tolyposporella*, *Tolyposporium*, *Ustilago*.

Mainly tropical (especially Old World and Gondwanan), extending to temperate.

Panicodae

Inflorescence usually espatheate, the axes persistent or condensed into deciduous spikelet clusters. Spikelets sometimes in long-and-short combinations, but usually all alike in form and sexuality. Female-fertile lemma firm or indurated, with a germination flap, usually awnless and entire; the palea well developed. C_3 or C_4 (occasionally intermediate), types PCK, NAD-ME and NADP-ME (the latter all XyMS−). Basic chromosome number usually $x = 9$.

Pantropical to temperate; diverse habitat and rainfall requirements.

Isachneae (126): *Coelachne*, *Cyrtococcum*, *Heteranthoecia*, *Hubbardia*, †*Isachne* (100), *Limnopoa*, *Sphaerocaryum*.

Usually with both florets female-fertile; C_3.

Paniceae: *Achlaena*, *Acostia*, *Acritochaete*, *Acroceras*, *Alloteropsis*, *Amphicarpum*, *Ancistrachne*, *Anthaenantia*, *Anthaenantiopsis*, *Anthephora*, *Arthragrostis*, *Arthropogon*, †*Axonopus* (110), *Baptorhachis*, *Beckeropsis*, *Boivinella*, †*Brachiaria* (100), *Calyptochloa*, *Camusiella*, *Cenchrus*, *Centrochloa*, *Chaetium*, *Chaetopoa*, *Chamaeraphis*, *Chasechloa*, *Chloachne*, *Chlorocalymma*, *Cleistochloa*, *Commelinidium*, *Cymbosetaria*, *Cyphochlaena*, †*Dichanthelium* (120), †*Digitaria* (220), *Digitariopsis*, *Dimorphochloa*, *Dissochondrus*, *Eccoptocarpha*, †*Echinochloa*, *Echinolaena*, *Entolasia*, *Eriochloa*, *Holcolemma*, *Homolepis*, *Homopholis*, *Hydrothauma*, *Hygrochloa*, *Hylebates*, *Hymenachne*, *Ichnanthus*, *Ixophorus*, *Lasiacis*, *Lecomtella*, *Leptocoryphium*, *Leptoloma*, *Leucophrys*, *Louisiella*, *Megaloprotachne*, *Melinis*, *Mesosetum*, *Microcalamus*, *Mildbraediochloa*, *Odontelytrum*, *Oplismenopsis*, *Oplismenus*, *Oryzidium*, *Otachyrium*, *Ottochloa*, †*Panicum* (370), *Paratheria*, *Parectenium*, †*Paspalidium*, †*Paspalum* (320), †*Pennisetum*, *Perulifera*, *Plagiantha*, *Plagiosetum*, *Poecilostachys*, *Pseudechinolaena*, *Pseudochaetochloa*, *Pseudoraphis*, *Reimarochloa*, *Reynaudia*, *Rhynchelytrum*, *Sacciolepis*, *Scutachne*, †*Setaria* (110), *Setariopsis*, *Snowdenia*, *Spheneria*, *Spinifex*, *Steinchisma*, *Stenotaphrum*, *Stereochlaena*, *Streptolophus*, *Streptostachys*, *Tarigidia*, *Tatianyx*, *Thrasya*, *Thrasyopsis*, *Thuarea*, *Thyridachne*, *Trachys*, *Tricholaena*, *Triscenia*, *Uranthoecium*, *Urochloa*, *Whiteochloa*, *Xerochloa*, *Yakirra*, *Yvesia*, *Zygochloa*.

C_3, or C_4 type PCK, NAD-ME and NADP-ME.

Neurachneae (10): *Neurachne*, *Paraneurachne*, *Thyriodolepis*.

Inflorescence a single spicate raceme, C_3, C_4 type NADP-ME (XyMS−) or intermediate.

Arundinelleae (184): †*Arundinella*, *Chandrasekharania*, *Danthoniopsis*, *Diandrostachya*, *Dilophotriche*, *Garnotia*, *Gilgiochloa*, *Isalus*, *Jansenella*, *Loudetia*, *Loudetiopsis*, *Trichopteryx*, *Tristachya*, *Zonotriche*.

Spikelets compressed laterally to terete, disarticulating above the glumes.

Lemma usually bilobed or cleft, usually with a geniculate awn. Hilum usually long-linear. C_4 type NADP-ME (XyMS−, often with isolated PCR cells), or rarely C_3 (?). Chromosome base number $x = 10$ or 12.

Andropogonodae

Inflorescence often spatheate and/or spatheolate, the axes usually disarticulating. Spikelets usually in long-and-short combinations, often heterogamous, the members of the combinations different in sexuality (the longer-pedicelled members usually male or sterile). Glumes usually very dissimilar. The female-fertile lemma usually reduced to a stipe, insubstantial or hyaline; often bifid, with a geniculate awn; without a germination flap. Palea commonly reduced, vestigeal or absent. Exclusively C_4, type NADP-ME (XyMS−). Chromosome base number, x = mostly 5 or 10.

Tropical/warm temperate; mostly requiring seasonal high rainfall.

Andropogoneae (967):

Plants usually bisexual with hermaphrodite (upper) florets. Lodicules usually present.

Andropogoninae: *Agenium*, *Anadelphia*, †*Andropogon* (100), *Andropterum*, *Apluda*, *Apocopis*, *Arthraxon*, *Asthenochloa*, *Bhidea*, *Bothriochloa*, *Capillipedium*, *Chrysopogon*, *Chumsriella*, *Cleistachne*, †*Cymbopogon*, *Dichanthium*, *Digastrium*, *Diheteropogon*, †*Dimeria*, *Dybowskia*, *Eccoilopus*, *Elymandra*, *Eremopogon*, *Erianthus*, *Eriochrysis*, *Euclasta*, *Eulalia*, *Eulaliopsis*, *Exotheca*, *Germainia*, *Hemisorghum*, *Heteropogon*, *Homozeugos*, †*Hyparrhenia*, *Hyperthelia*, *Hypogynium*, *Imperata*, †*Ischaemum*, *Ischnochloa*, *Iseilema*, *Kerriochloa*, *Lasiorrachis*, *Leptosaccharum*, *Lophopogon*, *Microstegium*, *Miscanthidium*, *Miscanthus*, *Monium*, *Monocymbium*, *Narenga*, *Parahyparrhenia*, *Pleiadelphia*, *Pogonachne*, *Pogonatherum*, *Polliniopsis*, *Polytrias*, *Pseudanthistiria*, *Pseudodichanthium*, *Pseudopogonatherum*, *Pseudosorghum*, *Saccharum*, †*Schizachyrium*, *Sclerostachya*, *Sehima*, *Sorghastrum*, *Sorghum*, *Spathia*, *Spodiopogon*, *Thelepogon*, *Themeda*, *Trachypogon*, *Triplopogon*, *Vetiveria*, *Ystia*.

Rottboelliinae: *Chasmopodium*, *Coelorhachis*, *Elionurus*, *Eremochloa*, *Glyphochloa*, *Hackelochloa*, *Hemarthria*, *Heteropholis*, *Jardinea*, *Lasiurus*, *Lepargochloa*, *Loxodera*, *Manisuris*, *Mnesithea*, *Ophiuros*, *Oxyrhachis*, *Phacelurus*, *Pseudovossia*, *Ratzeburgia*, *Rhytachne*, *Robynsiochloa*, *Rottboellia*, *Thaumastochloa*, *Thyrsia*, *Urelytrum*, *Vossia*.

Maydeae (32): *Chionachne*, *Coix*, *Euchlaena*, *Polytoca*, *Sclerachne*, *Trilobachne*, *Tripsacum*, *Zea*.

Plants monoecious, with all the fertile spikelets unisexual; without hermaphrodite florets. The male and female-fertile spikelets in different inflorescences, on different branches of the same inflorescence, or on different parts of the same branch. Lodicules absent.

Note added in proof

Flowering of bamboo precociously *in vitro*

While this book was in press, a report appeared of bamboos flowering in culture (Nagauda, Parasharami & Mascarenhas, 1990). Seeds of *Bambusa arundinacea* and *Dendrocalamus brandisii* were germinated *in vitro*. Tissue pieces were subcultured in various media. Using Murashige and Skoog basal medium supplemented with 2% sucrose, 0.5 ppm benzylaminopurine (BAP) and 5% coconut milk, after three subcultures 70% of *B. arundinacea* and 40% of *D. brandisii* cultures developed panicles of normal spikelets. Seed was obtained for both species.

The authors point out that such work should lead to an understanding of the shift from a vegetative to a monocarpic floral state, the specific role of cytokinins and the causes of gregarious (mast) flowering. This report was greeted with enthusiasm by Hanke (1990).

Many questions remain to be answered. Would it be possible to separate precocious flowering from precocious monocarpy? Would sterile hybrids be less prone to monocarpy? After intensive *in vitro* breeding, if this proved possible, how far might the new types differ from their long generation progenitors?

Clearly, the study of bamboos seems likely to gather considerable momentum.

References

Hanke, D. E. (1990). Seeding the bamboo revolution. *Nature*, **334**, 291–2.

Nadgauda, R. S., Parasharami, V. A. & Mascarenhas, A. F. (1990). Precocious flowering and seeding behaviour in tissue cultured bamboos. *Nature*, **334**, 335–6.

Author index

Aarssen, L. W., 146, 150
Abdullah, R., 200, 209
Abe, T., 186, 209
Abouguendia, Z. M., 168, 174
Ackerman, M. C., 207, 210
Adams, W. T., 132, 150
Akagi, H., 211
Ali, O. M. M., 157, 175
Allard, R. W., 132, 150, 152
Allred, K. W., 32, 49
Altman, P. L., 139, 150
Alwen, A., 209
Anderson, A. M., 63, 71
Anderson, E. G., 87, 95
Anderson, P. C., 197, 216
Anderson, W. W., 138, 153
Andersson, L., 46, 49
Anon., 9, 240, 247, 254
Ansari, A. A., 225, 235, 237
Antonovics, J. A., 137, 138, 139, 143, 144, 146, 147, 150, 152
Appels, R., 7, 27
Arber, A., 240, 241, 254
Armstrong, C. L., 209
Armstrong, T. A., 184, 185, 210
Arrillaga de Maffei, B. R., 43, 50
Artschwager, E., 56, 71
Ashida, E. R., 222, 235
Asker, S., 136, 150
Avdulov, N. P., 27
Aviv, D., 201, 219
Avjioglu, A., 238

Babcock, E. B., 125, 128
Bailey, E. T., 167, 174
Baker, B., 214
Baldo, B. A., 225, 226, 236
Band, S. R., 26, 29
Barker, R., 218
Barkworth, M. E., 30
Barnard, C., 53, 72
Barnes, S. H., 81, 95
Barrett, S. C. H., 133, 150
Bashaw, E. C., 54, 75, 100, 101, 105, 106, 107, 108, 111, 116, 117, 119, 124, 126, 128, 129, 130, 244
Bates, M. R., 238
Batygina, T. B., 94, 95
Baum, B. R., 19, 28
Baumer, M., 165, 174
Bayliss, M. W., 95
Beaudry, J. R., 54, 66, 68, 70, 72
Beck, R., 60, 72
Becwar, M. R., 212
Beetle, A. A., 9, 28, 36, 49
Bell, T. J., 28, 49
Belousova, V. I., 130
Benito Moreno, R. M., 192, 209
Bennet, A. G., 175
Bennett, M. D., 26, 28, 60, 72, 95, 245, 255, 256
Benzion, G., 183, 209
Berg, C. C., 129
Berg, R. Y., 249, 254
Bergquist, R. R., 211, 255
Bhalla, P. L., 235
Bhaskaran, S., 197, 209
Bhatnager, S. P., 83, 96
Blackmore, S., 81, 95
Blaikrie, P., 157, 175
Blaser, H. W., 32, 49
Blau, A., 239
Bobin, M., 216
Bogdan, A. V., 34, 49, 164, 174, 249, 254
Bohm, L. R., 53, 56, 57, 60, 68, 70, 73
Bokareva, L. I., 241, 255
Bolt, D. J., 256
Bonnett, O. T., 72
Booth, A., 155, 175
Booth, T. A., 132, 150
Bor, N. L., 10, 28, 162, 163, 175
Borgaonkar, D. S., 129
Borromeo, E. S., 214
Bosch, O. J. H., 161, 175
Bottermans, J., 210
Boulton, M. I., 218

Boyes, C. J., 196, 209
Bradley, J., 238
Bradshaw, A. D., 145, 146, 152, 153
Brar, D. S., 200, 209
Bremen, I. H., 156, 175
Bremer, G., 244, 255
Brettell, R. I. S., 193, 209
Brewbaker, J. L., 79, 96
Bright, S. W. J., 184, 191, 196, 197, 209
Bright, S. W. S., 214
Brightwell, B., 212
Brink, D. E., 211, 255
Brink, R. A., 68, 70, 72
Brinster, R. L., 256
Brookes, M. R., 129
Brookfield, H. H., 157, 174
Brown, A. H. D., 133, 150
Brown, P. T. H., 183, 202, 214, 215
Brown, W. V., 3, 28, 128, 244, 255
Buckner, R. C., 129
Burman, A. G., 50
Burson, B. L., 114, 118, 124, 128, 130
Burton, G. W., 110, 120, 129
Byrne, M., 212

Cai, Q., 200, 209
Calderon, C. E., 34, 50, 241, 257
Cameron, J. W., 120, 130
Campbell, D. H., 226, 237
Campbell, S. C., 2, 11, 12, 15, 18, 28, 29, 30, 35, 49
Cao, J., 218
Capra, J. D., 232, 237
Carman, J. G., 101, 107, 123, 129
Carroll, T. W., 62, 72
Carswell, G. K., 216
Cartwright, P. M., 53, 74
Cass, D. D., 57, 63, 66, 67, 68, 69, 72, 78, 80, 81, 83, 85, 93, 95, 98, 221, 238
Chakravarty, A. K., 168, 175
Chaleff, R. J., 191, 197, 209
Chaleff, R. S., 209
Chan, F., 175
Chandler, S. F., 187, 209
Chandra, N., 55, 65, 72, 83, 96
Chapman, G. D., 7, 27
Chapman, G. P., 170, 171, 175, 240
Chapman, V., 52, 60, 72
Charles, A. H., 145, 150
Chase, A., 10, 29
Chawla, H. S., 194, 210
Chen, C. H., 187, 214
Chen, D., 186, 210
Chen, S., 252, 255
Chen, W. H., 204, 210
Cheplick, G. P., 28, 49
Chikkannaiah, P. S., 68, 71, 72
Cho, M. S., 189, 210
Chourey, P. S., 199, 210
Christie, E. K., 171, 175
Christou, P., 207, 210
Chua, K. Y., 232, 235
Clarke, A. E., 232, 236
Clarke, A. J., 252, 255
Clausen, Jens, 126, 128
Clayton, W. D., 1, 10, 11, 12, 13, 27, 28, 91, 158, 160, 165, 175, 242, 244, 251, 255
Clifford, H. T., 2, 9, 10, 28, 30, 32, 33, 46, 49, 50, 240, 245, 255, 257
Cocking, E. C., 202, 209, 210, 217
Coeckel, P., 255
Cohen, C. E., 211, 255
Coleman, A. W., 92, 96
Conde, M. F., 80, 87, 96
Conert, H. J., 11, 28
Conger, B. V., 212, 213
Connor, H. E., 15, 28, 35, 49
Constabel, F., 209, 211
Cooper, D. C., 61, 68, 70, 72
Cooper-Bland, S., 217
Coras, G., 165, 175
Corriveau, J. L., 92, 96
Cottam, G. P., 225, 235
Coumans, M. P., 189, 210
Crampton, B., 9, 30
Crane, C. F., 19, 29, 101, 107, 123, 125, 129
Crawford, T. J., 137, 150
Creemers-Molenaar, J., 217
Cronquist, A., 2, 28
Crosby, W. L., 216
Cunningham, G. A., 211

Dabadghao, P. M., 159, 175
Dahlgren, R. M. T., 2, 9, 28, 46, 50
Dale, H. M., 54, 56, 65, 71, 74
Dale, P. J., 186, 187, 210, 213
Dallwitz, M. J., 1, 10, 13, 15, 17, 25, 26, 30, 31, 245, 257
Dalton, S. J., 185, 186, 187, 200, 210
Darlington, C. D., 125, 129
Darwin, C., 251, 255
Datta, S. K., 189, 210
Daub, M. E., 191, 192, 195, 210
Davey, M. R., 202, 210
David, B., 226, 238
David, C. C., 175
Davidse, G., 8, 28, 43, 44, 50
Davies, J. W., 212
Davies, S., 238
Davies, M. S., 138, 152
Davis, T. A. W., 241, 255
Davy, A. J., 143, 144, 151
Day, A., 190, 210
Deaton, W. R., 188, 210
de Block, M., 210

de Cleene, M., 207, 210
de Greef, W., 197, 210
de la Pena, A., 206, 211
Delon, R., 210
DeMason, D. A., 60, 74
Detmar, J. J., 216
De Triquell, A. A., 53, 66, 72
de Wet, A. E., 211, 255
de Wet, J. M. J., 110, 118, 122, 127, 129, 136, 152, 205, 211, 244, 246, 247, 250, 252, 255, 256
Dewey, D. R., 129
Diallo, A. K., 164, 180
Diboll, A. G., 62, 63, 64, 66, 67, 68, 69, 70, 71, 72, 78, 93, 94, 96
Dilworth, R. J., 235
Dimaio, J. S., 216
Dirk, J., 216
Dittmer, D. S., 139, 150
Djitaye, M. A., 159, 179
Dodson, R. K., 137, 151
Donn, G., 197, 211
Donovan, C. M., 193, 212
Dover, G. A., 7, 28
Dover, M., 154, 156, 175
Dow Mongkolsmai, 175
Dowd, J. M., 7, 29
Draz, O., 167, 175
Duan, X., 252, 255
Dubois, M., 257
Dudits, D., 202, 211, 217
Dujardin, M., 121, 122, 123, 129, 172, 175
Dumas, C., 74, 76, 81, 84, 96, 97, 221, 235, 236, 237, 255, 257
Duncan, D. R., 185, 191, 196, 211, 215
Dunn, G. M., 129
Dunstan, D. I., 183, 211
Dunwell, J. M., 187, 188, 189, 211
Dupuis, I., 222, 236, 253, 255

Ebert, K. M., 256
Eckes, P., 239
Ehrlich-Kantzy, E., 236
Ekramoddoullah, A. K. M., 225, 226, 236
El Moghraby, A. I., 157, 175
El Shourbagui, M. N., 168, 178
Ellis, R. P., 8, 12, 28, 30, 248, 255
Ellis, T. H. N., 190, 210
Ellstrand, N. C., 143, 144, 146, 150
Emery, W. H. P., 128
Engel, L. M., 129
Engell, K., 78, 93, 94, 96
Ennos, R. A., 137, 151
Epstein, E., 197, 211
Esau, K., 96
Esch, R. E., 225, 226, 233, 236
Esen, A., 7, 28
Estes, J. R., 19, 28
Evans, D. A., 189, 190, 215

FAO, 163, 166, 175
Fabina, E. S., 130
Fairbrothers, D. E., 7, 29
Fan Weihong, 160, 181
Farias, R. M., 247, 257
Farooq, S., 246, 255
Fatahallah, M. M., 163, 175
Feldman, M., 243, 257
Ferl, R. J., 200, 215, 217
Filgueiras, T. S., 50
Fisher, D. B., 86, 91, 97
Flavell, R. B., 243, 255
Forbes, I., 110, 128
Ford, S. A., 225, 226, 236
Forgeois, P., 216
Forlani, G., 216
Foroughi-Wehr, B., 187, 188, 190, 211, 214, 217, 218
Foury, A., 167, 175
Fowke, L. C., 216
Fraley, R. T., 207, 211, 212
Freidhoff, L. R., 225, 236
French, N. H., 168, 176
Friedt, W., 188, 211, 217
Fromm, M. E., 204, 211, 213, 218
Fu Yikun, 162, 179
Fujimoto, H., 217
Fujimura, T., 186, 200, 211
Fukui, K., 184, 211
Funk, C. R., 117, 120, 130
Futsuhara, Y., 186, 209

Gaff, D. R., 248, 255
Gallagher, R., 167, 176
Galun, E., 201, 219
Gamborg, O. L., 186, 209, 211, 213
Gartland, J. S., 210
Gartland, K. M. A., 210
Gaude, T., 76, 96, 235
Gebrehiwot, L., 162, 176
Geltz, N. R., 238
Gendloff, E. H., 195, 212
Gengenbach, B. D., 193, 194, 212, 216
Gerlach, W. L., 235, 236
Gibbon, C. V., 171, 177
Gleaves, J. T., 138, 151
Gleich, G. J., 226, 237
Gobbe, J., 60, 72
Gobel, E., 202, 212, 214
Godard, M., 257
Goldman, S. L., 207, 212
Gong, Z., 252, 255
Goodfriend, L., 232, 237
Goodman, H. M., 211
Gorst, J., 182, 218
Gottlieb, L. D., 133, 151

Goudie, A., 156, 176
Gould, F. W., 62, 73
Goumandakoye, M., 165, 179
Gradziel, T., 213
Grandbacher, F. J., 38, 50
Granier, F., 216
Grant, C. A., 129
Grant, J. H., 236
Graves, A. C. F., 207, 212
Gray, A. J., 141, 142, 145, 148, 151, 152
Gray, D. J., 186, 212
Green, C. E., 185, 186, 193, 209, 212
Greenham, J., 52
Griffiths, D. J., 138, 151
Grime, J. P., 26, 29, 135, 145, 151
Grimsley, N., 207, 212
Gronenborn, B., 217
Grubb, P. J., 135, 151
Guest, E., 158, 162, 163, 175, 176
Gunn, H., 218
Gupta, P. K., 19, 28
Gustafsson, A., 100, 109, 129

Hackel, E., 9, 29
Haddad, Z. H., 226, 237
Hadlackzy, G., 202, 217
Hagemann, R., 80, 81, 88, 96
Hair, J. B., 101, 123, 129
Hakansson, A., 60, 68, 71, 73
Hamby, K. R., 7, 29
Hammer, R. E., 252, 256
Hamrick, J. L., 134, 135, 137, 151, 152
Handel, S. N., 139, 151
Hanna, W. W., 100, 101, 111, 120, 121, 122, 123, 124, 129, 172, 175, 244
Hanning, G. E., 212
Hanson, A. A., 120, 129
Harberd, D. J., 140, 141, 142, 151, 248, 256
Harlan, J. R., 100, 112, 118, 127, 129, 169, 176, 211, 244, 247, 249, 250, 256
Harms, C. T., 213, 216
Harper, J. L., 43, 50, 141, 142, 143, 146, 153
Harris, D., 143, 144, 151
Harris, R., 185, 212
Hartke, S., 215
Hartley, W., 164, 176
Harvey, B. L., 218
Haseloff, J., 235, 236
Hattersley, P. W., 3, 29, 244, 256
Hauptmann, R. M., 204, 212
Hause, G., 86, 96
Hayashi, Y., 201, 212
Hayes, J. V., 138, 152
Heberle-Bors, E., 209
Heddle, R. J., 238
Heizmann, P., 257
Hemming, C. F., 158, 161, 176
Hepburn, A. G., 255
Herr, J. M., 54, 73
Heslop-Harrison, J., 76, 77, 78, 96, 97, 99, 114, 130, 223, 236, 237, 247, 252, 256
Heslop-Harrison, Y., 26, 96, 97, 99, 223, 236
Hess, D., 205, 212
Heyn, C. C., 163, 176
Heyser, J. W., 200, 212
Hignight, K. W., 119, 128
Hill, D. J., 226, 236
Hilu, K. W., 7, 28
Hinata, K., 217
Hirose, A., 211
Hitchcock, A. S., 9, 29
Ho, W., 186, 212
Hodges, T. K., 184, 186, 198, 212, 213
Hodgkin, T., 192, 212
Hoff, B. J., 120, 128
Hohn, B., 212
Hohn, T., 212
Holden, L. R., 137, 151
Holt, E. C., 106, 126, 128
Hooykaas, P. J. J., 207, 213
Horn, M. E., 200, 213
Horsch, R. B., 207, 211, 212
Horton, J. S., 60, 72
Hough, T., 81, 97, 222, 236, 237, 238
Hoveland, C. S., 129
Hovin, A. W., 114, 126, 128, 129
Howlett, B. J., 223, 224, 226, 232, 236, 239
Hu Zizhui, 162, 179
Hu, S-V., 81, 97, 99
Huang, J., 255, 257
Hubbard, C. E., 9, 29
Hughes, J., 135, 151
Huiskes, A. H. L., 142, 144, 151
Hunter, C. P., 189, 213
Hutchinson, J., 255

IBPGR/Kew, 162, 169, 171, 176
Ibrahim, K. N., 161, 168, 176
Imbrie, C. W., 212
Imbrie-Milligan, C. W., 198, 213
Ingrouille, M. J., 5, 29
Izawa, T., 217

Jackson, G. C., 241, 256
Jackson, J. E., 120, 128
Jacobs, S. W. L., 11, 29
Jacquemin, J. M., 216
Jacques-Felix, H., 9, 29
Jain, S. B., 133, 136, 137, 139, 151
Jain, S. K., 152, 153
Janzen, D. H., 35, 50, 241, 256
Jauhar, P. P., 19, 29
Jeffery, C., 97

Jensen, W. A., 57, 66, 67, 68, 69, 71, 72, 78, 79, 85, 86, 91, 93, 96
Jerling, L., 142, 152
Jianming, S., 194, 214
Jiong, R., 209
Jobson, S., 237, 239
Johnson, M. A., 7, 13, 29, 31
Johnson, B. B., 186, 213
Johnson, C. M., 216
Johnson, C. R., 31
Johnson, P., 225, 232, 237
Jones, M. G. K., 184, 186, 209, 213
Jones, T. J., 69, 73
Joysey, K. A., 7, 29
Julen, G., 120, 129
Juska F. V., 120, 129

Kabak, A., 167, 178
Kabuye, C. H. S., 161, 176
Kackar, M. L., 168, 175
Kamo, K. K., 186, 212, 213
Kannenberg, L. W., 132, 152
Kao, K. N., 200, 211, 213, 218
Karas, I., 80, 81, 83, 96
Kartha, K. K., 187, 213
Kasha, K. J., 186, 213
Kassas, M. A. F., 155, 176
Kavi Kishor, P. B., 184, 213
Keijzer, C. J., 81, 97
Kelley, D. B., 211
Kellogg, E. A., 2, 11, 12, 19, 28, 29
Kernick, M. D., 91, 154, 157, 163, 166, 167, 169, 171, 172, 176, 249, 254
Khan, C. M. A., 168, 177
Kiedrowski, ?. ?., 206
Kihara, T. K., 225, 235
King, J., 216
King, P. J., 185, 213
King, G. J., 5, 29
Kingsbury, R. W., 211
Kisil, F. T., 226, 236
Kitamura, S., 173, 177
Klapper, D. G., 225, 226, 232, 233, 236, 237, 238
Klein, T. M., 205, 206, 213, 218
Kleinhofs, A., 217
Knight, R., 169, 171, 177
Knobloch, I. W., 19, 29
Knox, R. B., 76, 78, 81, 96, 97, 114, 130, 136, 152, 220, 221, 222, 223, 224, 225, 226, 230, 235, 236, 237, 238, 239, 247, 253, 256
Koblitz, H., 199, 213
Kohler, F., 189, 213
Kolesniker, V., 215
Kollmorgen, J., 182, 245
Koop, H.-U., 216
Koren, E., 238
Kott, L. S., 186, 213
Kranz, E., 205, 213, 216
Krishnamurthy, K. V., 57, 74
Kroskey, C. S., 197, 215
Kruse, A., 245, 256
Kueh, J. S. H., 196, 214
Kuhlmann, K., 188, 190, 214
Kuo, C., 209
Kuraoiwa, T., 92, 97
Kyozuka, J., 185, 198, 201, 212, 214, 217

Ladyman, J. A. R., 77, 78, 97
Laguardia, A., 43, 50
Laikova, L. I., 130
Lamb, H. H., 156, 177
Langer, R. H. M., 36, 50
Lapeyre, J., 226, 238
Larkin, P. J., 193, 214
Larson, D. A., 63, 64, 66, 67, 68, 69, 70, 71, 72
Latz, P. K., 248, 249, 255
Laurie, D. A., 245, 256
Law, R., 145, 146, 147, 152
Lawrence, P., 155, 177
Lazarides, M., 249, 256
Lazzeri, P. A., 182, 185, 186, 200, 204, 214, 245, 253
Le Houerou, H. N., 158, 161, 166, 168, 177
Leemans, J., 210
Leiferman, K. M., 226, 237
Lersten, N. R., 67, 74
Leu, L. S., 187, 214
Levings, C. S., 80, 96
Li Chonghao, 168, 177
Li Jiandong, 160, 162, 181
Li, H. R., 99
Lichfield, J., 155, 177
Lihua, S., 194, 195, 214
Lin, Shu-Chang, 69, 70, 71, 73
Linder, H. P., 46, 50
Ling, D. H., 193, 214
Linhart, Y. B., 151
Liu, G., 257
Lo, P. F., 187, 214
Longly, B., 60, 71, 72, 73
Lord, E. M., 252, 257
Lorz, H., 182, 183, 185, 186, 200, 204, 205, 206, 211, 212, 213, 214, 215, 217, 219, 245
Louant, B. P., 60, 72, 73
Love, A., 19, 29
Loveless, M. D., 133, 134, 135, 152
Lowe, K. I. S., 199, 216
Lowenstein, H., 226, 237
Loxton, A. E., 245, 256
Lu Lian, 169, 177
Lu, C. Y., 185, 186, 199, 200, 214, 215
Luhrs, R., 186, 200, 214, 215
Luk, S. H., 156, 177

Luke, H. H., 193, 216
Lundquist, A., 76, 97
Luo, Z. X., 205, 215, 256
Lupotto, E., 197, 215
Lusardi, M. C., 197, 215
Lynch, N. R., 225, 237

Ma Zhi Guang, 168, 177
Mabbutt, J. A., 156, 177
MacArthur, R. H., 145, 152
McClure, F. A., 48, 50
McConchie, C. A., 81, 83, 96, 97, 235, 237
Macfarlane, T. D., 5, 11, 29
McGuire, R. C., 56, 71
McHugh, D. M., 197, 215
Macke, F., 209
McNeilly, T., 137, 138, 139, 146, 152
McWilliam, J. R., 171, 177
Maddock, S. E., 186, 215
Mahalingappa, M. S., 68, 71, 72
Maliga, P., 191, 215
Malik, C. P., 93, 99
Maltby, S., 257
Mares, D. J., 95, 97
Marsh, D. G., 225, 226, 232, 233, 235, 236, 237
Marshall, D. R., 133, 150
Martin, P. G., 7, 29
Mascarenhas, D., 216, 228
Mascarenhas, J. P., 239
Mascia, P. N., 210
Matthys-Rochon, E., 74, 236, 255
Mauldon, R. G., 175
May, R. M., 2, 29
Mayasmi, Y., 185, 214
Mayhew, D. E., 62, 72
Mayo, O., 190, 215
Maze, J. 53, 56, 57, 60, 63, 65, 68, 69, 70, 71, 73
Meher-Homji, V. M., 173, 178
Mehlenbacher, L. E., 53, 55, 56, 73
Mendel, R. R., 206, 215
Mensching, H. G., 157, 178
Mettler, I. J., 216
Metz, S. G., 210
Mew, T. W., 214
Meyanathan, S., 175
Meyers, D. A., 236
Miao, S., 196, 215
Michayluk, M. R., 200, 213
Migahid, A., 168, 178
Mii, M., 216
Milkovits, L., 182, 218
Miller, R. A., 211
Miller, T. E., 243, 256
Millot, J. C., 129
Mitton, J. B., 133, 151
Miyamura, S., 92, 97
Mogensen, H. L., 61, 67, 73, 77, 78, 79, 80, 81, 82, 83, 84, 85, 86, 87, 90, 221, 222, 237, 249
Mohamed, A. H., 62, 73
Mongodi, M., 197, 215
Moran, D. M., 225, 235
Mori, K., 218
Mornan, K., 257
Morrison, I. N., 58, 69, 73
Morrison, J. W., 85, 98
Morrison, R. A., 189, 190, 215
Mortimore, M. J., 159, 178
Morton, E., 44, 50
Muller, B., 184, 215
Muller, E., 215
Mulligan, B. J., 210
Mullineaux, P. M., 218
Muniyamma, M., 60, 63, 68, 73
Muntzing, A., 101, 130, 136, 152
Myles, D. G., 222, 238

Nabors, M. W., 197, 215
Naegele, A., 165, 171, 178
Nagata, T., 92, 97
Naqvi, S. H. M., 246, 255
Narayanaswami, S., 56, 60, 68, 73
Negishi, T., 211
Nelson, O. E., 196, 215
Nelson, R., 156, 178
Nemati, N., 158, 166, 178
Nettencourt, D., de 76, 98
Neuhaus, G., 216
Nevo, E., 172, 178
Newell, C. A., 211, 255
Nicholson, S. E., 156, 178
Nielsen, E. L., 60, 73, 216
Niklas, K. J., 34, 50
Nishibayashi, S., 217
Nishiyama, I., 67, 73
Nitsch, C., 246, 257
Nitzsche, W., 117, 130
Nogler, G. A., 100, 130
Norlyn, J. D., 211
Norstog, K., 66, 74, 93, 94, 95, 98
Numata, M., 160, 164, 178
Nygren, A., 60, 74, 100, 130

O'Neill, P., 226, 238
Ogden, E. C., 138, 152
Ohaliwal, H., 211
Ohta, Y., 205, 215, 252, 257
Olson, J., 225, 238
Olsson, K., 156, 178
Omar, B., 167, 178
Oswald, T. H., 197, 215
Otoo, E., 185, 214
Ozias-Akins, P., 200, 201, 202, 212, 215, 216

Pabot, H., 158, 162, 163, 164, 178
Paliwal, M. K., 168, 179
Palmiter, R. D., 256
Parker, B. B., 217
Parlevliet, J. E., 120, 130
Parodi, L. R., 9, 30
Paterniani, E., 139, 152
Pauly, M. H., 194, 216
Pearce, C. K., 157, 178
Peart, M. H., 46, 50
Pelanda, R., 216
Peltre, G., 226, 238
Pennell, R. I., 222, 238
Penning de Vries, F. W. T., 159, 179
Pepin, G. W., 117, 130
Perry, S., 256
Peteya, D. J., 63, 66, 72, 77, 96
Petolino, J. F., 184, 188, 216
Petrov, D. F., 101, 121, 122, 130
Petruska, J., 216
Peyre de Fabreque, M. B., 159, 179
Philipson, M. N., 52, 57, 74
Philipson, W. R., 74
Phillips, D. V., 197, 215
Phillips, R. C., 184, 209
Phillips, R. L., 183, 209
Picard, E., 205, 216
Pierce, D. A., 216
Pilger, R., 9, 30
Pissarek, H. P., 35, 50
Platt, S. G., 207, 210
Plozza, T. M., 235
Pohl, R. W., 241, 257
Polhill, R. M., 8, 30
Pope, M., 66, 74, 78, 85, 86, 94, 98
Potrykus, I., 204, 216
Powell, J. B., 120, 129
Power, J. B., 210
Prat, H., 9, 30
Pratt, D., 159, 179
Praznovsky, T., 202, 211
Price, E. G., 250, 256
Primakoff, P., 222, 238
Pring, D. R., 80, 96, 201, 215, 217
Prioli, L. M., 199, 200, 216
Puri, D. C., 168, 179
Pursel, V. G., 256
Putwain, P. D., 145, 152

Qia, Y., 209
Qian, S., 255, 257
Quarin, C. L., 247, 257
Quinn, J. A., 28, 49, 147, 152

Rabau, T., 60, 73
Racchi, M. L., 198, 216
Raghavan, V., 95, 98
Rai, K. N., 137, 139, 152
Rajasekaran, K., 183, 187, 216
Rambold, S., 209
Randolph, L. F., 74
Rangan, T. S., 197, 216
Rao, M. K., 95
Rasmusson, E. M., 156, 179
Rattray, J. M., 157, 161, 179
Raven, P. H., 8, 30
Ray, J., 250, 257
Ray, T. B., 197, 209
Raynor, G. S., 138, 152
Read, J. C., 124, 130
Reddy, G. M., 184, 213
Reddy, P. J., 197, 216
Redman, R. E., 38, 50
Reekie, E. G., 38, 50
Reger, B. J., 76, 97, 98
Ren Jizhou, 162, 179
Renvoize, S. A., 1, 10, 11, 12, 13, 28, 33, 49, 175, 242, 244, 251, 255
Rexroad, C. G. Jr, 256
Rhodes, C. A., 186, 199, 200, 204, 212, 216
Richards, A. J., 91, 131, 132, 135, 136, 138, 139, 142, 150, 151, 152, 248
Richards, P. W., 241, 255
Richardson, B. J., 133, 150
Rihan, J. R., 141, 142, 152
Riley, R., 60, 72
Rineker, C. M., 129
Rines, H. W., 183, 193, 209, 216
Robertson, B. L., 63, 66, 72
Robertson, D. W., 80, 87, 90, 98
Rochon, E., 257
Rodkiewicz, B., 61, 74
Roeckel, P., 236, 252, 257
Rogers, S. G., 207, 211, 212
Roksandie, Z., 256
Roman, H., 81, 98
Roos, F. H., 147, 152
Roose, M. L., 146, 152
Rosengurtt, B., 43, 50
Ross, J. G., 187, 214
Rossetti, C., 158, 161, 179
Rossnagel, B. G., 218
Rost, T. L., 69, 73
Ruby, K. L., 199, 216
Rusche, M., 80, 81, 83, 84, 98
Rush, D. W., 211
Russell, S. D., 58, 61, 66, 74, 81, 86, 96, 97, 98, 221, 222, 235, 237, 238
Rutishauser, A., 118, 130
Ryle, G. J. A., 36, 50

Sager, R., 88, 98
Said, M., 167, 179
Sakurai, M., 211
Salinger, A. P., 88, 99
Sanders, D. M., 238
Sanders, L. C., 252, 257

Sanford, J. C., 213, 218
Sangenberg, G., 216
Sankary, M. N., 179
Saran, S., 136, 152
Sarayanan, K. R., 217
Sasaki, T., 217
Saul, M. W., 216
Savidan, Y. H., 111, 118, 130
Saxena, P. K., 202, 216
Saxena, S. K., 160, 179
Schacht, W., 71
Schank, S. C., 216
Scheffler, R. P., 195, 212
Schell, J., 206, 211, 214, 217, 239
Schertz, K. F., 101, 129, 130, 197, 209
Schilperoort, R. P., 207, 213
Schnarf, K., 60, 74
Schoenenberger, A., 168, 179
Scholes, G., 7, 27
Scholtz, H., 163, 180
Schroder, M. B., 80, 81, 86, 88, 96
Schubert, K., 212
Schulze, J., 215
Schwab, C. A., 61, 74
Schweiger, H.-G., 204, 205, 216
Scofield, V. L., 222, 235
Scowcroft, W. R., 193, 214
Seal, A. G., 255
Sears, B. B., 80, 98
Sehon, A. H., 226, 236
Sen, J. H., 99
Sendulsky, T., 50
Shacklock, J. M. L., 6, 29
Shane, W. W., 194, 216
Shaner, D. L., 197, 216
Shankarnaryan, K. A., 159, 160, 179
Sharman, B. C., 53, 74
Shaw, C. H., 233, 238
Shaw, G. H., 238
Shen, W., 255
Shenker, A., 165, 179
Sherwood, R. T., 54, 75
Shikanai, T., 218
Shillito, R. D., 184, 185, 199, 200, 216
Shimamoto, K., 185, 201, 204, 212, 214, 216, 217
Shivanna, K. R., 76, 99
Short, A. C., 139, 152
Short, K. C., 211
Shu Jiang, S., 162, 179
Sikora, I., 165, 179
Silander, J. A., 139, 147, 152
Silberbauer-Gottsberger, I., 43, 50
Simpson, C. E., 108, 124, 130
Simpson, E., 217
Simpson, R. J., 235
Singh, M. B., 93, 99, 220, 221, 226, 230, 231, 232, 235, 236, 237, 238, 239, 253, 256
Sivalingam, G., 175
Sjors, H., 142
Skovlin, F. M., 158, 180
Smart, I. J., 225, 236, 238
Smith, A. E., 197, 215
Smith, G. S., 184, 217
Smith, J. A., 211
Smith, J. B., 26, 28, 95
Smith, P. 238
Smith, R. H., 197, 209
Snape, J. N., 187, 188, 217
Snaydon, R. W., 138, 140, 152
Sneath, P. H., 5, 30
Soane, I. D., 145, 153
Soderstrom, T. R., 8, 10, 12, 30, 34, 49, 50, 241, 257
Sohota, S., 210
Somers, D. A., 202, 217
Somerville, S. C., 195, 212
Sondahl, M. R., 199, 200, 216
Song, Y. C., 74, 257
Songqiao, Z., 156, 180
Soriano, A., 163, 180
Sorokina, T. P., 130
Sotak, R., 210
Southworth, D., 221, 237, 253, 256
Sprague, G. F., 85, 86, 99
Srinivasan, C., 199, 200, 217
Staff, I. A., 226, 229, 239
Standring, R., 225, 235
Stapf, O., 33, 47, 50
Stebbins, G. L., 9, 27, 30, 100, 109, 125, 128, 130, 243, 257
Steinbiss, H.-H., 217
Stephens, J. C., 138, 153
Stewart, G. A., 235
Stockhaus, J., 233, 239
Stolarz, A., 186, 217
Stone, B. A., 97
Stover, E. L., 66, 74
Sukopp, H., 163, 180
Swamy, B. G. L., 57, 74
Swanson, E. B., 192, 210, 217
Syarifuddin Baharsyah, 175
Szabados, L., 202, 217

Tabaeizadeh, Z., 200, 201, 212, 217
Tadesse, A., 162, 176
Takeoka, T., 9, 30
Takhtajan, A., 18, 30
Talbot, L. M., 154, 156, 175
Taliaferro, C. M., 111, 116, 117, 130
Tang, C. Y., 101, 130
Tang, D. T., 218
Tasi, T.-K., 257
Tatintseva, S. S., 80, 99
Taylor, L. P., 211
Taylor, P. E., 239
Terada, R., 201, 216, 217

Terrell, E. E., 43, 51
Theerakulpisut, P., 237, 238
Thomas, E., 193, 209, 211
Thomas, W. R., 235
Thomasson, J. R., 27, 30
Thompson, J. A., 209
Thompson, S. A., 184, 188, 216
Thurman, D. A., 146, 153
Tilney-Basset, R. A. E., 88, 99
Tilton, V. R., 67, 74
Tinney, F. W., 60, 74, 101, 130
Tischner, E., 211
Todorov, A. V., 156, 180
Tolba, M. K., 156, 180
Tomes, D. T., 184, 217
Topfer, R., 206, 217
Toriyama, K., 200, 217
Toutain, B., 158, 180
Tovey, E. R., 226, 236
Tran, V. N., 32, 51
True, R. H., 56, 74
Tsai, L.-K., 246
Tsvelev, N. N., 9, 30
Tuchy, M., 97
Tuddenham, W. G., 225, 238
Tuohy, M., 237
Turkington, R., 146, 150
Turner, K. J., 235, 237
Turner, M. E., 138, 153, 225

UNESCO/UNEP/UNDP, 157, 180
Uithol, P. N. J., 156, 175
Usberti, J. A., 136, 153

Vaid, K. M., 51
Vaidyanath, K., 197, 216
Valenza, J., 164, 180
Van der Valk, P., 200, 217
Van Lammeren, A. A. M., 78, 93, 94, 99
Van Went, J. L., 91, 99
Van Winkle-Swift, K. P., 88, 99
Varnum, J., 212
Vasil, I. K., 182, 183, 184, 185, 186, 187, 209, 212, 214, 215, 216, 217, 218
Vasil, V., 185, 186, 199, 200, 212, 215, 218
Vesey-Fitzgerald, D. F., 163, 180
Vidhyaseharan, P., 214
Vilmorin, P. L. F. L. de, 251
Vithanage, H. I. M. V., 226, 239
Voigt, P. W., 107, 114, 118, 130

Waddington, S. R., 53, 74
Wagner, V. T., 57, 67, 69, 74, 81, 82, 83, 85, 90, 98, 99, 221, 237, 253, 257
Wakasa, K., 196, 218
Walbot, V., 211
Wall, R. J., 256
Wang, D., 218
Wang, Y. C., 206, 218
Watkinson, A. R., 141, 142, 143, 145, 146, 153
Watson, L., 1, 3, 5, 10, 12, 13, 15, 17, 25, 26, 29, 30, 31, 91, 244, 245, 248, 251, 257
Weatherwax, P., 78, 99
Webb, K. J., 187, 210
Webster, R. D., 249, 256
Weir, C. E., 54, 56, 65, 71, 74
Weng, J., 257
Went, F. W., 18, 31
Wenzel, G., 187, 188, 189, 191, 192, 194, 210, 211, 213, 218
Wernicke, W., 182, 218
Whatley, J. J., 80, 99
White, L. P., 158, 180
Whitehead, D. R., 153
Whyte, R. O., 159, 164, 180
Wickens, G. W., 158, 180
Widholm, J. M., 191, 196, 211, 215, 218
Willemse, M. T. M., 91, 99
Williams, G. C., 147, 153
Williams, J. T., 247, 257
Williams, M. E., 211
Willing, R. P., 228, 239
Willmitzer, L., 234, 239
Wilms, H. J., 81, 97, 99
Wilson, E. O., 145, 152
Withers, L. A., 187, 218
Woldu, Z., 162, 180
Wolf, E. D., 213
Wolff, D., 216
Wong, K. M., 241, 257
Wong, S.-C., 256
Wood, G. M., 129
Woodbury, R. O., 241, 256
Woods, J. A., 255
Woolston, C. J., 208, 218
World Bank, 155, 180
World Resources, 156, 158
Worthington, M., 186, 213
Wright, M., 212
Wright, S., 137, 153
Wrona, A. F., 211
Wu, L., 146, 153
Wu, R., 204, 205, 213, 215, 218, 219, 246, 256
Wu, S. H., 257

Xia, C.-A., 186, 210
Xiang-Yuan, Xi, 60, 74
Xu, L. V., 99
Xuefeng, L., 194, 214

Yabuno, T., 67, 73
Yamada, Y., 200, 218
Yamura, A., 68, 75
Yang Dianchen, 160, 181

Yang, M. Y., 189, 190, 218
Yang, W., 257
Yang, Y. H., 222, 239
Yang, Z. Q., 202, 218
Yatsenko, R. M., 130
Ye Juxin, 177
Ye, J. M., 189, 192, 218
Yeo, P. F., 2, 28, 46, 50
Yeoh, Hock-Hin, 3, 31
You, R., 78, 79, 99
Young, B. A., 54, 75
Yu, H.-S., 81, 97

Zaal, M. A. C. M., 217
Zapata, F. J., 189, 210, 214
Zaroug, M. G., 157, 181
Zehr, B. E., 211
Zelcer, A., 201, 219
Zelenin, A., 215
Zen, Y., 257
Zeng, Y., 257
Zhang, W., 204, 219
Zhao Ji, 160, 162, 181
Zhao Kuiyi, 177
Zheng Xuanfeng, 177
Zhou Giming, 160, 181
Zhou, C., 189, 190, 218, 222, 239
Zhou, G.-Y., 246, 252, 257
Zhou, G., 255
Zhou, Y., 209
Zhu Ting-Cheng, 160, 162, 169, 181
Zhu, C., 83, 99
Zimmer, E. A., 7, 29
Zimny, J., 186, 219
Zohary, D., 243, 244, 257
Zola, M., 238
Zu Yuangang, 162, 181
Zuwarski, D. B., 199, 210
Zwollo, P., 225, 237

Organism index

Poaceae

Note: Intergeneric hybrids are set out on pages 20–5 and 246. For protoplast fusion hybrids see pages 200–1.

Authorities differ on higher taxa and Cenchrinae mentioned on page 248 is among those that do not occur in the Appendix. See Watson (this volume) and Clayton & Renvoize (1986) for discussion.

Acamptoclados, 263
Achlaena, 264
Aciachne, 14, 261
Acidosasa, 261
Acostia. 264
Acrachne, 14, 263
Acritochaete, 264
Acroceras, 264
Actinocladum, 42, 261
Aegilops, 20, 21, 23, 24, 26, 37, 41, 243, 244, 258
  biuncialis, 243
  ovata, 243
  squarrosa, 243
  umbellulata, 243
  variabilis, 243
Aegopogon, 37, 263
Aeluropus, 163, 263
  littoralis, 243
  repens, 263
Afrachneria, 261
Afrotrichloris, 263
Agenium, 265
Agropyron, 20–5, 68, 164–8, 226, 258
  caespitosum, 164
  cristatum, 165–6, 226
  desertorum, 165–6
  libanoticum, 167
  repens, 68
Agropyropsis, 259
Agrostis, 14, 19–21, 23, 26, 40, 43, 57, 60, 63, 68, 133, 139, 145–6, 226, 242, 259
  alba, 226
  capillaris, 133, 139
  curtisii, 145–6
  interrupta, 57, 68
  pilosula, 60
  stolonifera, 146
Aira, 14, 26, 259
Airopsis, 242, 259
Alloeochaete, 39, 261
Allolepis, 15, 263
Alloteropsis, 264
Alopecurus, 38, 40, 259
Alvimia, 42, 261
Amblyopyrum, 258
Ammunochloa, 259
  arenaria, 142, 144
Ammophila, 1, 20, 21, 142, 144, 259
  arenaria, 142, 144
Ampelodesmos, 1, 34
Amphibromus, 14, 259
Amphicarpum, 14, 43, 264
Amphipogon, 39, 41, 261
Anadelphia, 265
Ancistrachne, 45, 264
Ancistragrostis, 259
Andropogon, 158–9, 161, 164–5, 172
  gayanus, 158–9, 164–5
  kelleri, 161, 172
Andropogoneae, 7, 12, 35, 37–41, 43, 47–8, 246, 265
Andropogonodae, 6, 16, 265
Andropterum, 265
Anemanthele, 261
Aniselytron, 14, 259
Anisopogon, 3, 34, 261
  acenaceus, 3
Anomochloa, 2, 33, 49, 260
Anomochloeae, 260

Anthaenantia, 264
Anthaenantiopsis, 264
Anthephora, 14, 24, 41, 245, 264
Anthochloa, 259
Anthoxanthum, 26, 39, 53, 139–40, 143, 146, 224, 226, 242, 259
odoratum, 53, 139, 143, 146, 224, 226
Antinoria, 259
Apera 259
Aphanelytrum, 39, 259
Apluda, 14, 40, 265
Apochiton, 45, 263
Apoclada, 261
Apocopis, 15, 265
Arberella, 15, 260
Arctagrostis, 14, 259
Arctophila, 20–1, 259
Aristida, 157–8, 161–3, 165, 168, 262
adscensionis, 161
caelorescens, 158
mutabilis, 159, 161, 165
pallida, 161
papposa, 161
plumosa, 158, 163
pungens, 161, 165, 168
sieberinia, 161
Aristideae, 27, 39–40, 45, 262
Arrhenatherum, 15, 20, 140, 259
elatius, 140
Arthragrostis, 264
Arthraxon, 265
Arthropogon, 264
Arthrostylidium, 261
Arundinaria, 240, 261
Arundineae, 18, 35, 39, 42, 261
Arundinella, 264
Arundinelleae, 11–12, 37–9, 264–5
Arundinoideae, 5–6, 11–12, 16, 261–2
Arundo, 261
Arundoclaytonia, 15, 261
Asthenochloa, 265
Astrebla, 14, 263
Athroostachys, 261
Atractantha, 261
Aulonemia, 261
Australopyrum, 14, 258
Austrochloris, 263
Austrofestuca, 259
Avellinia, 259
Avena, 8, 14, 19, 20, 26, 57, 67, 137, 139, 226, 245, 259
barbata, 137, 139
sativa, 226
Aveneae, 39, 43, 45, 259
Axonopus, 264

Bambusa, 5, 20, 23, 38, 48, 60, 65, 241, 261, 266
arundinacea, 266
polymorpha, 241
Bambuseae, 8, 18, 38, 42, 47, 261
Bambusodae, 6, 16–17, 260–1
Bambusoidae, 5–7, 11–12, 16–17, 19, 26–7, 33, 246, 259–61
Baptorhacis, 264
Bealia, 263
Beckeropsis, 264
Beckmannia, 259
Bellardiochloa, 259
Bewsia, 263
Bhidea, 265
Blepharidachne, 39, 263
Blepharoneuron, 263
Boissiera, 26, 259
Boivinella, 264
Bothriochloa, 14, 20–1, 113–14, 118, 136, 186, 247, 265
decipiens, 247
ischaemum, 186
Bouteloua, 14, 18, 62, 263
curtipendula, 62
Boutelouinae, 37, 41, 45
Brachiaria, 14, 43–4, 60, 91, 264
decumbens, 91
cruciformis, 43
nigropedata, 44
ruziziensis, 60
Brachyachne, 14, 263
Brachychloa, 263
Brachylytreae, 260
Brachyellytrum, 12, 39, 260
Brachypodium, 11, 14, 259
Brachypodieae, 11, 259
Briza, 14, 26, 133, 259
maxima, 133
media, 133
Bromuniola, 39, 260
Bromeae, 11, 259
Bromus, 3, 5, 7, 20, 26, 60, 164, 166, 168, 173, 186, 200, 226, 259
cappodocicus, 166
inermis, 186, 200, 226
persicus, 164
tomentellus, 164, 166
Brylkinia, 39, 40, 259
Buchloe, 15, 35, 45, 263
Buchlomimus, 15, 263
Buergersiochloa, 15, 260

Calamagrostis, 14, 19, 20–1, 107, 160, 172, 259
squarrosa, 160, 172
Calamovilfa, 263
Calderonella, 15, 260
Calosteca, 14, 259
Calyptochloa, 14, 36, 264
Camusiella, 264
Capillipedium,, 14, 20, 21, 265

Castellia, 259
Catabrosa, 259
Catabrosella, 259
Catalepis, 43, 263
Catapodium, 14, 26, 259
Cathestechum, 15, 263
Cenchrinae, 37, 245, 246
Cenchrus, 14, 26, 42, 45, 101, 106, 110–13, 116, 119, 124, 158–61, 164–5, 167–9, 171, 250, 264
  biflorus, 159–61
  ciliaris, 101, 106, 111, 113, 116, 119, 124, 159, 164–5, 167–9, 171
  setigerus, 112, 119, 159, 164, 168
Centotheca, 45, 260
Centotheceae, 12, 35, 260
Centrochloa, 264
Centropodia, 261
Cephalostachyum, 261
Chaetium, 264
Chaetobromus, 261
Chaetopoa, 41, 264
Chaetopogon, 40, 259
Chaetostichium, 263
Chamaeraphis, 15, 264
Chandrasekharania, 264
Chasechloa, 264
Chasmanthium, 14, 39, 260
Chasmopodium, 265
Chevalierella, 260
Chikusichloa, 14, 260
Chimonobambusa, 48, 261
Chionochloa, 14, 261
Chloachne, 264
Chlorideae, 7, 12, 263
Chloridoideae, 5–7, 11, 16–17, 26, 246, 262–3
Chloris, 14, 21, 26, 68, 164, 249, 263
  gayana, 68, 164, 249
Chlorocalymma, 42, 264
Chrysochloa, 263
Chrysopogon, 37, 44, 158–9, 161, 164, 172–3, 265
  asciculatus, 44
  aucheri, 158–9, 161, 172
  fulvus, 159
  gryllus, 158, 164, 172
  montanus, 164, 172
Chumsriella, 15, 265
Chusquea, 39, 241, 261
  abietifolia, 241
Cinna, 259
Cladoraphis, 42, 263
Cleistachne, 265
Cleistochloa, 14, 36, 264
Clinelymus, 168
  dahuricus, 168
Cockaynea, 258
Coelachne, 264
Coelachyropsis, 263
Coelachyrum, 263
Coelorhachis, 14, 265
Coix, 14, 15, 35, 42, 265
Colanthelia, 261
Coleanthus, 39, 60, 259
Colpodium, 259
Commelinidium, 264
Cornucopiae, 43, 259
Cortaderia, 14, 15, 35, 57, 261
Corynephorus, 26, 259
Cottea, 14, 36, 262
Craspedorhachis, 262
Criciuma, 48
Crinipes, 261
Crithopsis, 258
Crypsis, 263
Cryptochloa, 15, 260
Ctenium, 39, 263
Ctenopsis, 259
Cutandia, 259
Cyathopus, 259
Cyclostachya, 15, 263
Cymbopogon, 265
Cymbosetaria, 264
Cynodon, 2, 21, 159, 164, 224, 226, 248, 249, 263
  dactylon, 159, 164, 224, 226, 249
Cynodonteae, 12, 35, 37–9, 41–3, 45
Cynosurus, 41, 137, 259
  cristatus, 137
Cyperochloa, 262
Cyperochloeae, 262
Cyphochlaena, 264
Cypholepis, 263
Cyrtococcum, 264

Dactylis, 26, 133, 139, 186, 200, 226, 259
  glomerata, 133, 139, 186, 200, 226
Dactyloctenium, 14, 165, 172, 263
  aegyptium, 165
  robecchii, 172
Daknopholis, 263
Danthonia, 12, 14, 21, 23, 36, 261
Danthoniastrum, 261
Danthonidium, 261
Danthonieae, 8, 261–2
Danthoniopsis, 264
Dasychloa, 263
Dasypoa, 259
Dasypyrum, 20, 21, 258
Decaryella, 263
Decaryochloa, 42, 261
Dendrocalamus, 261
Dendrochloa, 261
Deschampsia, 14, 26, 133, 242, 259
  caespitosa, 133
Desmazeria, 259
Desmostachya, 263

Deyeuxia, 14, 259
Diandrochloa, 14, 263
Diandrolyra, 15, 42, 260
Diandrostachya, 264
Diarrhena, 12, 61, 260
Diarrheneae, 260
Dichaetaria, 262
Dichanthelium, 14, 264
Dichanthium, 14, 20–1, 113–14, 118, 136, 159, 164, 169, 174, 244, 247, 265
  annulatum, 159, 164, 169
  aristatum, 247
  intermedium, 244
Dichelachne, 14, 259
Dielsiochloa, 259
Digastrium, 265
Digitaria, 14, 19, 24, 43, 167, 242–3, 245, 264
  commutatum, 167
  divaricatissima, 43
Digitariinae, 245
Digitariopsis, 264
Dignathia, 263
Diheteropogon, 159, 265
  hagerupii, 159
Dilophotriche, 264
Dimeria, 265
Dimorphochloa, 14, 264
Dinebra, 37, 263
Dinochloa, 38, 42, 43, 261
Diplachne, 14, 263
Diplopogon, 262
Dissanthelium, 259
Dissochondrus, 39, 264
Distichlis, 15, 35, 263
Drake-Brockmania, 263
Dregeochloa, 42, 262
Dryopoa, 259
Dupontia, 20, 21, 259
Duthiea, 262
Dybowskia, 265

Eccoilopus, 265
Eccoptocarpha, 40, 264
Echinaria, 259
Echinochloa, 2, 14, 26, 44, 57, 60, 201, 264
  callopus, 44
  colonum, 160
  frumentacea, 57, 60
  oryzicola, 201
Echinolaena, 264
Echinopogon, 259
Ectrosia, 14, 263
Ectrosiopsis, 263
Ehrharta, 3, 12, 39, 260
Ehrharteae, 12, 260
Ekmanochloa, 15, 260
Eleusine, 2, 14, 26, 56, 159, 248, 263
  compressa, 159
Eleusiniae, 246
Elionurus, 265
Elymandra, 265
Elymus, 14, 21, 24–5, 54, 67–8, 101, 107, 123, 133, 143–4, 165–6, 167–9, 259
  cinereus, 167
  elongatus, 166–7
  hispidus, 165–6
  nutans, 169
  rectisetus, 101, 107
  sibiricus, 169
  tauri, 166
  virginicus, 54, 67–8
Elytrigia, 19, 20–5, 258
Elytrophorus, 41, 262
Elytrostachys, 261
Enneapogon, 14, 36, 159, 262
  persicus, 159
Enteropogon, 26, 263
Enterolasia, 264
Entoplocamia, 263
Eragrosteae, 12
Eragrostideae, 35, 37, 42, 246
Eragrostiella, 248, 263
  bifaria, 248
Eragrostis, 3, 8, 14, 19, 35, 39, 43–4, 60, 66, 101, 107, 114, 118, 160–1, 164–5, 169, 242, 248, 263
  cilianensis, 60, 66
  curvula, 101, 114, 164–5
  reptans, 35
  tef, 60
  tremula, 161, 165
  viscosa, 44
Eremochloa, 265
Eremopoa, 259
Eremopogon, 265
Eremopyrum, 26, 258
Eriachne, 14, 262
Eriachneae, 39, 262
Erianthecium, 14, 259
Erianthus, 14, 21, 23, 265
Eriochloa, 14, 44, 91, 264
Eriochrysis, 265
Erioneuron, 263
Erythranthera, 262
Euchlaena, 15, 21, 25, 61, 265
Euclasa, 265
Eulalia, 265
Eulaliopsis, 265
Eustachys, 263
Euthryptochloa, 259
Exotheca, 41, 265

Fargesia, 261
Farrago, 263
Festuca, 14, 19, 20, 22, 25–6, 60, 70, 131,

133, 140–2, 186, 200, 226, 228, 242–3, 248, 259
arundinacea, 166–7, 186, 200
elatior, 226, 228
microstachys, 131, 133
ovina, 133, 136, 142, 248
rubra, 133, 140–2, 248
Festucella, 259
Festucopsis, 258
Fingerhuthia, 14, 39, 263
Froesiochloa, 260

Garnotia, 14, 264
Gastridium, 259
Gaudinia, 37, 41, 259
Gaudiniopsis, 259
Germainia, 15, 41, 265
Gigantochloa, 34, 261
Gilgiochloa, 40, 264
Glaziophyton, 48, 261
Glyceria, 14, 259
Glyphochloa, 265
Gouinia, 263
Gouldochloa, 260
Greslania, 261
Griffithsochloa, 15, 263
Guadiella, 49, 261
Guaduelleae, 261
Gymnachne, 14, 259
Gymnopogon, 14, 263
Gynerium, 15, 35, 262

Habrochloa, 14, 262
Hackelochloa, 265
Hainardia, 259
Hainardoeae, 37, 39, 259
Hakonechloa, 262
Halopyrum, 263
Harpachne, 40, 263
Harpochloa, 263
Helictotrichon, 14, 26, 259
Helleria, 259
Hemarthria, 14, 265
Hemisorghum, 265
Henrardia, 258
Heterachne, 14, 263
Heteranthelium, 41, 258
Heteranthoecia, 264
Heterocarpha, 263
Heteropholis, 265
Heteropogon, 265
Hickelia, 261
Hierochloe, 13, 14, 17, 39, 242, 259
Hilaria, 14, 263
Hitchcockella, 42, 261
Holcolemma, 40, 264
Holcus, 133, 135, 146, 226, 259
lanatus, 146, 224, 226
mollis, 133, 135

Homolepis, 44, 264
Homopholis, 264
Homozeugos, 265
Hookerochloa, 259
Hordelymus, 258
Hordeum, 8, 14, 20–2, 24, 26, 37, 41, 57, 60, 66–9, 132–3, 163–4, 166–7, 169, 171–2, 174, 186, 200, 203–4, 206, 221, 242, 258
brevisubulatum, 169
bulbosum, 163, 166–7, 172
fragile, 164, 166, 172
jubatum, 68
murinum, 132–3
spontaneum, 163, 171–2
violaceum, 172
vulgare, 60, 186, 200, 203, 206, 221
Hubbardia, 40, 264
Hubbardochloa, 14, 51, 260
Hyalopoa, 259
Hydrochloa, 15, 260
Hydrothauma, 264
Hygrochloa, 15, 35, 264
Hygroryza, 260
Hylebates, 264
Hymenachne, 264
Hyparrhenia, 14, 38, 41, 158, 163, 167, 168–9, 248–9, 265
dissoluta, 158
hirta, 158, 163, 167–9, 248
Hyperthelia, 3, 265
edulis, 3
Hypogynium, 15, 265
Hypseochloa, 14, 259
Hystrix, 258

Ichnanhus, 44, 264
Imperata, 2, 22, 23, 265
Indocalamus, 261
Indopoa, 263
Indosasa, 261
Isachne, 15, 264
Isachneae, 18, 39, 264
Isalus, 264
Ischaemum, 265
Ischnochloa, 265
Ischnurus, 263
Iseilema, 15, 37, 41, 42, 242, 265
vaginiflorum, 42
Ixophorus, 15, 40, 264

Jansenella, 264
Jardinea, 265
Jouvea, 15, 35, 39–40, 45, 263

Kampochloa, 263
Kaokochloa, 262
Karroohloa, 262
Kengia, 14, 263

Kerriochloa, 265
Koeleria, 14, 22, 24, 259

Lagurus, 14, 259
Lamarckia, 41, 259
Lamprothyrsus, 14–15, 35, 262
Lasiacis, 44, 265
Lasiorrachis, 265
Lasiurus, 37, 159, 161, 163–4, 168, 248, 265
  hirsutus, 161, 163–4, 168
  sindicus, 168
Lecomtella, 264
Leersia, 14, 260
Lepargochloa, 265
Leptagrostis, 259
Leptaspis, 15, 39, 45, 260
Leptocarydion, 263
Leptochloa, 14, 171, 246, 263
  fusca, 171
Leptochloopsis, 263
Leptocoryphium, 264
Leptoloma, 264
Leptosaccharum, 265
Leptothrium, 263
Lepturella, 263
Lepturidium, 15, 263
Lepturopetium, 20, 263
Lepturus, 20, 37, 41, 263
Leucophrys, 264
Leucopoa, 259
Leymus, 20–4, 160, 164, 168–9, 258
  chinense, 160, 164, 168
  dasystachys, 169
Libyella, 14, 39, 259
Limnas, 259
Limnodea, 259
Limnopoa, 15, 264
Lindbergella, 259
Lintonia, 263
Lithachne, 15, 260
Littledalea, 259
Loliolum, 259
Lolium, 14, 22, 26, 40, 133, 138–40, 146, 186, 200, 203–4, 224, 226, 259
  multiflorum, 186, 203–4, 226
  perenne, 133, 138–9, 146, 186, 200, 224, 226
Lombardochloa, 14, 25
Lophachme, 263
Lophatherum, 45, 260
Lopholepis, 263
Lophopogon, 265
Lophopyrum, 21–2, 24, 26, 259
Lorenzochloa, 261
Loudetia, 3, 264
  esculenta, 3
Loudetiopsis, 37, 264
Louisiella, 264
Loxodera, 265
Luziola, 15, 34–5, 42, 260
Lycochloa, 259
Lycurus, 37, 263
Lygeneae, 261
Lygeum, 1, 12, 34, 39, 261

Maclurolyra, 15, 260
Maillea, 259
Malacurus, 259
Maltebrunia, 260
Manisuris, 265
Maydeae, 18, 265
Megalachne, 34, 259
Megaloprotachne, 264
Megastachya, 260
Melanocenchris, 263
Melica, 14, 259
Meliceae, 259
Melinis, 264
Melocalamus, 42–3, 46, 261
Melocanna, 42, 46, 241, 261
Merostachys, 42, 261
Merxmuellera, 262
Mesosetum, 264
Metasasa, 261
Metcalfia, 34, 262
Mibora, 26, 259
Micraira, 2, 49, 248, 262
Micrairieae, 262
Microbriza, 14, 259
Microcalamus, 264
Microchloa, 14, 263
Microlaena, 3, 14, 260
  stipoides, 3
Micropyropsis, 259
Micropyrum, 259
Microstegium, 14, 40, 265
  vagans, 40
Mildbraediochloa, 264
Milium, 259
Miscanthidium, 22–3, 265
Miscanthus, 22–3, 265
Mnesithea, 265
Mniochloa, 15, 260
Molinia, 262
Monachather, 262
Monanthochloe, 15, 39, 263
Monelytrum, 263
Monium, 265
Monocymbium, 265
Monodia, 262
Monostachya, 262
Mosdenia, 263
Muhlenbergia, 14, 36, 263
Munroa, 263
Myriocladus, 261
Myriostachya, 40, 263

Nardeae, 261
Narduroides, 259
Nardus, 12, 14, 34, 39, 261
Narenga, 22, 23, 265
Nassella, 14, 43, 261
  trichotoma, 43
Nastus, 39, 261
Neeragrostis, 15, 263
Neesiochloa, 263
Nematopoa, 262
Neobouteloua, 263
Neohouzeaua, 261
Neostapfia, 2, 262
Neostapfiella, 263
Nephelochloa, 259
Neurachne, 34, 244, 264
  minor, 244
Neurachneae, 264
Neurolepis, 261
Neyraudia, 263
Notochloe, 262

Ochlandra, 34, 42, 49, 241, 261
Ochthochloa, 263
Odontelytrum, 34, 264
Odyssea, 263
Olmeca, 42, 261
Olyra, 15, 241, 260
Olyreae, 18, 35, 39–40, 260
Ophiuros, 265
Opizia, 15, 263
Oplismenopsis, 264
Oplismenus, 44, 91, 264
  aemulus, 91
Orcuttia, 262
Orcuttieae, 18, 262
Oreobambos, 34, 261
Oreochloa, 259
Orinus, 263
Oropetium, 37, 40, 248, 263
  capense, 248
Orthachne, 261
Orthoclada, 260
Oryza, 5, 14, 19, 23–4, 26, 65, 186, 200–1, 203–4, 206, 242, 246, 260
  brachyantha, 201
  eichingeri, 201
  officinalis, 201
  perrieri, 201
  sativa, 186, 201, 203, 206
Oryzeae, 7, 9, 35, 39–40, 246, 260
Oryzidium, 15, 264
Oryzodae, 6, 16, 19, 26, 260
Oryzopsis, 14, 23–4, 53, 55–6, 63, 71, 159, 166, 261
  aequiglumis, 159
  hendersoni, 53, 55–6
  holciformis, 166
  miliacea, 56, 63, 71

Otachyrium, 15, 40, 264
Otatea, 261
Ottochloa, 264
Oxychloris, 263
Oxyrhachis, 264
Oxytenanthera, 261

Paniceae, 7, 27, 35, 37, 39–40, 43, 245–6, 264
Panicodae, 6, 16, 264–5
Panicoideae, 5–6, 7, 10–13, 16–17, 26–7, 35, 39, 245–6, 263–5
Panicum, 2, 7–8, 19, 43–4, 60, 91, 101, 106, 110–11, 116, 118, 133, 136, 159–61, 163–5, 167, 169–70, 186, 199–200, 203–4, 242–3, 247–9, 264
  antidotale, 159–60, 164, 167, 169, 248
  cervicatum, 44
  coloratum, 91, 249
  maximum, 101, 111, 116, 118, 133, 136, 186, 200, 203–4
  miliaceum, 60, 200
Pappophoreae, 18, 39, 45, 262
Pappophorum, 14, 36, 262
Parafestuca, 259
Parahyparrhenia, 265
Paraneurachne, 264
Parapholis, 259
Paratheria, 14, 264
Parectenium, 264
Pariana, 14–15, 34, 40, 241, 260
Parianeae, 35, 37
Parodiolyra, 15, 260
Pascopyrum, 259
Paspalidium, 264
Paspalum, 14, 19, 26, 36, 56, 101, 106, 110, 120, 247, 264
  amphicarpon, 36
  cromyorrhizon, 247
  dilatum, 101, 106, 120
  notatum, 101, 110
Pennisetum, 7, 14, 34, 36, 41, 43, 57, 68, 108–9, 121–4, 139, 159, 162–3, 169, 172, 174, 186, 199–203, 242, 246, 248, 250, 264
  americanum, 172, 186, 200–1, 203, 250
  clandestinum, 36
  dichotomum, 163
  flaccidum, 108
  glaucum, 109
  mezianum, 109
  mollissimum, 250
  polystachion, 43
  purpureum, 122–3, 186, 200
  schimperi, 162, 172
  setaceum, 121–2, 124
  squamulatum, 121–3, 172
  typhoideum, 57
  violaceum, 250

Pentameris, 42, 262
Pentapogon, 259
Pentarrhaphis, 263
Pentaschistis, 262
Pereilema, 14, 263
Periballia, 26, 242, 259
Perotis, 263
Perrierbambus, 42, 261
Perulifera, 264
Petriella, 260
Peyritschia, 259
Phacelurus, 265
Phaenanthoecium, 262
Phaenosperma, 12, 260
Phaenospermateae, 260
Phalaris, 14, 26, 39, 41, 133, 167, 171, 223, 226, 259
  aquatica, 167, 171, 223
  arundinacea, 133, 226
  paradoxa, 41
Phareae, 35, 260
Pharus, 15, 45, 260
Pheidochloa, 14, 262
Phippsia, 14, 23, 39, 259
Phleum, 14, 26, 138, 224, 226, 259
  pratense, 38, 224, 226
Pholiurus, 259
Phragmites, 5, 38–9, 91, 135, 142, 248, 261
  australis, 91, 135, 142, 248
Phyllorachideae, 35, 37, 40–41, 260
Phyllostachys, 38, 241, 261
Pilgerochloa, 259
Piptatherum, 261
Piptochaetium, 14, 261
Piptophyllum, 263
Piresia, 15, 260
Plagiantha, 40, 264
Plagiosetum, 264
Planichloa, 263
Plectrachne, 262
Pleiadelphia, 265
Pleuropogon, 259
Plinthanthesis, 262
Poa, 14–15, 19, 26, 35, 60, 63, 68, 71, 91, 101–2, 105–7, 114, 116–17, 120, 133, 145–7, 155–6, 167, 172–4, 200, 224, 226, 242–3, 259
  alpina, 68, 71
  annua, 133, 145–7
  bulbosa, 91, 172–3
  compressa, 226
  × *jemtlandica*, 133, 135
  nervosa, 133
  pratensis, 60, 101–2, 105–6, 114, 116–17, 120, 136, 200, 224
  siniaica, 167, 172–3
Poagrostis, 262
Poales, 2
Podophorus, 259
Poeae, 7, 35, 259
Poecilostachys, 264
Pogonachne, 265
Pogonarthria, 38, 263
Pogonatherum, 265
Pogoneura, 263
Pogonochloa, 263
Pohlidium, 15, 35, 39, 260
Poidium, 14, 259
Polevansia, 263
Polliniopsis, 265
Polypogon, 14, 20, 23, 40, 186, 259
  fugax, 186
Polytoca, 15, 265
Polytrias, 265
Pommereulla, 263
Poodae, 6–7, 16, 26, 259
Pooideae, 5–7, 10, 12–13, 16, 18, 26–7, 36, 43, 258, 259
Porteresia, 260
Potamophila, 260
Pringleochloa, 15, 263
Prionanthium, 262
Prosphytochloa, 260
Psammagrostis, 263
Psammochloa, 261
Psathyrostachys, 22, 23, 259
Pseudanthistiria, 265
Pseudarrhenatherum, 259
Pseudechinolaena, 45, 264
Pseudobromus, 259
Pseudochaetochloa, 15, 264
Pseudocoix, 261
Pseudodanthonia, 34, 262
Pseudodichanthium, 265
Pseudopentameris, 262
Pseudophleum, 259
Pseudopogonatherum, 265
Pseudoraphis, 15, 264
Pseudoroegneria, 259
Pseudosasa, 261
Pseudosorghum, 265
Pseudostachyum, 261
Pseudovossia, 265
Pseudozoysia, 263
Psilathera, 259
Psilolemma, 263
Psilurus, 39, 259
Pterochloris, 263
Puccinellia, 14, 23, 145, 148, 167, 259
  capillaris, 167
  distans, 167
  maritima, 145, 148
Puelia, 15, 17, 44, 261
Puelieae, 261
Pyrrhanthera, 42, 262

Racemobambos, 261

Raddia, 15, 260
Raddiella, 15, 260
Ratzeburgia, 265
Redfieldia, 263
Reederochloa, 15, 263
Rehia, 15, 260
Reimarochloa, 264
Reitzia, 15, 260
Relchela, 14, 259
Rendlia, 263
Reynaudia, 264
Rhipidocladum, 261
Rhizocephalus, 259
Rhomboelytrum, 14, 259
Rhynchelytrum, 264
Rhynchoryza, 45, 260
Rhytachne, 265
Richardsiella, 263
Robynsiochloa, 265
Rottboellia, 14, 247, 265
  exaltata, 247
Rottboelliinae, 44, 265
Rytidosperma, 14, 262

Saccharum, 1, 5, 9, 14, 19–25, 186, 200–1, 204, 242, 244, 265
  officinarum, 186, 200, 244
  spontaneum, 244
Sacciolepis, 264
Sartidia, 262
Sasa, 23, 33, 68, 261
  paniculata, 68
Saugetia, 263
Schaffnerella, 263
Schedonnardus, 263
Schismus, 262
Schizachne, 259
Schizachyrium, 14, 265
Schizostachyum, 241, 261
  zollingeri, 241
Schmidtia, 262
Schoenefeldia, 159, 161, 165, 248, 263
  gracilis, 159, 161, 165
Sclerachne, 15, 265
Sclerochloa, 259
Sclerodactylon, 263
Scleropogon, 15, 35, 263
Sclerostachya, 22–3, 265
Scolochloa, 259
Scribneria, 259
Scrotochloa, 15, 260
Scutachne, 264
Secale, 1, 14, 20–4, 26, 60, 164–6, 206, 226, 259
  cereale, 206, 226
  montanum, 164–6
Sehima, 265
Semiarundinaria, 23, 261
Sesleria, 41, 259
Seslerieae, 259
Sesleriella, 259
Setaria, 14, 37, 41, 45, 57, 264
  italica, 57
Setariopsis, 264
Shibataea, 261
Sieglingia, 14, 21, 23, 262
Silentvalleya, 263
Simplicia, 259
Sinarundinaria, 261
Sinobambusa, 261
Sinochasea, 259
Sitanion, 20–2, 24, 41, 259
Snowdenia, 264
Soderstromia, 15, 263
Sohnsia, 15, 35, 263
Sorghastrum, 265
Sorghinae, 246
Sorghum, 2, 5, 14, 23–4, 56, 226, 242, 246, 265
  halepensis, 236
Spartina, 5, 64, 133, 135, 139, 147, 263
  patens, 139, 147
  × *townsendii*, 133, 135
Spartochloa, 2, 262
Spartochloeae, 262
Spathia, 14, 265
Sphaerocaryum, 264
Sphaerobambos, 261
Spheneria, 264
Sphenopholis, 24, 259
Sphenopus, 259
Spinifex, 15, 35, 42, 45, 264
Spodiopogon, 265
Sporobolus, 14, 43, 159–61, 171, 248, 263
  helvolus, 161, 171
  iocladus, 161, 171
  marginatus, 159
  portobolus, 248
  ruspolinus, 161, 171
Steinchisma, 40, 264
Steirachne, 263
Stenotaphrum, 37, 40, 41, 45, 264
Stephanachne, 259
Stereochlaena, 264
Steyermarkochloa, 45, 261
Steyermarkochloeae, 8, 261
Stiburus, 263
Stilpnophleum, 259
Stipa, 1, 3, 8, 12, 14, 19, 23–4, 36, 44, 56, 64–5, 69, 70–1, 158–60, 162, 166–8, 170, 173, 243, 261
  baicalensis, 162
  barbata, 158, 162, 166–7
  caucasica, 162
  chrysophylla, 162
  elmeri, 69
  gobica, 162
  grandis, 160, 162

Stipa (*cont.*)
humilis, 162
krylovii, 162
lagascae, 158, 162, 166–8
linearis, 159
pariflora, 162
pulcherrima, 44
speciosa, 162
szowitziana, 159
tenacissima, 158
tortilis, 56, 64
Stipeae, 27, 34, 39–40, 261
Stipagrostis, 161, 169–70, 262
plumosa, 169–70
pungens, 161
uniplumis, 161
Streblochaete, 45, 259
Streptochaeta, 33, 45, 48, 260
Streptochaeteae, 260
Streptogyna, 45, 260
Streptogyneae, 260
Streptolophus, 42, 45, 264
Streptostachys, 264
Styppeiochloa, 262
Sucrea, 15, 260
Suddia, 260
Swallenia, 263
Swallenochloa, 241, 261
Symplectrodia, 262

Taeniatherum, 26, 259
Tarigidia, 24, 245, 264
aequiglumis, 245
Tatianyx, 264
Teinostachyum, 261
Tetrachaete, 263
Tetrachne, 39, 263
Tetrapogon, 14, 161, 263
villosus, 161
Tetrarrhena, 260
Thamnocalamus, 261
Thaumastochloa, 37, 265
Thelepogon, 265
Thellungia, 14, 263
Themeda, 14, 41, 265
Thinopyrum, 14, 21, 22, 24, 259
Thrasya, 40, 264
Thrasyopsis, 40, 264
Thuarea, 15, 37, 41, 45, 264
Thyridachne, 264
Thyridolepis, 14, 264
Thyrsia, 265
Thyrsostachys, 261
Thysanolaena, 3, 40, 261
Torreyochloa, 259
Tovarochloa, 259
Trachypogon, 41, 265
Trachys, 40, 42, 264
Tragus, 45, 263
Tribolium, 262
Tricholaena, 14, 158, 163, 172, 264
teneriffae, 158, 163, 172
Trichoneura, 263
Trichopteryx, 264
Tridens, 14, 263
Trikeraia, 261
Trilobachne, 15, 265
Triniochloa, 259
Triodia, 248, 262
Triodieae, 262
Triplachne, 259
Triplasis, 14, 36, 263
Triplopogon, 15, 265
Tripogon, 248, 263
loliiformis, 248
minimus, 248
Tripsacinae, 246
Tripsacum, 7, 14–15, 24–5, 101, 107, 122, 242, 246, 252, 265
dactyloides, 101, 107, 122
Triraphis, 263
Triscenia, 264
Trisetum, 14, 22, 24, 259
Tristachya, 37, 264
Triticeae, 1, 7–8, 11, 19, 37, 39, 41, 43, 251, 258–9
Triticodea, 6–7, 16, 26, 258–9
Triticosecale, 186
Triticum, 7–8, 14, 19, 20–4, 26, 38, 60, 123, 186, 203–4, 206, 242–3, 245, 259
aestivum, 60, 123, 186, 206
boeticum, 243
monococcum, 203–4, 206
Tsvelevia, 259
Tuctoria, 262

Uniola, 14, 39, 263
Uranthoecium, 37, 264
Urelytrum, 265
Urochlaena, 262
Urochloa, 14, 91, 264
Urochondra, 263

Vahlodea, 259
Vaseyochloa, 263
Ventenata, 259
Vetiveria, 1, 64, 265
zizanioides, 164
Viguierella, 40, 263
Vossia, 265
Vulpia, 14, 22, 25, 141–3, 146, 259
fasciculata, 141–3, 146
Vulpiella, 259

Wangenheimia, 259
Whiteochloa, 264
Willkommia, 14, 263

Xerochloa, 15, 264

Yakirra, 44, 249, 264
Ystia, 265
Yushania, 261
Yvesia, 264

Zea, 5, 15, 21–6, 34–5, 56–8, 61, 65–71, 78, 101, 122, 139, 186, 200, 203–4, 206–7, 226, 242, 245–6, 253, 265
  mays, 56–8, 67–8, 71, 78, 101, 122, 139, 186, 200, 203–4, 206–7, 226, 242
Zenkeria, 262
Zeugites, 15, 260
Zingeria, 242, 259
Zizania, 14–15, 26, 35, 54, 56, 65, 260
  aquatica, 54, 65
Zizaniopsis, 15, 35, 42, 260
Zonotriche, 37, 264
Zoysia, 263
Zoysiinae, 37, 39, 41
Zygochloa, 15, 35, 42, 264

Other organisms
Note: Rusts and smuts associated with grass taxa are set out in the Appendix.

Agrobacterium, 203, 205, 207–8, 234, 253
  tumefaciens, 203, 234
Alternaria brassicola, 192
Anarthriaceae, 2
Apis, 241, 249
  dorsata, 241
  florea, 241
  mellifera, 241
Apoidea, 241
Arales, 207

Brassica, 81, 192, 221, 222
  campestris, 221
  napus, 192
  oleracea, 221

Capsella, 62
Centrolepidaceae, 2
Claviceps paspali, 124
Commelinaceae, 47
Commeliniflorae, 2
Compositae, 1
Cyperaceae, 2, 46

Datura, 202
Drechslera maydis, 193

Ecdeiocolaceae, 2
Elegia vulgaris, 46
Escherichia coli, 228

Flagellariaceae, 2, 46, 47

Gaeumannomyces graminis, 195

Helminthosporium maydis, 193
  sacchari, 193
  sativum, 194
  victoriae, 193, 195
Hemerocallis, 252
Hymenoptera, 241

Ischnosiphon arouma, 47
Ixeris, 107

Joinvilleaceae, 2, 46, 47
Juncaceae, 46

Leguminosae, 1
Liliales, 207

Marantaceae, 46, 47
Melipona, 241

Nicotiana, 205
Nomia, 249

'Oenothera', 56, 62
Orchidaceae, 1

Periconia circinata, 101
Petunia, 205
Phrynium, 47–8
Plumbago, 81, 86, 221–2
  zeylanicum, 221
'Polygonum', 59–62, 103, 105
Pseudomonas syringae, 194
Puccinia
  coronata, 193
  sorghi, 252

Raphanus, 252–3
Restionaceae, 2, 46–7
Restio
  complanatus, 46
  schoenoides, 48
Rosaceae, 100

Taraxacum, 107
Thalia geniculata, 47
Trigona, 241

Vicia, 252–3

Xanthomonas oryzae, 194

Subject index

abortion, 102
abscission, 36–7, 41, 43
adventitious embryony, 103
agamic complex, 119, 126–7, 163
agamospermy, 132–3
agroinfection, 207
albinism, 189, 190, 200
alfalfa, 197
alien genomes, 118
allergens, 27, 220, 223–34
 actinidin, 232
 *Agr a*I, 228
 *Ant o*I, 228
 Bromelain, 232
 *Fes e*I, 233
 *Lol p*I, 225–34
 *Poa p*I, 228
alternation of generations, 62
amino acid
 analogues, 197
 lysine, 196
 overproduction, 196–7
 phenyl alakine, 196
 proline, 196, 232
 threonine, 196
 tryptophan, 196
aneuploidy, 114, 117
angiosperms, 80
aniline-blue fluorescence, 61, 77
animal–flower interaction, 13
 transport, 44
annualism, 243, 249
anther culture, 187–90, 208
anthesis, 34, 38, 79, 95, 105, 109, 125, 131
anthoecium, 33
antipodals, 54, 57, 62–70, 86, 103–5
ants, 44, 249
apical pocket, 64, 67
apomixis, 14, 16, 18, 36, 52, 55, 62, 100–28, 169–70, 243, 249–50
 cytology, 102
 environment, 110, 114, 126
 evolutionary role, 125–7
 facultative, 101–2, 106, 108–10, 112–13, 116–18, 127
 genetics, 109–14
 obligate, 100, 102, 108, 110, 112–13, 115–16, 118–19, 121–2, 124, 126–7
apospory, 58–9, 101, 103, 110–11, 132, 136
archesporial cells, 55–6, 58
asexual reproduction, 3, 132, 135–6
associate cell, 221–2
assortative mating, 137
asthma, 220, 223–4
awn, 38, 41, 44–5, 48, 163, 168
awnless mutants, 170–71
axenic cultures, 187

B chromosomes, 81
BIII hybrids, 118–19
bamboos, 1–2, 17–18, 27, 44, 68, 240–2, 247, 266
bandicoots, 249
bees, 34, 241, 249
binding specificities, 77–8
birds, 44–5
bract, 32, 46–7
branch, 41
 -let, 32
breeding barriers, 138
bridging species, 122–3
bulbils, 132, 163, 173
bulbs, 132
burs, 45

C_3, 244, 258, 261, 263–4
C_{3-4}, 244, 264
$C_3 \times C_4$, 246, 261–5
C_4, 3, 47, 240, 244
callose, 61, 63, 67
callus (floral), 44
callus (*in vitro*), 183, 185, 188
carbohydrate, 43, 77, 189
carpel, 52–4

caryopsis, 2, 42–3, 55
central cell, 64, 66, 69, 84, 86, 88
cereal, 154–5, 174, 184, 188–9, 196–7, 202–3, 208, 221–2, 234, 240, 242, 247, 249
chalaza, 53, 55, 108
'chloridoids', 9
chromosome
 aberrations, 100
 basic number, 242, 248
 chromatin, 86
 colchicine, 188
 dihaploids, 188–9, 192
 doubling, 187–8
 haploids, 108, 187–8, 245
 polyploidy, 3, 27, 102, 109–11, 114–19, 121, 127, 132, 135, 242–4
citrus, 103
cladistics, 5
classification, 1–27, 33, 258–65
clearing techniques, 132
climate, 132
cloning, 128
collection, 172, 174
common names
 arctic poa, 132
 bahia grass, 110–11
 bamboos, 1–2, 17–18, 27, 44, 52, 68, 240–2, 247
 barley, 53, 66–7, 69, 79, 80, 82–94, 171, 184, 189, 194, 196, 202, 206, 208, 253
 barnyard grass, 201
 Bermuda (couch), 224, 226
 birdwood grass, 112
 buffel grass, 111, 116–17
 bulbous poa, 91
 cocksfoot (orchard grass), 224, 226
 dallis grass, 124
 fountain grass, 124
 Kentucky blue, 224
 maize (corn), 27, 64, 66–7, 76–8, 83–4, 93–4, 101, 140, 184–5, 189, 192, 196–9, 202–3, 205–8, 222, 228, 249, 250–1
 napier grass, 250
 oats, 53, 184, 195, 197, 206, 226
 pearl millet, 120, 122–4, 196–7, 249–50
 purple wire grass, 91
 rice, 27, 69, 184, 189–90, 193–4, 196, 199, 201–3, 206–8, 246, 251
 rye, 53, 60, 76, 184, 249
 ryegrass, 221, 224–6, 228–9
 sorghum, 101, 197, 250
 sugar cane, 193, 199, 201
 surinam grass, 91
 sweet vernal grass, 224, 226
 triticale, 184, 206
 velvet grass, 224, 226
 wheat, 27, 53, 58, 69, 78, 80, 81, 83, 94–5, 101, 123, 184, 189, 190, 194–5, 202, 205–7, 222, 243, 250–1
'compilo-species', 233
competitive interaction, 131
computer-assisted quantitation, 81
condensation (ontogenetic), 36
coolamon, 249
cotton, 86, 252
Cretaceous, 27
cryopreservation, 184
cytoplasmic body, 88–91
cytoplasmic extensions, 80, 83–4

DNA
 alien, 245, 251, 253
 amplification, 243
 cDNA, 228, 230–1, 233–4
 2*C* values, 26
 estimations, 13
 infection, 54
 non-recombinant, 202
 plasmid, 206
 rDNA 200–1
 recombinant, 220
 single copy,
 total genome, 205
deforestation, 155
desert, 163, 173
desertification, 154–6, 169, 254
DAPI, 92
diaspore, 36, 43, 45–6
dicotyledons, 58
dictyosomes, 67, 69, 85, 89, 91, 93–4
differentiation, 106, 183, 245
diplospory, 58–9, 103, 107–8, 114
disease resistance, 101, 120, 190, 194, 201, 203
 bacterial blight, 194
 brownspot, 193–4
 crown rust, 193
 ergot, 124
 eyespot disease, 193
 maize rust, 252
 southern corn leaf blight, 193
 take-all, 195
 (*see also* virus)
diversity, 248
domestication, 249–51
dormancy, 247, 250
double fertilization, 79–92, 220–3
drought, 3, 18, 155–60, 166–7
dryland farming, 158
dyads, 59, 108

ecotypes, 115, 117, 126, 148, 169, 172, 174, 240
ecotypic adaptation, 148
egg, 63, 64, 60, 66, 78, 84–5, 87, 89–92,

105–6, 108, 118–19, 221, 127, 251, 253
apparatus, 60–7, 69–70
elaiosomes, 39, 44, 249
electronmicroscopy, 81–9, 220, 229
electroporation, 125, 203
embryo, 2, 9, 43, 63, 67, 71, 94–5, 104, 109, 188
rescue, 246
embryogenesis, 93–5
embryogenic suspensions, 199
embryo sac, 54, 56, 58–60, 62–3, 65–7, 78, 85–90, 102–8, 112, 119, 253
Adoxa-type, 62
Antennaria-type, 107–8
Polygonum-type, 61–2, 103, 105, 107
pseudo-*Oenothera* type, 62
pseudo-*Polygonum* type, 62
wall, 52–3, 66
endomitosis, 68
endoplasmic reticulum, 57, 67, 69, 71, 85, 93–4
endosperm, 23, 43, 63, 67–9, 70–1, 93–4, 103–4, 222
endothelium, 57
enzyme activation, 76
epistasis, 111–12
'eternal triangle', 132–3
ethological mechanisms, 131
evolution, 19, 33, 38–9, 47, 125–7, 131–2, 232, 241, 251

farming, 33
fecundity, 142–3
'feedback', 132, 144, 149
fertilization, 35, 55–7, 65, 68–71, 76–94, 100, 115, 119, 127, 221–3, 234–5, 251, 253
filiform apparatus, 64, 66, 70–1
fire, 3, 18, 33, 145, 154, 158, 164
fitness, 131
flag leaf, 38, 42
flies, 241
flooding, 157
floral apex, 53
floral biology
andromonoecism, 35
anemophily, 34
apomixis, 14 (*and see* apomixis–main entry)
autogamy, 14, 60
breeding systems, 132–6, 144
chasmogamy, 36, 131, 247
cleistogamy, 14, 16, 18, 35–6, 132, 133, 135, 247
cleistogenes, 14, 16, 18, 36
dichogamy, 131
dicliny, 35
dioecy, 15–18, 34–5, 131
environmental effects, 132, 247
female fertility, 121, 123
gynodioecy, 131
inbreeding, 14
male sterility, 123, 221, 235
monoecy, 15–18, 34–5, 131
outbreeding, 35, 76, 131–5, 148
panmixis, 132
protandry, 52
protogyny, 35, 52
rhizanogenes, 14, 18, 36
self- and cross-compatibility, 35, 76–7, 101, 109, 131–2, 134–5, 140, 147, 220, 241
wind, 34, 138, 157
floral formulae, 33, 46
florescence, 47
floret, 13, 32–40, 43, 52, 117, 131, 137
floristic Kingdoms, 18
forage, 100, 104–5, 110, 114, 116, 120, 127–8, 160, 164, 166, 169, 172, 247
fluorescence, 61, 77
founders, 146
fruit, 33, 42, 58, 131, 146, 241
baccoid, 42, 241
cistoid, 42
false, 43
follicoid, 42
mimic, 44
mucoid, 42
fungal pathogens, 9, 193–5, 258, 260–2, 264
fusibility factors, 222
fusion of gametes, 88–92, 253
fusoid cells, 48

gametocide, 119
'gametoclonal variation', 189
gametophyte, 52, 58–9, 62, 106–7, 112, 120
dispersal, 131
geitonogamy, 131, 139
gene
clustering, 139
dispersal, 133, 139
exchange, 140
flow, 137
transfer, 115, 122–3, 125
transport, 137
generative cell, 80, 81, 131–3, 137, 139, 141–6, 220–1
genet
density, 133, 139, 143–4, 146
distribution, 144
mobility, 142
replacement, 137
turnover, 145
genetic
diversity, 132–6, 140, 147, 169, 183–4

genetic (*cont.*)
erosion, 169
fidelity, 184
fixation index, 140
renewal, 135
'turbulence', 242
genera, 1, 19, 127, 201, 203–4, 207
genome relationships, 122, 243–4
genotype, 135, 141–3
geographic distribution, 125–6
germination, 38
glumes, 32, 39, 44, 48–9
glycoprotein, 77, 225
grain, 120, 127
grazing, 3, 33, 36, 154, 156–8, 160, 164
'Green Revolution', 155
growth circles, 248
growth factor, 112
growth regulators (including herbicides)
auxin, 183
chlorsulphuron, 192, 197
2,4-D, 182, 189
Dicamba, 182
dichlorphenoxy acetic acid, 197
glufosinate, 197
glyphosphate, 198
imidazolinones, 197
Picloram, 182
sulphometuron, 197
gynoecium, 52

habitats
alkaline soil, 171
arid/semi-arid, 154–74
closed, 135
closed swards, 150
disclimax grassland, 157, 159
heathland, 145
lawn, 1, 146
forest, 34, 157
mine spoil, 137, 140
moor, 140–2
mountain grassland pasture, 117–18, 140, 146, 156, 169, 240
plains, 34, 163
rangelands, 156, 165–9, 171–2
reed beds, 142, 144, 147
saline soils, 171
saltmarsh, 145, 147–8
sand dune, 141–2, 144, 147–8, 161, 163, 165, 170, 247
savannah, 157, 164
seral communities, 147
stable, 135
steppe, 158, 162–4
thorn scrub, 157
turf, 101, 118, 120, 145
wadis, 163
Hardy–Weinberg equilibrium, 134, 136

haustoria, 57
hay, 1
herbivores, 44
heterofertilization, 84
heterokaryons, 200–1
grass–dicot, 200
grass–grass, 200
heteromorphy, 131
heterosis, 115, 126
'hitch hiking', 150
homogamy, 135
hormones, 109
'hump-backed' model, 145
hyaline layer, 58
hybridization, 3, 19–25, 70, 101, 112, 115–18, 122–4, 243–6, 251–2
asymmetric, 201
BIII, 118
'counterfeit', 246, 252
cybridization, 201
intergeneric, 20–5, 246
interspecies, 109, 115, 117–18, 222
natural, 124, 245
permanent, 115
somatic, 199, 200–2
true breeding, 116
wide crosses, 244–6
hygroscopic flexion, 25
hypodermal cells, 54

inflorescence, 13, 32, 38, 41, 43, 45–6, 132, 187
capitate, 41
culture, 187
false fruits, 43
iterauctant, 38
panicle, 32, 46, 47
raceme, 32
spike, 32
immunoglobulin IgE, 224–6, 230–3
inbreeding, 14, 16, 18
drepression, 110, 116
indigenous (native) species, 160, 164, 166, 168–9
induration, 39
insect pollination, 34, 131, 241
insect tolerance, 201
irradiation, 120, 200–1
integuments, 54–6, 59, 78–9, 103
in vitro technology, 182–209
involucre, 41–2
iso-zymes (including allozymes), 132–5, 202, 242

karyogamy, 86
karyology, 9
'K' versus 'r' strategists, 145, 149

'large' genera, 242–3

latex particles, 252–3
leaf anatomy, 9, Appendix
  blade, 42
  sheath, 3
lectin, 77
legumes, 8
lemma, 32, 34, 39–40, 45, 49, 53
lipids, 43, 67, 69
livestock 157
locule, 78
lodicules, 22, 33–6, 48–9
long cells, 1
longevity, 91, 135, 140–3, 248

male cytoplasm, 87–9
male fertility, 121, 123
male gamete, 76–9, 84, 221
  gametoplasts, 222, 253
  organelles, 88
male germ unit, 81–5, 221
marker genes, 200–6, 234, 252
  amino-ethyl cystein (AEC), 200
  beta-galactosidase, 235
  chloramphenicol transferase (CAT), 234, 252
  glucuronidase (GUS), 206, 234
  hygromycin, 203
  Kanamycin, 203
  morphological, 140
  neomycin phosphotransferase (NPT II), 205, 234, 252
  nitrate reductase, 202
  nopaline synthase, 234
  octopine synthase, 234
mast cell, 224
mast crop, 241
mate choice, 131
maternal inheritance, 80, 87, 90
megagametogenesis, 59, 62, 102–8
megagametophyte, 54, 57, 63–5, 69
  lamellae 66
  wall, 52–3, 66
megasporangium, 54
megaspore, 58, 60, 103–4, 108
megaspore mother cells, 55, 103–4, 108
megasporogenesis, 53, 59, 102–4, 106, 108, 113
meiosis, 58, 60, 63, 77, 102–4, 107–9, 114, 118, 124
meristems, 183
metal tolerance, 137, 139
5 methyl-DL-tryptophan, 196
microgametogenesis, 87
microinjection, 189, 202–5, 208
microprojectiles, 204, 209
micropyle, 53, 54–7, 63, 66, 79, 255
  associated hypertrophy, 56
microsporocytes, 61, 77
microspore, 192, 204
  culture, 189, 208
microtillering, 183
microtubules, 71
mitochondria, 62, 67–71, 80–1, 85–7, 89, 91, 93–4, 221
mitochondrial genomes, 220–2
mitosis, 62, 69, 94, 103, 105, 107, 108
mixed strategy, 134–5, 149
monoclonal antibodies, 220, 226–71, 229–33
monocotyledons, 32, 58
morphogenetic capacity, 183–4
mowing, 3
mutation, 7, 119–20, 136, 196, 234, 243

natural selection, 126, 172
'neighbourhood' concept, 137–8
'nobilization', 244
nucellus, 53–9, 63–4, 66, 71, 78–9, 95–5, 103, 106
  crassinucellate, 55–6
  embryosacs, 112
  protoderm, 65
  sporophyte, 62
  tenuinucellate, 55–6
nuclear fusion, 86
  genes, 184, 235
  membranes, 86

oedema, 224
oil, 34, 44
ontogeny, 32, 52, 183
organelles, 65, 68, 80, 86–7, 90, 92
organogenesis, 183
ovary, 32, 78
  culture, 187, 189
overgrazing, 158–60, 167
ovule, 52–71, 78, 83, 106
  orientation, 53
  outgrowths, 56
  vascularization, 54
  ventral suture, 54

palea, 32, 39–40, 46, 48
panicle, 32, 46, 47
panicoids, 55
paper, 1
parietal cells, 58, 65
parthenogenesis, 108, 111
partial genome transfer, 202
particle bombardment, 189
pectin, 77
pedicel, 37, 40, 41
peduncle, 42
perennialism, 243, 248
perianth, 32, 46
pericarp, 42–3, 58
'peripheral' grasses, 242

pesticide, 188
phenetics, 5
phenotypic evaluation, 114
phosphorus, 171
photosynthesis, 3, 9, 33, 38, 47, 240, 244, 258, 260–5
phyllotaxy, 2
phylogeny, 46
pioneer, 147–8, 150, 156, 161, 169
'pivotal' genomes, 243–4
placenta, 53–5
plant breeding, 114, 125, 127, 169, 187, 221, 245, 253
plasma membrane, 79–80, 85, 91–2
plasmodesmata, 61–3, 67, 69, 71, 94
plasticity, 136, 147, 247
plastid genomes, 190
plastids, 62, 69, 80, 85, 87, 89, 91, 93, 221
Pleistocene, 157, 165
polarity, 60, 61, 65
polar nuclei, 62–4, 66–7, 84, 86–8, 103–8, 119
  fusion, 66–7, 86–8
pollen, 13, 27, 34, 58, 76–80, 83–4, 109, 118–19, 137–40, 147, 188, 204–6, 220, 223, 225, 228–30, 234, 241
  adhesion, 76
  dispersal, 138–40, 147
  entrapment, 34
  enzyme activation, 76
  germination aperture, 76
  genes, 228
  hydration, 76
  hydrolases, 78
  hypersensitivity, 233
  mediated transformation, 203
  quality, 109
  transferases, 78
  tricellular, 79
  tube, 70, 76–80, 83–119, 221, 252
pollination, 38, 82–3, 95, 100, 103, 108, 119, 139, 249
pollinator, 33
  specialization, 131
polyembryony, 106, 136
polymorphism, 125–6, 133–6
population structure, 131–50
pre-pollination isolation, 131
prophyl, 32, 46–8
protection/regeneration, 158–60, 165
protoplasts, 125, 185, 191, 198, 200–1, 203, 208
pseudogamy, 59, 108–9, 126, 245
pseudospikelet, 47–9

quarantine, 187

raceme, 32, 37–8, 47, 49
racemelets, 41–3
radioallergo-sorbent test (RAST), 226
ramets, 141, 144
rapeseed, 192
recognition, 82, 220–35, 245, 251–2
  molecules, 82, 222, 232,
recombination, 131
regions
  Afghanistan, 167
  Africa, 17
  Alaska, 17
  Algeria, 157, 166–7
  Andean Highlands, 163
  Argentina, 165
  Asia, 159–60, 162, 170
  Atlas Mountains, 157
  Australasia, 17, 171, 225, 248–9
  Baluchistan, 159, 173
  Barbary, 250
  Burkina Fasso, 165
  Canada, 17
  China, 17, 156–7, 160, 162, 164, 168, 172
  Denmark, 250
  Eastern Asia, 17
  East Indies, 250
  Egypt, 250
  Elburg Mountains, 157, 164
  Ethiopia, 161–2
  Eurasia, 17
  Europe, 174
  Guinea, 250
  Himalayas, 157, 164, 172
  India, 159, 168, 171
  Indonesia, 17
  Iran, 157, 164, 166–7, 172
  Irano-Turan, 162
  Iraq, 157, 166, 172
  Jamaica, 249
  Japan, 17
  Jordan, 167
  Karakoram Mountains, 173
  Kenya, 158, 249
  Madagascar, 17, 250
  Malaysia, 17
  Mali, 159, 161, 165
  Mauritania, 161, 165, 250
  Mediterranean, 162–3
  Middle East, 162
  Mexico, 17
  Morocco, 157, 166–7
  Near East, 160, 162–3, 165–6, 169–72
  Niger, 157, 165
  Nigeria, 159
  North Africa, 162, 165, 169, 172
  North America, 17
  Pacific, 17
  Pakistan, 159, 167, 173
  Patagonia, 163, 165

Punjab, 171
Rajasthan, 159, 168
Sahara, 157, 163, 169, 172, 174
Sahel, 156–7, 159–62, 165, 169, 171
Samarkand, 167
Saudi Arabia, 17
Scotland, 250
Senegal, 161, 165
Sind, 168
Somalia, 161
Somali-Masai, 162
Southern Africa, 160, 165
sub-Saharan Africa, 156–7
Sudan, 157, 161, 165
Syria, 167
Thar desert, 159–60
Tibet, 163
Tunisia, 157, 166–7
Turkey, 157, 164
USA, 17, 166, 174
Zagros Mountains, 157, 164
regeneration, 183, 188, 199
relic species, 125
reseeding, 154, 164–7, 169
'resurrection' plants, 248
rhachilla, 32, 40
rhachis, 39–40, 42–3
rhizanthogenes, 14, 18, 36
rhizome, 36, 132, 241
ribosomes, 62, 71, 85, 93–4
RNA, 7, 69–70, 80, 228, 230, 233
rodents, 44
'r' strategists, 145, 149
rust, 13, 252, 258, 260–2, 264

'safe site', 131
salinization, 155, 254
salt tolerance, 166, 171, 197, 201, 247
seed (cf. caryopsis, fruit), 3, 43–5, 69, 100, 109, 126, 128, 131, 137, 139, 145, 159–60, 165, 168, 170, 186, 240–1, 171, 249
burs, 45
dispersal, 44–5
multiplication, 165, 186
production, 109, 128
scutellum, 43, 205, 207
setting, 170
testa, 55, 58
segregation, 109, 112–14, 120
selection, 145, 148, 190–1, 195–6, 247
sexuality, 52, 67, 111–12, 118, 126, 147
short cells, 2
silica bodies, 2
smut, 13, 258, 260–2, 264
soil erosion, 154, 157
somoclonal variation, 199
somatic cell genetics, 198–202
embryogenesis, 208
hybridization, 199, 200–2
spatheole, 38, 42, 48, 49
speciation, 126–7
sperm, 79, 81–3, 85–92, 221–2
dimorphism, 81–2, 85
discharge, 86
nuclei, 80, 81, 87–8, 90
pair, 84
wall, 83
spikelet, 2, 9, 13, 32–49, 52, 131
anomalous, 48
origin, 46
sporophyte, 52, 58–9
dispersal, 131
sporopollenin, 223
stamens, 32, 34, 35, 241
starch, 43, 69–70, 93
stigma, 32, 34, 77, 223, 251
papilla, 78
secretion, 76
'strawberry coral', model 147
style, 52, 251–3
stylodium, 78
succession, 146, 154–74
survivorship, 141, 148
suspension culture, 183, 185
suspensor, 93–4
synchronous (gregarious, mast) flowering, 241, 247
syncytium, 68
synergids, 57, 63–4, 70–1, 84–6, 88–90, 92, 103–4, 106, 251
synflorescence, 46–7
syngamy, 88, 91, 245
synonomy, 8
syphonogamy, 251–3
S,Z self-incompatibility, 76, 220, 241 (*see also* floral biology)

teratology, 32, 36
Tertiary, 27
tetrad, 60, 103, 108, 113
tiller, 146, 187, 206, 247
Ti-plasmid, 207
tobacco, 192, 197, 202
totipotency, 182, 198, 245, 252
toxins, 192–5
victorin, 193, 195
syringomycin, 194
transformation, 189, 202–8, 252
transplant performance, 143
trichomes, 78
tumbleweeds, 44

ultra-thin sectioning, 81, 85

vasoactive amines, 224
vegetative cell, 79–84, 221
virus, 9, 13, 187

virus (*cont.*)
  barley stripe mosaic, 62
  maize streak, 207
  wheat dwarf, 208
vivipary, 46, 132–2, 136

Wallacean isolation, 137

'yandying', 249
yield, 190, 202

zygote, 67, 79, 86, 93–5, 100, 208, 222, 245, 251–3